T0139945

Hobbes and Galileo: Method, Matter and the Science of Motion

INTERNATIONAL ARCHIVES OF THE HISTORY OF IDEAS

ARCHIVES INTERNATIONALES D'HISTOIRE DES IDÉES

230

HOBBES AND GALILEO: METHOD, MATTER AND THE SCIENCE OF MOTION

Gregorio Baldin

More information about this series at http://www.springer.com/series/5640

Gregorio Baldin

Hobbes and Galileo: Method, Matter and the Science of Motion

 Springer

Gregorio Baldin
University of Piemonte Orientale
Vercelli, Italy

ISSN 0066-6610 ISSN 2215-0307 (electronic)
International Archives of the History of Ideas Archives internationales d'histoire des idées
ISBN 978-3-030-41416-0 ISBN 978-3-030-41414-6 (eBook)
https://doi.org/10.1007/978-3-030-41414-6

This Springer imprint is published by the registered company Springer Nature Switzerland AG.
The registered company address is: Gewerbestrasse 11, 6330 Cham, Switzerland

Preface and Acknowledgements

This book is the English translation of my work, *Hobbes e Galileo. Metodo, Materia e Scienza del Moto*, published in Italian in 2017 by Leo S. Olschki Editore. I chose to not alter the original text, therefore making no substantial changes. I have only added some considerations in the introduction, which seek to clarify the central thesis that I intend to support in this book. I would like to thank in particular Gianni Paganini, whose support made this edition possible. Special thanks also go to the anonymous referee for his insight comments and remarks and to Guido Giglioni, who gave me important and useful advice.

I started the research leading to this book during my doctoral studies in Philosophy and the History of Philosophy at the University of Piemonte Orientale (Facoltà di Lettere e Filosofia, Vercelli), under the guidance of Professor Gianni Paganini, and continued at the École Normale Supérieure de Lyon, within the LabEx Comod, directed by Professor Pierre-François Moreau.

I would also to thank Catherine Bolton, Miriam Hurley, and Sonia Hill for translating my book into English. The last revision of the text has been made by Valerie Beavers, whom I thank warmly. Sadly, Catherine Bolton passed away suddenly in 2017 before completing the translation. We dedicate this translation to the memory of Catherine.

Oulx (Turin), Italy   Gregorio Baldin
January 2020

Contents

Abbreviations

OG *Edizione nazionale delle opere di Galileo Galilei*, 20 vols., Antonio Favaro (ed.) (Barbera: Rome, 1968 [original ed. 1890–1909]).

AT *Œuvres de René Descartes*, 11 vols., 13 tomes, Charles Adam and Paul Tannery (eds.) (Vrin: Paris, 1982–1992 [original ed. 1887–1913]).

CM *Correspondance du P. Marin Mersenne*, 16 vols., Paul Tannery, Cornelis de Waard, René Pintard, and Bernard Rochot (eds.) (Paris: Éditions Universitaires de France et Centre National de la Recherche Scientifique, 1932–1986).

OL *Thomae Hobbes Malmesburiensis opera philosophica quae Latine scripsit omnia*, 5 vols., William Molesworth (ed.) (London: Johannem Bohn, 1839–1845).

EW *The collected English works of Thomas Hobbes*, 12 vols., William Molesworth (ed.), 1839–1845 (reprint: Routledge/Thoemmes Press: London, 1997).

CH *The correspondence of Thomas Hobbes*, 2 vols., Noel Malcolm (ed.) (Oxford: Clarendon Press, 1994).

EL *The elements of law natural and politic*, Ferdinand Tönnies (ed.) (London: Frank Cass, 1984 [original ed. 1889]).

MLT *Critique du De mundo de Thomas White*, Jean Jacquot and Harold Whitmore Jones (eds.) (Paris: Vrin, 1973); Eng. Trans. by Harold Whitmore Jones (London: Bradford University Press, 1976).

TO I *Tractatus opticus I*, in *OL*, V, 215–248.

TO II *Tractatus opticus II*, MS Harley 6796, fols. 193–266 r–v / first complete edition by Franco Alessio (1963) in *Rivista critica di storia della filosofia*, 18/2, 147–228, without illustrations[1].

FD   *First draught of the Optiques*, MS Harley 3360, fols. VI + 193 r–v, available
         online and searchable at the website: http://www.bl.uk/manuscripts/
         FullDisplay.aspx?ref=Harley_MS_3360 *Thomas Hobbes' A minute of first
         draught of the Optiques: A critical edition*, by Elaine C. Stroud PhD
         Dissertation, (Madison: University of Wisconsin-Madison, 1983[1]).

---

[1] I have decided to refer to these texts in both the manuscript form and printed edition; first, I mention the manuscript folio, followed by the page in the printed version.

Introduction

In his autobiography in verses completed in about 1673 at the venerable age of 84,[1] Thomas Hobbes underscores the ties between his third and last Grand Tour of the European continent and the first reflections he developed in the field of natural philosophy.[2] In these passages, two elements capture the reader's attention. The first is the image—presenting notable Galilean echoes[3]—of the world as a book (*Nec tamen hoc tempus libris consumpsimus omne, Ni mundum libri dixeris esse loco*).[4] This metaphor invites us to dwell on the sole principle, identified by Hobbes, that holds up and governs all of nature. This is a truth that, while hidden from our eyes (*Et mihi visa quidem est toto res unica mundo Vera, licet multis falsificata modis*), underpins every phenomenon present in the natural world (*Unica vera quidem, sed quae sit basis earum rerum*): 'simply the movement that inhabits our inner parts', and these movements are the cause of our concepts, the 'children of our brain' (*Phantasiae, nostri soboles cerebri, nihil extra; Partibus internis nil nisi motus inest*).

The second interesting element is represented by the central role of Marin Mersenne, who was Hobbes' main counterpart, approved his reflections, and introduced him to the 'intellectual community' (*Is probat, et multis commendat*). Until then, Hobbes essentially considered himself to be a scholar of classical Greek and Latin.[5] It was Mersenne's appreciation and approval that facilitated his rise as a

---

[1] See Schuhmann 1998, 213. Hobbes started writing his autobiography in 1655 (ibid., 135). See also Tricaud 1985, 277–286, especially 281. Without further references except specific indications, biographical information about Hobbes is discussed by Pacchi 1971; Bernhardt 1989a; Tuck 1989; Malcolm 1996; Schuhmann 1998; Martinich 1999; Skinner, 'Hobbes's life in philosophy', in Skinner 2002, III, 1–37; Terrel 2008, 11–124.

[2] See Hobbes, 'Vita carmine expressa', in *OL*, I, lxxxix–xc.

[3] See Jesseph 2004, 191–211 and, above all, Paganini 2010a, 11 ff.

[4] On the image of the *liber mundi* in modern science, see Garin, 'La nuova scienza e il simbolo del libro,' in Garin 1961, 451–65, as well as Blumemberg's classic: Blumenberg 1979, *passim*. To limit ourselves to two authors known to Hobbes, this image can be found in Gassendi (see Magnard 1997, 21–29, esp. 21–22) and Campanella (see Ernst 2010, 8–9).

[5] Regarding the years after his education in Oxford, when he became the tutor of William Cavendish, Hobbes wrote: 'Ille (*i.e. the future second earl of Devonshire*) per hoc tempus mihi praebuit otia,

philosopher (*tempore ab illo Inter philosophos et numerabar ego*).[6] Hobbes even arrived at establishing a tie between his reflections developed during the Grand Tour and the idea of developing the triptych of the *Elementa philosophiae*.[7] A few years later—in 1640—he finished the *Elements of Law, Natural and Politic*. This was the first attempt to construct a philosophical system in which physical, anthropological, and political problems were connected in the wake of a demonstrative and deductive method, shaped according to the model of geometric sciences.[8]

Several elements in Hobbes' correspondence lead us to believe that he had already sketched out some of the subjects he explored in his *Elements*, just after his Grand Tour.[9] A letter from Sir Kenelm Digby, dated January 1637, cites a 'Logike'[10] by Hobbes (initially—and erroneously—identified as a manuscript at the National Library of Wales and now known as *De principiis*).[11] Unfortunately, this treatise of logic has been lost. Nevertheless, based on Digby's request, we can surmise the subject, which represents a cornerstone of the *Elements* and, in general, Hobbesian philosophy: the study of the formation of the concepts of the mind, starting from the action outside the senses. Knowing that the principles of nature were contained 'in the nature and variety of motions' (*in natura et varietate motuum*),[12] Hobbes started

---

libros / Omnimodos studiis praebuit ille meis. / Vertor ego ad nostras, ad Graecas, atque Latinas / Historias; etiam carmina saepe lego. Flaccus, Virgilius, fuit et mihi notus Homerus, / Euripides, Sophocles, Plautus, Aristophanes, / Pluresque; et multi Scriptores Historiarum: / Sed mihi prae reliquis Thucydides placuit'. Vita Carmine Expressa', in *OL*, I, lxxxviii. On the humanist education of Hobbes, see Skinner, 'Hobbes and the *studia humanitatis*', in Skinner 2002, III, 38–65; Skinner 1996, 215–244; Iori 2015; and Raylor. 2018.

[6] This aspect was already emphasized by Robertson 1886, 37–8 and 56–9.

[7] On the connection between the third Grand Tour and the genesis of Hobbes' philosophy, see Brandt 1928, 143 ff.; Pacchi 1971, 17–18; Bernhardt 1989a, 33 ff.; Terrel 2008, 17 ff.

[8] See Hobbes, *EL*, The Epistle Dedicatory, xv. See also the dedicatory letter of *De cive*, *Epistola dedicatoria*, *OL*, II, 137. The letter is dated November 1, 1641, in the first version of *De cive*. See the reproduction in Hobbes 1983b, 76.

[9] Baumgold emphasized—rightly in my opinion—that the composition of much of the *Elements of Law* is probably earlier than 1640 (although Hobbes underscored the occasional nature of the text, inspired by the outbreak of civil war). According to Baumgold, Hobbes devoted himself simultaneously to natural philosophy and political problems in the second half of the 1630s. See Baumgold 2004, 16–43, in part. 25–26. Indeed, Sommerville also thought that *Elements* presents the 'stamp of Mersenne's group' (Sommerville 1992, 15).

[10] See Sir Kenelm Digby to Hobbes, from Paris, January 17–27, 1637, *CH*, I, 42–43. See also Digby's three letters to Hobbes dated September 11–21, 1637, *CH*, I, 50, and October 4–14, 1637, in which he sends Hobbes 'Monsieur des Cartes… his book', i.e. the *Discours de la méthode*. *CH*, I, 51

[11] MS 5297, National Library of Wales, discovered by Mario Manlio Rossi (see Rossi 1942, 104–119) and reproduced as *Appendix II*, in Hobbes, *MLT*, 448–460. Now, we know that the text was unquestionably later (after 1643). See Minerbi Belgrado 1993, 165–170. It is likely that the 'Logic' referred to by Digby has been lost (see Malcolm 1996, 29–30).

[12] See 'Thomae Hobbes Malmesburiensis vita authore seipso', in *OL*, I, xiv.

to investigate which movements produced sensation, intellection, and *phantasmata*, i.e. the images[13] and concepts of external things.[14]

In fact, even prior to his departure for the Grand Tour, Hobbes had started to focus on scientific matters,[15] as suggested by a letter dated October 17, 1634, which the elderly mathematician Walter Warner (ca. 1557–1643)[16] sent to Robert Payne (1596–1649). In it, Warner mentioned a geometrical demonstration by Hobbes.[17] In another letter from the same period, which Hobbes sent to William Cavendish, Earl of Newcastle, he informed Cavendish that he was heading to London to get a copy of Galileo's *Dialogo sopra i due massimi sistemi* (*Dialogue Concerning the Two Chief World Systems*)[18]—and that he was probably aware of Joseph Webbe's desire to translate the work.[19]

Several autobiographical references seem to backdate Hobbes' interest in science to the early 1630s,[20] though these reflections are the outcome of mere informal discussions with the habitués of the so-called 'Newcastle Group'.[21] The same can be said of the scanty references to the so-called Euclidean Illumination in the

[13] On the formation of images in Hobbes' optics, see Stroud 1983, 41 ff., and Malet 2001, 303–333.

[14] As emphasized by Paganini, the Latin term *phantasma*, used by Hobbes, 'alludes first of all to the sensible representation of an object. Starting with the *Elements*, however (see Hobbes, *EL*, Part I, chaps. II–IV, 3–17), Hobbes establishes a close correlation between sensation, imagination and conception (all reducible, from an ontological standpoint, to the movement of the internal parts of the sentient being). As a result, the term *phantasma* takes on a larger meaning to signify any mental representation'. Hobbes 2010, 147–148, note. The idea is shared by the author of *Short Tract*. See Hobbes 1988a, 40 and 45–46.

[15] See also Hobbes to ? from Paris, October 21–31, 1634, *CH*, I, 22–23.

[16] On Warner, see Aubrey 1962, 365–366, and Clucas 2004. Seth Ward accused Hobbes of plagiarizing Warner's theories (also those of Descartes, Gassendi, Digby, and Roberval; see *Six Lessons*, *EW*, VII, 340–343), but Prins emphasized that Hobbes' claim of originality was not unfounded. See Prins 1993, 195–224, in particular 210 ff.

[17] See Walter Warner to Robert Payne, Westminster, October 17, 1634, in Halliwell 1841, 65 (which I have quoted), partially reproduced in *CM*, IV, 380–381. The complete text is also in Schuhmann 1998, 43–44. See also Médina 2013a, 99. The letter refers to an 'analogy' of Hobbes, and among the papers of Walter Warner now at the British Museum, there is an autograph letter from Hobbes dealing with this problem, at the bottom of which Warner added: 'Mr Hobbes analogy'. British Museum, Birch MS 4395, fol. 131. See Jacquot and Jones, 1973, 17–18. On Robert Payne, see below, *Appendix*.

[18] Hobbes to William Cavendish, Earl of Newcastle, from London, 26 Jan. [5 Feb.] 1634, in *CH*, I, 19.

[19] *CH*, I, 19: 'I doubt not but the Translation of it will here be publiquely embraced, and therefore wish extreamely that Dr Webbe would hasten it'. The translation, almost complete but never published, is at the British Library (MS Harl. 6320). See note 10 in *CH*, I, 20.

[20] In a 1641 letter, Hobbes recalled that he had developed his first reflections on the nature of light with William and Charles Cavendish of Newcastle (see Hobbes to Mersenne, Paris, March 30, 1641, *AT*, III, 342–343), and he emphasized this a few years later in *First Draught* (see Hobbes, *FD*, n. pag., fol. 3r/76–77).

[21] See Halliwell 1841, 65 ff. See also François Derand to William Oughtred (commencement d'), October 1634, *CM*, IV, 365–367.

Examinatio et emendatio mathematicae hodiernae (1660)[22] and in Hobbes' autobiography in prose.[23]

These clues tell us that Hobbes' main interests already included natural philosophy and, in particular, phenomena connected with optics,[24] though the most interesting documents from this period include two letters he sent to William Cavendish in 1636. In the first, Hobbes explains the fundamental difference he notes between *mathematics* and *natural philosophy*:

> In *Thinges that are not demonstrable*, of which kind is the *greatest part of Natural Philosophy*, as dependinge upon the motion of bodies so subtile as they are invisibile, much as are ayre and spirits, the most that can be atteyned unto is to have such *opinions*, as no certayne experience can confute, and from wich can be deduced by lawfull argumentation, no absurdity, and such are your opinions in your letter of the 3rd July wich I had the honour to receave the last weeke; namely that *the variety of thinges is but the variety of locall motion* in the spirits or invisibles parties of bodies. And That such motion is heate.[25]

The letter reveals three especially relevant aspects. First, Hobbes maintains that natural philosophy is 'not demonstrable'[26] and that thus only probabilistic arguments are possible in this area. Second, he indicates the reason why the methodology to be applied diverges in the two areas of study: *mathematics* and *physics*. Physical science presents an inevitably hypothetical nature because it is applied to phenomena that are not entirely knowable nor can be mastered by the observer. They belong to a world of singular bodies, with which we come into contact only through our sensory

---

[22] See Hobbes 1660, Prop. 41, 154–155. See also Aubrey 1962, 230; Bernhardt 1986, 281–282; Bernhardt 1988, 61–87. On the English university system and its relationship with the scientific culture, see Feingold 1984b, esp. 45 ff.

[23] See 'Thomae Hobbes Malmesburiensis Vita, Authore Seipso', *OL*, I, xiv.

[24] See Charles Cavendish to Walter Warner, Welbeck, 2 May 1636, in Halliwell 1841, 66. (See also *CM*, VI, 66, and Schuhmann 1998, 48–49.)

[25] Hobbes to William Cavendish, Earl of Newcastle, from Paris, July 29–August 8 1636, *CH*, I, 33 (my italic). See Kargon 1966, 56.

[26] On this point, there is the interesting analogy noted by Watkins 1965, 52 ff., and analyzed more extensively by Gargani between Hobbes' concept of science and reflections developed by several representatives of Paduan Aristotelianism in the late sixteenth century (see Gargani 1971, 32–51). In particular, Gargani emphasized a number of parallels between the works of Agostino Nifo and Jacopo Zabarella, on the one hand, and Hobbes' philosophical-scientific methodology, on the other. On the contrary, Jan Prins noted profound differences between the epistemological orientations of Hobbes and the Paduan Aristotelians (in particular Zabarella), shedding doubt on the idea of any direct influence of Zabarella over Hobbes. Prins preferred to suggest a possible convergence with Ramism and the Aristotelian humanist current tied to Philipp Melanchthon. See Prins 1990, 26–46. It should be said, however, that at Oxford University, where Hobbes studied, Nifo and Zabarella were known through the teachings of John Case. Furthermore, the Hobbes E.1 A manuscript in Chatsworth (Derbyshire), consisting of a list of volumes at the Hardwick Library, drawn up by Hobbes, mentions both *Logica* and *Fisica* by Zabarella (Chatsworth, MS Hobbes E.1A, 120, now printed in Talaska 2013). In general, on Hobbes' relationships with Aristotelianism, see the interesting and well-documented study by Leijenhorst 2002. The contribution made by the School of Padua to the birth of modern science has been emphasized above all by Randall 1961, 15–68. See also Crombie 1953, 274–403. On certain analogies between the conception of method in Hobbes (and Sarpi) and Paduan Aristotelianism, see also Baldin 2013.

organs and that we must thus codify, by applying a method that allows us to translate the empirical data offered to us by the senses into a conventional scientific language.[27] By contrast, in the mathematical sciences, the process is essentially a priori and entirely deductive.[28] This dual orientation of Hobbes' philosophical-scientific thought—which has led some interpreters of his works to talk about Hobbes' *conventionalism*[29]—characterizes his entire intellectual production, but it would acquire a different meaning and greater epistemological awareness over the years.[30]

In the letter mentioned above, however, there was another particularly significant element: Hobbes said he agreed with the opinion held by William Cavendish that 'the variety of thinges is but the variety of locall motion in the spirits or invisibles parties of bodies'.[31] The hypothesis that the variety of phenomena is solely the result of changes produced by matter and local motion is the cornerstone of Hobbes' natural philosophy. This point reveals the profound influence of Galileo's philosophy.

Nevertheless, there was another equally interesting letter to William Cavendish, dated October 1636, that is, the period immediately after Hobbes returned to England.[32] Here, he sketches out unique elements of the theory of optics that would appear later in various works: Hobbes explains that his conception does not turn to the visible *species*[33] and thinks that motion is propagated exclusively through a

---

[27] Hobbes argued that sensation is the first and only vehicle of knowledge, reproposing Aristotle's adage: 'There is nothing in the intellect that was not first in the senses'. See *MLT*, XXX, 349.

[28] Hobbes to Sir Charles Cavendish, from Paris, 29 Jan. [8 Feb.] 1641, *CH*, I, 83: 'In *mathematicall sciences* wee come at last to definition which is a beginning or Principle, made true by pact and consent among our selves'.

[29] The probabilistic aspect of Hobbes' natural philosophy was emphasized above all by Arrigo Pacchi, who considered Hobbes one of the exponents of 'mitigated' or 'constructive' scepticism, according to the expression coined by Richard Popkin. See Popkin 1979, 129–150, and Pacchi 1965, 12 ff.

[30] This aspect of Hobbes' thought was closely analysed by Gargani 1971, 152.

[31] Hobbes to William Cavendish, Earl of Newcastle, from Paris, July 29–August 8, 1636, in *CH*, I, 33: 'The variety of thinges is but the variety of locall motion in ye spirits or inuisibles partes of bodies'.

[32] Hobbes to William Cavendish, Earl of Newcastle, from Byfleet (Surrey), 16 [/26] October 1636, in *CH*, I, 37–38: 'For the reason of the species passing through a hole to a white paper, my opinion is this. The lucide body, as for example, the sunne, lighting on an object, as for example, the side of a house doth illuminate it, that is, to say giue it the same vertue, though not in the same degree, of diffusing light euery way, and illuminating other objects wth a lesse light, but the light that commeth from the house [sight *deleted*] side, is not pure light, but light mingled, that is to say, color, This light mingled, or colour, passing through the hole [is it *deleted*] there crosses, and goes wth ye figure inuerted to the white paper, and giues the paper in that part where it falles a power to diffuse light euery way, and so it comes to ye eye wheresoeuer they stand (if a direct line may thence come to ye eye) And it is not as Galileo sayes; that illumination is made by reflexion, and that the asperity of the obiect makes it be seene euery way wch otherwise would be seene onely in one point, where angles of incidence and refraction were equall. But whereas I vse the phrases, the light passes or the colour passes or diffuseth it selfe, my meaning is that the motion is onely in ye medium, and light and colour are but the effects of that motion in ye brayne'.

[33] The intromission theory of the species (medieval in origin, see Lindberg 1976, esp. 122 ff.), which envisages the passage of particles from the light source to the sentient's receptor organ, is

medium. Furthermore, he argues that light and colors consist solely of modifications produced by movement in the brain of the sentient being.[34] However, the letter
contains another significant point. Hobbes expresses his disagreement with Galileo's
opinion, who in his *Dialogue Concerning the Two Chief World Systems* maintained
that 'illumination is made by reflexion'[35] and that the roughness of the surfaces of
reflecting objects triggers the reflection of light in every direction.[36] The passage
suggests that, by this time, Hobbes had attentively read Galileo's *Dialogue*, carried
out some experimental observations, and also formulated personal reflections about
the nature of *light* on a notable scientific and systematic level.

It seems uncoincidental that, during the years of his third Grand Tour, Hobbes
had the chance to meet Mersenne and, probably, reflect attentively on Galileo's
Dialogue, which—like other works by Galileo—was well known among the habitués of Mersenne's circle.[37] Therefore, it comes as no surprise that the most significant figures in the autobiographies and correspondence of Hobbes are Marin
Mersenne and Galileo, whose image dominates Hobbesian natural philosophy, as
testified by the very laudatory references to Galileo in *De Motu, Loco et Tempore*
and *De Corpore*. In *De Motu*, Hobbes cites him as 'the greatest philosopher not only
of our century, but of all time'.[38] Likewise, in *De Corpore*, Galileo is described as
'the first that opened to us the gate of natural philosophy universal, which is the
knowledge of the nature of motion. So that neither can the age of natural Philosophy
be reckoned higher than to him'.[39]

---

instead present in the *Short Tract*. Hobbes 1988a, 24. An extensive and detailed analysis of Hobbes'
relationship with medieval optics can be found in Gargani 1971, 97–123, in which he also compares the *Short Tract* with texts by Roger Bacon and the metaphysics of light of Robert Grosseteste.
As opposed to Gargani, Bernhardt does not identify the presence of Oxford Franciscans in the
Short Tract, but in Hobbes' later works. See Bernhardt 1977, 3–24 and 11–13.

[34] As we know, optics represented one of Hobbes' main scientific interests. The most significant
studies on the subject are Alessio 1962, 393–410; Shapiro 1973, esp. 134–171; Bernhardt 1977;
Bernhardt 1990, 245–268; Stroud 1983, 1–71; Stroud 1990, 269–277; Prins 1996; Médina 1997;
Médina 2015a; Médina 2015b; Médina 2016b; and esp. Giudice 1999. See also Giudice 2015 and
Giudice 2016. On Hobbes' theory of refraction, see Horstmann 2000.

[35] Hobbes to William Cavendish, Earl of Newcastle, from Byfleet (Surrey), 16 [/26] October 1636,
CH, I, 38

[36] Galileo examined the subject in the first day of his *Dialogue* (Galilei, *Dialogo sopra i due massimi sistemi, OG*, VII, p. 102. Eng. trans. Galileo 2001, 77), but did not affirm exactly what Hobbes
attributed to him in the examined letter. On the contrary, he argued that '[i]f the surface, being
spherical, were as smooth as a mirror, it would be entirely invisible, seeing that that very small part
of it which can reflect the image of the sun to the eyes of any individual would remain invisible
because of the great distance'. See also *OG*, VII, 123.

[37] Regarding the presence of some of Galileo's works in Mersenne's library and in particular *Il
Saggiatore* and *Discorsi,* see Buccolini 1998.

[38] *MLT*, X, 9, 178, Eng. trans. Hobbes 1976, 123, modified: 'Galilaeus, non modo nostri, sed
omnium saecolorum philosophus maximus'

[39] *De corpore*, Epistola dedicatoria, *OL*, I, n. pag., Engl. trans. *EW*, I, viii

In the preface to his *Ballistica* (1644), Mersenne described Hobbes' philosophy as being characteristically defined by 'mechanical' features.[40] The documents mentioned above lead us to believe that the seeds of this explanation of natural phenomena date to the 2-year period of 1634–1636.[41] During this time, Hobbes came up with the key idea of his philosophy: 'in nature everything is made mechanically, and all phenomena are generated from one matter, agitated by different kinds of motion'.[42] To understand the meaning of this *mechanical world picture*,[43] we must turn to two of his points of reference, namely, Mersenne and, above all, Galileo. In fact, there was another 'privileged interlocutor' of Hobbes, Pierre Gassendi, with whom he not only had a deep friendship[44] but also shared many aspects of his philosophy. Although a study of the relationship between these two authors, which has been examined extensively by Paganini,[45] is beyond the scope of this work, I will refer to Gassendi every time it is necessary to mention his work.

Nevertheless, before addressing in full of Hobbes' mechanical philosophy, we should briefly discuss the meaning of the key concept of this work: *mechanism*. Much has been said about the true scope and essence of mechanical philosophy, the formulation of which—as we know—goes back to Robert Boyle.[46] By contrast, mechanism as a historiographic category acquired some importance during the

---

[40] Mersenne, 'Ballistica et anticosmologia'. In Mersenne 1644a, *Praefatio utilis ad lectorem* (n. pag.) (reproduced in Hobbes, *OL*, V, 309): 'CUM (in prop. xxiv BALLISTICAE) plura juxta subtilissimi philosophi Thomae Hobbes attulerimus, et quasdam philosophiae quam exornat partes legerim, *quae omnia fere per motum localem explicant*, velim etiam addere modum quo nostrarum facultatum operationes ex eodem motu concludit' (my italic)

[41] On the importance of discussions in Mersenne's circle, for the genesis of Hobbes' philosophy, see Baldin 2016b; Baldin 2017. See also Médina 2016a and Bernhardt 1989a, 34–36.

[42] *Vitae Hobbianae auctarium, authore R. Blackbourne*. In *OL*, I, xxviii: '…in natura omnia mechanice fieri, et ex una materia, variis motuum generibus et mensuris agitata, universa rerum phaenomena exurgere'.

[43] On the development of a mechanistic conception of the universe in modern philosophy, see Dijksterhuis 1961.

[44] See Paganini 2011, 23–38.

[45] On the relations between Hobbes and Gassendi, see Paganini 1990, 351–445; Paganini 2004b; Paganini 2006a; Paganini 2008b.

[46] The definition of 'mechanical philosophy' first comes up with Boyle in 1661, but, above all, in *The Origin of Forms and Qualities*, dated 1666, which contributed to circulating the concept. See *The Origine of Formes and Qualities (According to Corpuscolar Philosophy) Illustrated by Considerations and Experiments*. In Boyle 1999–2000, V, 302: 'That then which I chiefly aime at, is to make it Probable to you by Experiments… That allmost all sorts of Qualities, most of which have been by the Schooles either left Unexplicated, or Generally referr'd, to I know not what Incomprehensible Substantiall Formes, may be produced Mechanically, I mean by such Corporeall Agents, as do not appear, either to Work otherwise, then by vertue of the Motion, Size, Figure, and Contrivance of their own Parts (which Attributes I call the Mechanicall Affections of Matter, because to Them men willingly Referre the various Operations of Mechanical Engines:) or to Produce the new Qualities exhibited by those Bodies, their Action changes, by any other way, then by changing the Texture, or Motion, or some other Mechanical Affection of the Body wrought upon'. See also Garber 2013, 5–7.

1950s (especially following publication of a work by the science historian Marie Boas Hall).[47] Subsequently, more recent and prudent historiography beckons us to reflect on the overly rigid and schematic use of historiographic labels, with the awareness that they cannot grasp the wealth and variety of different philosophies and systems of thought, because of their inherently multifaceted nature. However, I argue that some of these categories can still be very meaningful and well still suited to describing a particular vision of the world or a philosophical and cultural movement.[48] This is even more the case with regard to concepts or definitions, such as that of mechanism, the roots of which date back to the historical period and cultural context examined here.

Here, it may be helpful to remind ourselves of the four defining characteristics of the mechanical philosophy, which were recently identified and discussed, in part, by Daniel Garber and Sophie Roux. With their necessarily indicative nature, they help us understand what we mean when we speak of mechanism: (a) the general program of replacing 'traditional philosophy' (i.e. Aristotelian-scholastic philosophy) with a 'new philosophy'; (b) rejection of Aristotelian hylomorphism and the adoption of an ontology, whereby all natural phenomena must be interpreted in terms of matter and the movement of the corpuscles composing macroscopic bodies; (c) the comparison of natural phenomena and, in particular, the world of animals, with the function and structure of machines, be they existent or imaginary; and (d) an ontology associated with mechanics, understood as the mathematical science of movement, the laws of which are described as the laws of nature in general.[49] Some of these elements, in fact, are 'more mechanical' than others, but the presence or absence of these indicators suggests if and when we are in the face of mechanical thought. As we will see ahead, Hobbes' philosophy meets all four criteria and thus can rightly be defined as strictly mechanical. This statement could even be considered obvious, since Hobbes and Descartes are often cited as examples of quintessential mechanical philosophers. However, I believe that Hobbes' mechanical philosophy must be interpreted in the light of the reflections and discoveries presented by Galileo in his most important and significant works: *The Assayer*, *Dialogue Concerning the Two Chief World Systems*, and *Discourses and Mathematical Demonstrations Relating to Two New Sciences*. Indeed, Hobbes' philosophy also differs in many respects from that of Galileo, in particular in consideration of its more hypothetical character, an element which was decidedly more attenuated in Galileo. And yet we must not forget that Hobbes' philosophy is based entirely on the concept of body and this notion of body

---

[47] See Boas Hall 1952, 412–541.

[48] Not incidentally, as also observed by Garber (Garber 2013, 4), mechanism was still considered a fundamental element of the scientific revolution, as demonstrated by several recent studies on the subject. See Shapin 1998, 30 ff; Dear 2001, 80–100; Dear 2006, 15–38.

[49] See Daniel Garber and Sophie Roux, *Introduction*, in Garber and Roux 2013, xi–xii. As a matter of fact, the authors are critical in considering the validity and correctness of these elements as principles of mechanism. Nonetheless, in my opinion, they can play the role of a litmus test, documenting the presence of a mechanical philosophy. A rather critical interpretation of the category can also be found in Gabbey 2004. In general, on mechanism, see also Lard and Roux 2008.

is axed on two strictly connected and essentially 'Galileian' elements: the quantitative and mathematical analysis of the body and the distinction between the real and the so-called subjective qualities of the bodies, as we will see in particular in Chap. 2, which contains the central thesis of this book. Obviously, the distinction between subjective and real qualities of the bodies is also present in Descartes and, later, in many other modern philosophers, but the traces present in Hobbes' works and correspondence lead me to believe that he had been particularly inspired by Galileo. Hobbes pushed his mechanism to its extreme consequences, coming to conceive even an anthropology which tried to be entirely mechanical and material. However, Hobbes' world is based on Galileo's notion of body and on the mathematical and mechanical laws of nature developed by Galileo, especially in his *Dialogues* and *Discourses*. For this reason, I think, Hobbes considered Galileo the greatest philosopher of all the centuries, as he refers to the Italian scientist and philosopher in his *De Motu, Loco et Tempore*. Hobbes uses the term *philosopher*, and this could seem obvious, since the conceptual distinction between philosophy and science is known to belong to a later era.[50] However, it is interesting to remark that Hobbes does not speak of Galileo simply as a *natural philosopher*, but he refers to him with the wider term of philosopher. In my opinion, this is not incidental, because Hobbes considered Galileo the founder of a new philosophy, on whose basis Hobbes built his entire philosophical thought.

In my book, I chose to refer to Galileo simultaneously as a natural philosopher and a scientist. Someone might consider this inappropriate or anachronistic, but it is necessary to keep in mind that modern science takes root precisely with Galileo and his mathematical conception of nature.[51] Therefore, it does not seem to me wrong to distinguish Galileo from Renaissance natural philosophers, attributing to him also the title of scientist, in addition to that of natural philosopher. Galileo certainly shares with the Renaissance philosophers the idea that natural phenomena should be explained by referring exclusively to principles inherent to nature. At the same time, however, Galileo stands out from his predecessors, because he intends to interpret the book of nature through mathematically and quantifiable laws.

This study starts with a chapter about Hobbes' relationship with Mersenne, and this could be surprising, since the book is dedicated to Hobbes and Galileo. However, we must remember that Mersenne played a key role in intermediating with and circulating Galileo's ideas. As we will see, Hobbes seems to propose a certain interpretation of Galileo in which some traces of Mersenne's influence are far from absent.

---

[50] See, in this respect, Gabbey 2001, for example.
[51] See, for example, Torrini 2013.

References

Manuscripts

British Library

Ms Harley 3360, ff. VI + 193 r et v (First Draught of the Optiques), available online at: http://www.
bl.uk/manuscripts/FullDisplay.aspx?ref=Harley_MS_3360
Ms Harley 6796, ff. 193–266 r et v (Tractatus Opticus II).
Ms Hobbes E.1 A (List of books kepts in the Hardwick Library, compiled by Hobbes).

Chatsworth House, Derbyshire

Ms Hobbes E.1 A (List of books kepts in the Hardwick Library, compiled by Hobbes).

Works of Thomas Hobbes

Hobbes, Thomas. 1839–1845a. Thomae Hobbes Malmesburiensis Opera Philosophica quae Latine
Scripsit Omnia, in unum corpus nunc primum collecta studio et labore Gulielmi Molesworth,
5 Vols. London: Johannem Bohn.
Hobbes, Thomas. 1839–1845b. The Collected English Works of Thomas Hobbes ed. William
Molesworth, 11 Vols. London: Bohn, Longman, Brown, Green, and Longmans.
Hobbes, Thomas. 1660. Examinatio et Emendatio Mathematicae Hodiernae. London: Crooke.
Hobbes, Thomas. 1976. Thomas White's De Mundo Examined, ed. Harold Whitmore Jones.
Bradford: Bradford University Press.
Hobbes, Thomas. 1983a. Thomas Hobbes' A Minute or First Draught of the Optiques, ed. Elaine
C. Stroud, PhD Dissertation. Madison: University of Wisconsin–Madison.
Hobbes, Thomas. 1983b. De cive. The Latin Version, ed. Howard Warrender. Oxford: Clarendon
Press.
Hobbes, Thomas. 1988a. (probably by Robert Payne), Court traité des premiers principes, ed. Jean
Bernhardt. Paris: Presses Universitaires de France.
Hobbes, Thomas. 1988b. Of Passions (British Library: Ms. Harl. 6083), ed. Anna Minerbi
Belgrado. Rivista di Storia della Filosofia (4): 729–738.
Hobbes, Thomas. 2010. Moto, luogo e tempo, ed. Gianni Paganini, Utet, Torino.

Other Works

Aubrey, John. 1962. Aubrey's Brief Lives, ed. Oliver Lawson Dick. Bristol: Penguin Book.
Boyle, Robert. 1999–2000. The Works of Robert Boyle, 14 Vols, ed. Michael Hunter and Edward
B. Davis. London: Pickering and Chatto.
Galilei, Galileo. 2001. Dialogue Concerning the Two Chief World Systems, Eng. Trans. Stillman
Drake. New York: The Modern Library (1st ed. Los-Angeles-Berkeley: University of California
Press, 1967).
Mersenne, Marin. 1644a. Cogitata physico matematica. In quibusdam naturae quàm artis effectus
admirandi artissimis demonstrationibus explicantur. Paris: Bertier.

Secondary Sources

Alessio, Franco. 1962. De Homine e A Minute of First Draught of the Optiques, di Thomas Hobbes. *Rivista critica di storia della filosofia* 17: 393–410.

Baldin, Gregorio. 2013. Hobbes and Sarpi: Method, Matter and Natural Philosophy. *Galilaeana* 10: 85–118.

―――. 2016b. 'La reflexion de l'arc' et le *conatus*: aux origines de la physique de Hobbes. *Philosophical Enquiries. Revue des philosophies anglophones* 7: 15–42.

―――. 2017. Archi, spiriti e *conatus*. Hobbes e Descartes sui principi della fisica. *Historia Philosophica* 15: 147–165.

Baumgold, Deborah. 2004. The Composition of Hobbes's *Elements of Law*. *History of Political Thought* 25 (Iss. 1): 16–43.

Bernhardt, Jean. 1977. Hobbes et le mouvement de la lumière. *Revue d'histoire des sciences* 30: 3–24.

―――. 1986. Témoignage direct de Hobbes sur son 'illumination euclidéenne'. *Revue philosophique de la France et de l'Etranger* 176 (n. 2): 281–282.

―――. 1988. Essai de commentaire. In *Thomas Hobbes, Court traité des premiers principes*, 61–274. France: Presses Universitaires de.

―――. 1989a. *Hobbes*. Paris: Presses Universitaires de France.

―――. 1989b. L'apport de l'aristotelisme à la pensée de Hobbes. In *Thomas Hobbes. De la métaphysique à la politique*, ed. Martin Bertman and Michel Malherbe, 9–15. Paris: Vrin.

―――. 1990a. L'œuvre de Hobbes en optique et en théorie de la vision. In *Hobbes Oggi*, ed. Bernard Willms et al., 245–268. Milan: Franco Angeli.

―――. 1990b. Grandeur, substance et accident: une difficulté du *De corpore*. In *Thomas Hobbes. Philosophie première, théorie de la science et politique*, ed. Yves-Charles Zarka and Jean Bernhardt, 39–46. Paris: Presses Universitaires de France.

Blumenberg, Hans. 1979. *Die Lesbarkeit der Welt*. Berlin: Surkhamp.

Boas Hall, M. 1952. The Establishment of the Mechanical Philosophy. *Osiris* 10: 412–541.

Brandt, Frithiof. 1928. *Thomas Hobbes' Mechanical Conception of Nature*. Copenhagen/London: Levin & Mungsgaard-Librairie Hachette (or. Ed. 1921).

Buccolini, Claudio. 1998. Opere di Galileo Galilei provenienti dalla biblioteca di Marin Mersenne: *Il saggiatore* e i *Discorsi e dimostrazioni matematiche intorno a due nuove scienze*. *Nouvelles de la Républiques des Lettres* (2): 139–142.

Clucas, Stephen. 2004. Walter Warner (c.1558–1643). In *Oxford Dictionnary of National Biographies*. Oxford: Oxford University Press. http://www.oxforddnb.com/view/article/38108.

Crombie, Alistair C. 1953. *Augustine to Galileo. The History of Science A.D. 400–1650*. Cambridge, MA: Harvard University Press.

Dear, Peter. 2001. *Revolutionizing the Sciences. European Knowledge and its Ambition, 1500–1700*. Basingstoke: MacMillan.

―――. 2006. *The Intelligibility of Nature*. Chicago/London: The University of Chicago Press.

Dijksterhuis, Eduard J. 1961. *The Mechanization of the World Picture: From Pythagoras to Newton*. Oxford: Clarendon Press.

Ernst, Germana. 2010. *Tommaso Campanella. Il libro e il corpo della natura*. Rome/Bari: Laterza.

Feingold, Mordechai. 1984a. Galileo in England: The First Phase. In *Novità celesti e crisi del sapere. Atti del Convegno di studi*, ed. Paolo Galluzzi, 411–420. Florence: Giunti-Barbera.

―――. 1984b. *The Mathematicians' Apprenticeship*. Cambridge: Cambridge University Press.

Gabbey, Alan. 2001. Disciplinary Transformation in the Age of Newton: The Case of Mathematics. In *Between Leibniz, Newton, and Kant. Philosophy and Science in the Eighteenth Century*, ed. Wolfgang Lefèvre. Dordrecht: Kluwer.

―――. 2004. What Was 'Mechanical' About 'The Mechanical Philosophy'. In *The Reception of Galilean Science of Motion in Seventeenth-Century Europe*, ed. Carla Rita Palmerino and J.M.M.H. Thijssen, 11–23. Dordrecht/Boston/London: Kluwer.

Garber, Daniel. 2013. Remarks on the Pre-History of Mechanical Philosophy. In *The Mechanization of Natural Philosophy*, ed. Daniel Garber and Sophie Roux, 3–26. Dordrecht: Springer.

Garber, Daniel, and Roux, Sophie. 2013. Introduction. In *The Mechanization of Natural Philosophy*, ed. Daniel Garber and Sophie Roux, xi–xviii. Dordrecht: Springer.

Gargani, Aldo G. 1971. *Hobbes e la scienza*. Einaudi: Turin.

Garin, Eugenio. 1961. *La cultura filosofica del Rinascimento italiano*. Florence: Sansoni.

Giudice, Franco. 1999. *Luce e visione. Thomas Hobbes e la scienza dell'ottica*. Florence: Olschki.

———. 2015. The Most Curious of Science: Hobbes's Optics. In *The Oxford Handbook of Hobbes*, ed. Aloysius P. Martinich and Kinch Hoekstra, 147–165. Oxford: Oxford University Press.

———. 2016. Optics in Hobbes's Natural Philosophy. *Hobbes Studies* 29 (1): 86–102.

Halliwell, James O. 1841. *Letters on the Progress of Science in England from Queen Elizabeth to Charles II*. London: Taylor.

Horstmann, Frank. 2000. Hobbes und Sinusgesetz der Refraktion. *Annals of Science* 57: 415–440.

Iori, Luca. 2015. *Thucydides Anglicus. Gli Eight Bookes di Thomas Hobbes e la Ricezione Inglese delle Storie di Tucicide (1450–1642)*. Roma: Edizioni di Storia e Letteratura.

Jacquot, Jean, and Harold Whitmore Jones. 1973. Introduction. In *Thomas Hobbes, Critique du De Mundo de Thomas White*, 9–102. Paris: Vrin.

Jesseph, Douglas M. 2004. Galileo, Hobbes and the Book of Nature. *Perspectives on Science* 12 (2): 191–211.

Kargon, Robert H. 1966. *Atomism in England from Hariot to Newton*. Oxford: Clarendon Press.

Laird, Walter Roy, and Sophie Roux, eds. 2008. *Mechanics and Natural Philosophy Before the Scientific Revolution*. Dordrecht: Springer.

Leijenhorst, Cees. 2002. *The Mechanisation of Aristotelianism. The Late Aristotelian Setting of Thomas Hobbes' Natural Philosophy*. Leiden/Boston/Köln: Brill.

Lindberg, David C. 1976. *Theories of Visions from Al-Kindi to Kepler*. Chicago: University of Chicago Press.

Magnard, Pierre. 1997. La mathématique mystique de Pierre Gassendi. In *Gassendi et l'Europe*, ed. Sylvia Murr, 21–29. Paris: Vrin.

Malcolm, Noel. 1996. A Summary Biography of Hobbes. In *The Cambridge Companion to Hobbes*, ed. Tom Sorell, 13–44. Cambridge: Cambridge University Press.

Malet, Antoni. 2001. The Power of Images: Mathematics and Metaphysics in Hobbes's Optics. *Studies in History and Philosophy of Science* 32 (2): 303–333.

Martinich, Aloysius P. 1999. *Hobbes. A Biography*. Cambridge: Cambridge University Press.

Médina, José. 1997. Nature de la lumière et science de l'optique chez Hobbes. In *Le siècle de la lumière 1600–1715*, ed. Christian Biet and Vincent Jullien, 33–48. Paris: ENS Éditions Fonteany-St. Cloud.

———. 2013a. Mathématique et philosophie chez Thomas Hobbes: une pensée du mouvement en mouvement. In *Lectures de Hobbes*, ed. Jauffrey Berthier, Nicolas Dubos, Arnaud Milanese, and Jean Terrel, 85–132. Paris: Ellipses.

———. 2013b. Physiologie mécaniste et mouvement cardiaque: Hobbes, Harvey et Descartes. In *Lectures de Hobbes*, ed. Jauffrey Berthier, Nicolas Dubos, Arnaud Milanese, and Jean Terrel, 133–162. Paris: Ellipses.

———. 2015a. La philosophie de l'optique du *De Homine*. In *Thomas Hobbes, De l'Homme*, ed. Jean Terrel et al., 31–82. Paris: Vrin.

———. 2015b. Le traité d'optique. In *Thomas Hobbes, De l'Homme*, ed. Jean Terrel et al., 87–146. Paris: Vrin.

———. 2016a. Matière, mémoire et mouvement chez Hobbes. In *Hobbes et le matérialisme*, ed. Jauffrey Berthier and Aranud Milanese, 33–74. Paris: Éditions matériologiques.

———. 2016b. Hobbes's Geometrical Optics. *Hobbes Studies* 29 (1): 39–65.

Minerbi Belgrado, Anna. 1993. *Linguaggio e mondo in Hobbes*. Rome: Editori Riuniti.

Pacchi, Arrigo. 1965. *Convenzione e ipotesi nella filosofia naturale di Thomas Hobbes*. Florence: La Nuova Italia.

———. 1971. *Introduzione a Hobbes*. Rome/Bari: Laterza.

Paganini, Gianni. 1990. Hobbes, Gassendi e la psicologia del meccanicismo. In *Hobbes Oggi*, ed. Bernard Willms et al., 351–445. Milan: Franco Angeli.

———. 2004b. Hobbes, Gassendi und die Hypothese der Weltvernichtung. In *Konstellationsforschung*, ed. Martin Mulsow and Marcelo Stamm, 258–339. Frankfurt am Mein: Surkhamp.

———. 2006a. Hobbes, Gassendi e l'ipotesi annichilatoria. *Giornale critico della filosofia italiana* 85 (1): 55–81.

———. 2008b. Le néant et le vide: les parcours croisés de Gassendi et Hobbes. In *Gassendi et la modernité*, ed. Sylvie Taussig, 177–214. Turnhout: Brepols.

———. 2010a. Introduzione. In *Moto, luogo e tempo*, ed. Thomas Hobbes, 9–104. Turin: Utet.

———. 2011. Il piacere dell'amicizia: Hobbes, Gassendi e il circolo neo-epicureo dell'Accademia di Montmor. *Rivista di Storia della Filosofia* (1): 23–38.

Popkin, Richard H. 1979. *The History of Scepticism from Erasmus to Spinoza*. Berkeley/Los Angeles/London: University of California Press (or. Ed. 1960).

Prins, Jan. 1990. Hobbes and the School of Padua: Two Incompatible Approaches of Science. *Archiv für Geschichte der Philosophie* 72: 26–46.

———. 1993. Ward's Polemic with Hobbes on the Sources of His Optical Theories. *Revue d'histoire des sciences* 46: 195–224.

———. 1996. Hobbes on Light and Vision. In *The Cambridge Companion to Hobbes*, ed. Tom Sorell, 129–156. Cambridge: Cambridge University Press.

Randall, John H. 1961. *The School of Padua and the Emergence of Modern Science*. Padua: Antenore.

Raylor, Timothy. 2018. *Philosophy, Rhetoric, and Thomas Hobbes*. Oxford: Oxford University Press.

Robertson Croom, George. 1886. *Hobbes*. Edinburgh/London: William Blackwood and Sons. (Reprint: Thoemmes Press, 1993).

Rossi, Mario Manlio. 1942. *Alle fonti del deismo e del materialismo moderno*. Florence: La Nuova Italia.

Schuhmann, Karl. 1998. *Hobbes. Une chronique*. Paris: Vrin.

Shapin, Steven. 1998. *Scientific Revolution*. Chicago: University of Chicago Press. (or. Ed. 1996).

Shapiro, Alan E. 1973. Kinematics Optics: A Study of the Wave Theory of Light in the Seventeenth Century. *Archive for History of Exact Sciences* 11: 134–266.

Skinner, Quentin. 1996. *Reason and Rhetoric in the Philosophy of Hobbes*. Cambridge: Cambridge University Press.

———. 2002. *Visions of Politics*, 3 Vols. Cambridge: Cambridge University Press.

Sommerville, Johann P. 1992. *Thomas Hobbes: Political Ideas in Historical Context*. New York: Plagrave MacMillan.

Stroud, Elaine C. 1983. *Thomas Hobbes' A Minute or First Draught of the Optiques: A Critical Edition*. PhD Dissertation. Madison: University of Winsconsin-Madison.

———. 1990. Light and Vision: Two Complementary Aspects of Optics in Hobbes' Unpublished Manuscript *A Minute or First Draught of the Optiques*. In *Hobbes Oggi*, ed. Bernard Willms et al., 269–277. Milan: Franco Angeli.

Talaska, Richard A. 2013. *The Hardwick Library and Hobbes's Early Intellectual Development*. Charlottesville: Philosophy Documentation Center.

Terrel, Jean. 2008. *Hobbes: vies d'un philosophe*. Rennes: Presses Universitaires de Rennes.

Torrini, Maurizio. 2013. Galilei, Galileo. In *Il Contributo italiano alla storia del pensiero*. Roma: Treccani. http://www.treccani.it/enciclopedia/galileo-galilei_%28Il-Contributo-italiano-alla-storia-del-Pensiero:-Scienze%29/.

Tricaud, François. 1985. Éclaircissements sur les six premières biographies de Hobbes. *Archives de Philosophie* 48: 277–286.

Tuck, Richard. 1989. *Hobbes. A Very Short Introduction*. Oxford: Oxford University Press.

Watkins, John W.N. 1965. *Hobbes's System of Ideas*. London: Hutchinson.

Chapter 1
Hobbes and Mersenne

1.1 The Philosophical Foundation of Science

Hobbes scholars have often remarked on the friendship between Hobbes and Marin Mersenne, sometimes emphasizing his role in promoting *De cive*. Yet little attention has been given to the intellectual relationship that formed between the two and that had growing importance during the most philosophically productive years of Hobbes' life.[1] Studies about Mersenne have tended to emphasize Hobbes' contribution to the development of Mersenne's thinking[2]; but, though there are several

---

[1] See Robertson 1886, 37–38. Fritjof Brandt noted 'that a philosopher with Mersenne's qualities as compiler could have exercised any decisive influence on such fundamental and daring thinker as Hobbes we consider little probable', though admitting Mersenne 'has been able to awaken interests in Hobbes, and certainly his criticism too has acted guidingly, a fact which is indicated by the autobiography'. (Brandt 1928, 160). Moreover, he noted some elements common to both authors, such as their interest in Galileo and speculations in the fields of optics and music (ibid., 154–160). See also Tönnies 1925, 19. Arrigo Pacchi argued that both could be counted, in a sense, as followers of '*constructive scepticism*'. See Pacchi 1965, 63 ff. and 179 ff. Aldo Gargani contrasted two different approaches to scientific knowledge: that of Galileo, Descartes, and Hobbes against that of Mersenne and Gassendi. According to Gargani, for the former group, the mechanism was the tool to reduce the sensible qualities of objects in their mathematically quantifiable objective reality, whereas Mersenne considered it only a descriptive model. In this interpretation 'An operation of ordering phenomena, with sensory appearances within the schemes of mathematic legality, is the limit beyond which man's powers of knowledge can legitimately operate'. Gargani 1971, 176–182).

[2] Robert Lenoble argued that Mersenne considered the Hobbesian system a valid alternative to Cartesian metaphysics. See Lenoble 1943, esp. 560 ff.; On the relationship between Mersenne and Hobbes see also Beaulieu 1990 and 1995, 210–214.

© Springer Nature Switzerland AG 2020
G. Baldin, *Hobbes and Galileo: Method, Matter and the Science of Motion*,
International Archives of the History of Ideas Archives internationales
d'histoire des idées 230, https://doi.org/10.1007/978-3-030-41414-6_1

studies about Hobbes in Mersenne's works,[3] little attention has been given to Mersenne's possible influence on Hobbes.[4]

Yet Hobbes considered his meeting with Mersenne of great import, so much so that he almost considered it the start of his philosophical career. He refers repeatedly in his writings to the discussions that Mersenne held at the Minim monastery at *Place Royale*,[5] and in Hobbes' writings, particularly in the 1640s, we see much influence from these discussions, especially on the topic of the 'return of the bow.'[6]

It is also worth exploring Hobbes' possible interest in the works that Mersenne had already published, or was in the midst of publishing, around 1635, considering the possible converging elements of their philosophies.

The first of Mersenne's many writings that attracts our attention is *La vérité des sciences* from 1625, in which he undertook a polemical, yet constructive, dialogue with radical scepticism, which continued to engage him for his entire life.[7] Throughout his discussion, Mersenne often reiterates his loyalty to the Aristotelian tradition, but nonetheless examines and shows appreciation for the writings of an innovative thinker like Francis Bacon.[8] He shows a particularly keen interest for mathematics, starting in the preface, for the contribution it can make to every branch of knowledge.[9] Mersenne sees knowledge of mathematics as essential for theology, Biblical exegesis, and even for jurisprudence.

---

[3] See Pacchi 1964; Schuhmann 1995a. See also Malcolm, 'Hobbes and Roberval.' In Malcolm 2002, 173 ff.

[4] Many scholars have assumed that Mersenne might have influenced Hobbes' thought, in particular regarding the foundation of epistemology of knowledge, against the sceptics. On this topic see, in particular, Tuck 1988.

[5] Hobbes makes special mention of the meetings in the Minims cloister in the 1640s. See, for example *Six Lessons*, *EW*, VII, 340–343. On Mersenne's circle, see Lenoble 1943, 590 ff. Armogathe 1992, esp. 135 ff.; Taton 1994; Beaulieu 1995, 63 ff. and 176–185; Maury 2003, 154–156; Lewis 2006, 110 ff. See also Jullien 2006. On the French academies in seventeenth century see also Fumaroli 2015.

[6] See Baldin 2016b and Baldin 2017.

[7] Mersenne's relationship with scepticism has been a point of debate, opened by an interesting and still relevant article by Popkin; see Popkin 1957.

[8] See Mersenne 1625, 206 ff. Bacon had been cited in the *Traité de l'harmonie universelle*, where Mersenne mentions the *De sapientia veterum* (see Mersenne 2003, 95) and in *Quaestiones in genesim*. See Buccolini 2014. On the *Verité*, 120 ff.

[9] Mersenne 1625, *Preface* (p. not numbered, fifth page of the preface): '*Je desire seulement pour toute satisfaction qu'un chacun fasse son profit de la verité, laquelle étant venüe de Dieu doit estre rapportée à son honneur, c'est pourquoy je treuve mauvais, de ce qu'il y en a qui ont si peu d'esprit, & de jugement, qu'ils croient que la verité des Mathematiques est inutile, & qu'elle ne peut servir à la pieté, & à la Religion: je m'assure que cette opinion ne vient que de l'ignorance, & qu'ils n'auront pas si tost compris ce que j'ai traité dans cette œuvre, qu'ils avouront librement: qu'il n'y a rien dans ces sciences qui ne soit tres-utile pur l'intelligence de l'écriture sainte, de saincts peres, de la Theologie, de la Philosophie, & de la Jurisprudence, & qui ne nous puisse servir pour nous élever à la connaisance, & à l'amour de Dieu. Car il n'y a point de sciences, apres la Theologie, qui nous proposent, & nous fassent voir tant des merveilles comme font les Mathematiques, lesquelles élevent l'esprit par dessus soy-mesme, & le forcent de reconoistre une*

The branch of mathematics that interests him most is that known as *mixed mathematics*, like hydraulics and pneumatics which take a prominent place in Mersenne's thinking.[10] In Mersenne, physics and metaphysics treat things *'absolûment,'* because 'they teach us the thing as it is in itself.'[11] Logic, instead, is metaphysic's *'cousine germaine'*[12] because 'as the first one has general principles of being and of the essence of creatures, the second has some propositions and discourses that fit to all things.'[13]

At the core of the discussion involving the three participants of Mersenne's dialogue, which is the work's main theme, there is the question of the veracity and reliability of human knowledge. Mersenne highlights the senses here, particularly vision and its science, optics.[14] Optics lets us demonstrate the real causes and specificities of a visual phenomenon, to know that which pertains to light and color, in contrast to the sceptical attitude that says—due to the fallibility of the sense—we cannot base incontrovertible knowledge on it.[15] Mersenne, on the contrary, believed that the natural knowledge of the external world was necessarily built on *'entendement,'* i.e. understanding. This aspect emerges clearly where Mersenne discusses Bacon's philosophy: he considered the *'Chancelier d'Angleterre'* worthy of esteem, because he had made an effort to 'find the means suited to knowing nature and its effects.'[16] Nonetheless, he sees as completely unfeasible Bacon's plan to make the human mind a 'mirror'—that is completely 'similar to the nature of things.'[17]

---

divinité; car la Statique, l'Hydraulique, & la Pneumatique produisent des effets si prodigieux, qu'il semble que les hommes puissent imiter les œuvres les plus admirables de Dieu...'.

[10] Scholars who particularly focused on the mathematics's dominance over physics in Mersenne's intellectual horizon included Dear (see Dear 1988, 48 ff., who maintains that Mersenne cleaved to a certain interpretation of Aristotelism, of which Mersenne's thinking kept the main concepts), and Daniel Garber (see Garber 2004, 148 ff.). Garber argues that the notion of *mixed mathematics* is fundamental to understanding Mersenne's relationship with Galileo and the new science, because—according to Garber—Mersenne considered Galileo a kind of 'mixed mathematicians', rather than a natural philosopher.

[11] Mersenne 1625, 51.

[12] Ibid., 52.

[13] Ibid. On this link between logic and metaphysics and the concept of metaphysics in Mersenne's thinking in general, see Marion 1994, esp. 139 ff.

[14] Mersenne 1625, 135 and 148–149. Optics covers about forty columns in *Quaestiones in Genesim*. See Beaulieu 1995, 28. Mersenne also links this to Claude Mydorge's experiments on parabolic mirrors (see Mersenne 1623, col. 488–538). However, as Lenoble has noted, Mersennes here again refers to light as a visible quality (*Ibid.*, col. 472), and it was not until later works that we would see a significant evolution in Mersenne's optics. See Lenoble 1943, 479. Mersenne was also concerned with establishing what *light* is, defining it as *accidens incorporeum*. (Mersenne 1623, col. 737–739). See also Beaulieu 1982b, 311; Cozzoli 2007, 31–3 and Cozzoli 2010, 11–14.

[15] Mersenne 1625, 148–149.

[16] Ibid., 206.

[17] Mersenne 1625, 212–213: 'Or quelques Phenomenes qu'on puisse proposer dans la Philosophie, il ne faut pas penser que nous puissions penétrer la nature des individus, ni ce qui se passe interieurement dans iceus, car nos sens, sans lesquels l'entendement ne peut rien conoître, ne voy-

Indeed, no matter how we try to reach the inner nature of objects—even through chemical and alchemical processes—we cannot fully control the phenomena that happen inside these bodies. On the contrary, Mersenne maintained that we could produce reasoning in this realm that was exclusively hypothetical in nature, 'as Ptolemy and Copernicus have done.'[18] We must not forget that the senses, by their very nature, do not lead us to the truth, which is the exclusive realm of the intellect:

> ... for even though the senses are the door of objects, they are not the door of conclusions, nor of counsels, which are taken in the secret cabinet of the understanding, which often makes fun of their suggestion, because it has a brighter, and more excellent light, by the means of which he discovers their error, when they are mistaken. For example the eyes are disappointed, when they judge that the straight stick seen in the water is broken, but the reason opposes itself because the light of dioptrics makes it recognize that it is straight.[19]

Mersenne is in no way diminishing the importance of the senses, but the goal of the senses is to be 'the couriers and messengers of reason,' so that reason can achieve certain knowledge 'which could not be called abuse nor deception nor vanity, when all precautions have been taken needed to reach some truth.'[20] The understanding

---

ent que ce qui est exterieur; qu'on anatomise, & qu'on dissolve les corps tant qu'on voudra soit par le feu, par l'eau, ou par la force de l'esprit, jamais nous n'arriverons à ce point que de rendre notre intellect pareil à la nature des choses, c'est pourquoy je croy que le dessein de Verulamius est impossible, & que ces instructions ne seront causes d'autre chose que de quelques nouvelles experiences, lesquelles on pourra facilement expliquer par la Philosophie ordinaire.' Mersenne spent an entire chapter of *Verité* on discussing Bacon's philosophy (chapt. 16), presenting the Baconian division of the *Eidola*, (ibid., 206 ff.). Mersenne's interest in Bacon's philosophy is also seen in his many mentions of his work (including in *Quaestiones in Genesim* and *Cogitata*), and, especially, in the translation that Mersenne did and never published (from 1626 to 1628) of the second century of *Sylva Sylvarum*. See Buccolini 2002. On Mersenne's relationships with Bacon, see also Fattori 2000.

[18] Mersenne 1625, 207. Popkin focused on this aspect, arguing that Mersenne's giving up the aspiration to master the inner nature of things opens a kind of a third way of a propositional nature that diverges from dogmatic realism and from radical scepticism alike. See Popkin 1957. Interesting observations by Crombie join Popkin's article. He argued that as Mersenne reflected on the sceptics, he gave up scientific realism and developed instead (especially after 1634) a complex dialectic between the theoretical element and experimental observation. See Crombie, 'Marin Mersenne (1588–1648) and the Seventeenth-Century Problem of Scientific Acceptability' (1975), now in Crombie 1990, 399–417, 402. However, on the concept of *hypotheses* in modern natural philosophy, it is key to keep in mind Sophie Roux's critical comments on Popkin's essay. See Roux 1998. Roux notes out that even with those writers often considered as 'dogmatic', such as Descartes, hypotheses have a fundamental role in the context of physics.

[19] Mersenne 1625, 221–222 (my trans.): '...car encore que les sens soient la porte des obiects, ils ne sont pas la porte des conclusions, ni des conseils, qui se prennent dans le cabinet secret de l'entendement, lequel se moque souvent de leur suggestion, parce qu'il a une plus vive, & plus excellente lumiere, par le moyen de laquelle il découvre leur erreur, quand ils se trompez: par exemple les yeus sont deçeus, quand ils jugent que le bâton droit veu dans l'eau est rompu, mais la raison s'y oppose, parce que la lumiere de la Dioptrique luy fait reconoistre qu'il est droit.'

[20] Ibid., 222.

(*Entendement*) indeed 'judges eventually, so that it recognizes, checks, and corrects errors and abuses that may arise from the senses' indisposition or inability.'[21]

Yet, Mersenne, the author of *La verité*, continuously stresses the need to constantly use both factors: *experience* and *ratio*, offering up an image of an indissoluble union of these two principles:

> ... it is necessary to consult the experience, in order to conjoin it with the reason, to avoid being disappointed by the imaginations of our mind, when the experience is missing. But when [the experience] is connected with [the reason], we no longer need to fear giving our consent to the truth. It is no longer necessary to say ἐπέχω, we must receive the truth in our understanding, as an ornament, and the greatest treasure it can receive. Otherwise it [our understanding] would be in perpetual darkness, and would have no consolation.[22]

Only in cases in which the phenomena are outside of experience are we required to rely on reason alone. On the contrary, when we have the chance to apply empirical and rational elements together, we are in a state of being able to receive the truth in our intellect.

But to fully understand the role reason plays in Mersenne's thinking, we cannot but look to mathematics. After his first book, introductory in nature, Mersenne went on to discuss mathematics specifically, especially arithmetic: he maintains that 'it serves as a general tool to explain all other sciences.'[23] The tools that serve to acquire new knowledge are those inherited from Aristotelian tradition, especially the *syllogism*[24] and the joint application of *analytical* and *synthetical* methods.[25]

As Buccolini emphasized, mathematics and geometry are inextricably bound to syllogistic reasoning in *La Verité des Sciences*.[26] Mersenne insists repeatedly on the need to follow the certainty of the Euclidean doctrine to provide a certain, incontrovertible basis to science that can dispel sceptical doubt and the *Isosthenia* of Pyrrhonism. Yet, though Euclidean geometry is the paradigm of certainty and evidence, it is based on modules of syllogistic demonstration, as emerges clearly from Chapter 4 of *La verité*, where he seeks to paraphrase some Euclidean demonstrations, actually applying a syllogistic process based on first figure syllogisms.[27]

---

[21] Ibid., 192. On the role of *entendement* in *La Verité,* in relationship to scepticism, see Paganini 2008, 144 ff.

[22] Mersenne 1625, 220 (my trans.): '...il faut consulter l'experience, afin de la conjoindre avec la raison, depeur que nous soions deceus par les imaginations de notre esprit, quand l'experience nous manque: mais quand l'un[e] est conjointe avec l'autre, il ne faut plus craindre de donner son consentement en faveur de la verité: il ne faut plus dire ἐπέχω, il faut recevoir la verité dans notre entendement, comme l'ornement, & le plus grand thresor, qu'il puisse recevoir, autrement il sera en des tenebres perpetuelles, & n'aura aucune consolation.'

[23] Ibid., 224. On the importance of mathematical and geometric elements regarding creation and universal harmony, see Fabbri 2003, 142 ff.

[24] Mersenne 1625, 199 ff. Popkin had already discussed the importance that Mersenne gave to syllogism (see Popkin 1957, 68), though it was Buccolini in particular who emphasized the centrality of syllogistic reasoning in *La Verité*. See Buccolini 1997.

[25] Mersenne 1625, 203.

[26] See Buccolini 1997, 10–11.

[27] See Mersenne 1625, 722–724. Cf. Clavius 1591, 20. *Scholivm*: '*Vt autem videas, plurès demonstrationes in vna propositione contineri, placuit primam hanc propositionem resolvere in prima*

Mersenne establishes a solid connection between syllogistic demonstration and Euclidean geometry, to the extent that, in the Mersennian perspective, the geometric proposition necessarily takes on the structure of the syllogism.[28] However, to understand the importance he gave to the geometric method, we should further investigate his interest in mathematics,[29] which emerges from the second book of *La Verité*.[30] He particularly insists on the idea of '*quantité intelligible*,' and an attempt emerges to apply the notions taken from mathematics and geometry to the realm of physics. For instance, Mersenne maintains that individual entities can be found in the mathematical figures that fill the physical reality around us, such as the cylindrical figure seen in trees, and that '*entendement*' extrapolates these figures from the phenomenal world 'by abstraction.'[31]

Nevertheless, mathematical disciplines' abstract and purely rational dimension is what sets them apart from physics, which cannot entirely escape the indeterminacy of empirical realities perceived through the senses.[32]

---

sua principia, initio facto ab vltimo syllogismo demonstrativo. Si quis igitur probare velit, triangulum A B C, constructum methodo praedicto, esse aequilaterum, utetur hoc syllogismo demonstrante...'. On the role given mathematics in Clavius and other sixteenth-century authors, such as Piccolomini, Barozzi, and Mazzoni, see Galluzzi 1973, esp. 53–54. See also Crapulli 1969. On the next century, Sergio 2006.

[28] See Clavius 1591, *Prolegomena*, not num.: '*Theorema autem appellant eam demostrationem, quae solùm passionem aliquam proprietatemve unius, vel plurium simul quantitatum perscrutatur. Vt si quis optet demonstrare, in omni triangulo tres angulos esse aequales duobus rectis, vocabunt talem demonstrationem Theorema, quia non iubet, aut docet triangulum, aut quippiam aliud construere, sed contemplatur tantummodo trianguli cuiuslibet constituti passionem hanc, quodanguli illius duobus sint rectis aequales. Unde a contemplatione ipsa, haec demonstratio theorema dicitur. In theoremate fieri nulla ratione potest, contradictionis utraque pars vera ut sit. Si enim quis demonstret, omnes angulos trianguli cuiuslibet duos esse rectis angulis aequales, nullo poterit modo fieri, ut inaequales quoque sint duobus rectis.*'

[29] On the conception of mathematics as a certain science in Mersenne's thought, see Lenoble 1943, 33 ff.

[30] Mersenne 1625, 225–226: 'Apres avoir discouru des sciences en general, & apres avoir montré que nous ne devons pas suspendre notre jugement à tout propos, ni sur toutes choses, ie veus maintenant vous faire voir que les mathematiques sont des sciences tres-certaines, & tres-veritables, es quelles la suspension ne treuue pas lieu: or auant que de vous apporter le demonstrations desquelles elles se servent, il faut que vous sçachiez qu'elles ont la quantité intelligible pour leur obiect, car elles ne considerent point le sensible que pour accident, & ce pour nous faire tumber en quelque façon sous les sens ce qui est releué par dessus l'incertitude de la matiere'.

[31] Buccolini has correctly pointed out a clear comparison between this passage and a corresponding one in the work of Giuseppe Biancani in *De mathematicarum natura dissertatio* (1615). See Buccolini 1997, 10–11.

[32] Mersenne 1625, 226–227: 'La Physiuc traite aussi de la quantité, mais entant qu'elle est sensible, & que ses proprietéz se peuvent cognoistre par quelque sorte de mouvement, selon lequel elle est sujette à divers changemens; mais la quantité Mathematique est invariable, car il ne peut se faire qu'un triangle ne sois compris par trois lignes, & par trois angles conjoints par trois points indivisibles: n'importe qu'il ni ayt aucun triangle parfait au monde, il suffit qu'il puisse estre pour établir la verité de cette science, & que la nature nous represente dans ses individus sensibles les figures de Mathematique le plus parfaictement qu'elle peut, comme la ronde dans le cieus, dans les

This makes mathematics the model of science that frees itself from any sceptical doubt. Its epistemological status lends certainty and the value of truth: addressing abstract entities, or entities of reason, it is applied to *invariable* quantities.[33] Mersenne seems to return to a theme whose roots are in late-medieval thought (such as in Roger Bacon, Robert Grosseteste, and William of Ockham[34]) and developed in depth by late sixteenth-century writers[35]: mathematics and geometry[36] produce *a priori* certain knowledge because they have a completely conventional constructive character. Starting from the first principles, established by convention, they build an internal logical system, which develops deductively, with inescapable necessity, starting from the same first principles.[37] Yet Mersenne does not consider the difference in epistemological status between the mathematical disciplines and physics as undermining or devaluing the latter over the former. On the contrary, Mersenne says that, though physics seems the 'most dubious,' we nonetheless cannot doubt the existence of physical entities, bodies, and physics pertains to the 'quantities, causes, and a thousand other things' that concern these objects. Mersenne essentially seeks to provide a foundation for physics as a science—in answer to the problems that scepticism raises—while maintaining its probabilistic nature and keeping the distance that separates it from the formal and deductive certainty of mathematics.

---

astres, & dans les élemens, sans mettre en ligne de conte tout ce qui est produit par les élemens ayant la figure Spherique, ni tous les artifices qui ont été inventé par les Mechaniques'.

[33] Popkin, in particular, stressed mathematics' character of conventional certainty in opposition to the hypothetical nature of physics in Mersennian thought. See Popkin 1979, 138.

[34] According to Grosseteste, physics is within the limits of the probable, whereas mathematics achieves certain, rigorous knowledge. Roger Bacon considered the primary knowledge from which all other knowledge came (see Bacon 1897–1900, pars IV, II, 1, Vol. 1, 109–111). However, it was William of Ockham who gave a great deal of attention to physical science, though he did consider it in probabilistic terms, based on restoring the validity and generality of scientific statements, adopting a principle of the uniformity of nature and the analysis of the linguistic matrix of scientific discourse, sometimes in terms consistent with the positions that Thomas Hobbes would take, as Aldo Gargani has noted. See Gargani 1971, 151 ff. See also Crombie 1953, 219 ff.

[35] Buccolini has correctly noted that Mersenne's thought follows in the tracks of speculations around *Quaestio de certidudine mathematicarum*, a considerable presence in the sixteenth-century Paduan philosophical milieu as well. Some authors have highlighted the prospect of establishing the value of mathematical demonstration specifically on their being able to be tied to the form of the scientific syllogism, which was considered a *demonstratio potissima*. See Buccolini 1997, 13. The author is mainly referring to *Opusculum de certitudine mathematicarum* (1560) by Francesco Barozzi and also noted Mersenne's privileged source in *De mathematicarum natura dissertatio* by Biancani.

[36] Mersenne discusses geometry in the work's fourth book and introduces the topic, restating that the validity of geometric demonstration is attested to by the fact that after almost 2000 years, the propositions in Euclid's *Elementa* still remain universally valid. See Mersenne 1625, 717 ff.

[37] Nonetheless, according to Buccolini, though 'the mathematical method is hypothetical, based on suppositions, on hypotheses that are the intellect's constructions, mathematics is applied to physics because it is in accordance with experience, and, indeed, learning mathematics comes from experience itself'. Buccolini 2015, 73. Buccolini also emphasizes some parallels between *Objectiones mersenniane e quelle hobbesiane alle Meditazioni* by Descartes (ibid., 61–62). The topic needs specific detailed analysis beyond the scope of this study.

Citing Bacon, Mersenne had chosen an anti-essentialist epistemological perspective, a choice which set him apart from the Aristotelian perspective. He insists on the idea of *intelligible quantity* and says that it is not necessary to delve into the inner essences of substances to establish a science. On the contrary, he believes that *external accidents of objects* are absolutely sufficient as a basis for scientific knowledge. From this perspective, the concept of *'ressemblance,'* takes on fundamental importance, as it emerges from the comparison with purely phenomenal data.[38]

Though the sceptic in Mersenne's text uses the Aristotelian adage: *'nihil in intellectu, qu'in prius in sensu fuerit'*[39] to infer that there could be no certain knowledge because the animal senses provided various and sometimes conflicting data[40]; the Christian philosopher—along with Father Mersenne whose opinions the *'philosophe'* took on—certainly does not stop at the limit set by the heterogeneous nature of sensory perception. Mersenne addresses the criteria at the basis of scepticism, clearly affirming that reason lets us overcome this obstacle. Optics plays a fundamental role in Mersenne's argument:

> ... Optics, in which we will make it appear that we know what belongs to colors, and to light. I will only say here that we do not rely on a single sense, nor even on all the senses taken together, because we use reason, which leaves nothing inexamined. It is not true that understanding knows only that which enters through the external senses, because it knows that air exists, and a thousand other things than the senses cannot perceive ....[41]

The issue of the sciences' epistemological status is addressed at length in all of Mersenne's subsequent works, including *Questions Inouyes* (1634), where he maintains that we do not know 'absolutely' much of natural phenomena.[42] Here he seems to make a concession to the sceptics: we do not have a certain demonstration of these phenomena,[43] because physics is not a certain science in terms of the object of its knowledge.[44] Indeed, though in *La Verité* he linked metaphysics to physics

---

[38] On this topic, see Buccolini 1997, 28 ff. Buccolini also pointed at the consistency between the concept of *ressemblance* and Gassendi's *similitudo* (ibid., 29).

[39] See Mersenne 1625, 149. See Aristotle, *De anima*, III (Γ) 8, 432a. For the Latin expression: *'Nihil est in intellectu quod prius non fuerit in sensu',* see Thomas Aquinas, *Quaestiones disputatae de veritate*, q. 2, art. 3, 19.

[40] Mersenne 1625, 141 ff.

[41] Ibid., 148–149 (my trans.): '...l'Optique, dans laquelle nous ferons paroistre que nous sçavons asseurément ce qui appartiens aus couleurs, & à la lumiere. Je diray seulement ici que nous ne nous fions pas a un seul sens, ni même à tous le sens pris ensemble, car nous nous servons de la raison, qui ne laisse rien qu'elle n'examine: or il n'est pas veritable que l'entendement ne comprenne rien que ce qui entre par les sens exterieurs, car il cognoît qu'il y a de l'air, & milles autres choses que le sens exterieurs ne sçauroient appercevoir.' In his analysis of Mersenne's optics, Cozzoli neglects *La verité des sciences*. See Cozzoli 2007, 2010.

[42] *Questions inouyes*, Question XVIII, now in Mersenne 1985, 53.

[43] Ibid.

[44] Regarding mathematics, Mersenne writes that '...si on leur oste la posibilité de la quantité, il semble qu'on leur oste le fondement, sur lequel elles establissent leurs demonstrations, et qu'elles ne peuvent tout au plus user que de la moindre demonstration que l'on appelle *à posteriori*; quoy que l'on puisse dire qu'il n'est pas necessaire que leur suject, ou leur object soit possible, d'autant

because they treated things '*absolument*,'[45] here metaphysics is instead equated with mathematics.[46] Mathematical sciences address 'the absolute or conditioned possible' and are, as such, sciences founded on imagination or on 'pure intelligence.' Mersenne notes that, precisely because of their aprioristic conventionalism, some prefer mathematics to physics, because of its 'certainty.' Nonetheless, once again, this idea in no way devalued physics. Indeed, if this science could show the same degree of certainty as mathematics, it would be enormously more fruitful and profitable for humanity.[47] Mersenne also wrote a few words about chemistry, as he had in *La Verité*: chemistry's positive side is in its experimental predisposition, though the speculations of chemists are so lacking in theory that they can be fairly called 'children of their discipline' (*enfans de la doctrine*).[48]

But in *Questions*, Mersenne seems to expand the hypothetical and probabilistic scope of physical inquiries, but still the attention he pays to Galileo[49] and the great effort he made to translate some of Galileo's works[50] lead him to notice the close relationship established between empirical observation and demonstrative reason.[51] Of course, the synthesis of *sensible experiences* and *necessary demonstrations* is

---

qu'elles peuvent proceder conditionnellement, et conclure absolument: par exemple, encore qu'il n'y eust point de quantité possibles, les Mathematiciens peuvent dire, s'il estoit possible de faire un triangle rectangle, c'est chose asseurée que l'hypotenuse ou la soustendante de l'angle droit seroit un quarré égal aux quarrez des deux autres costez: de là vient que l'on peut dire que la pure Mathematique est une science de l'immagination, ou de la pure intelligence, comme la Metaphysique, qui ne se soucie pas d'outre object que du possible absolu, ou conditionné'. (Ibid., 54).

[45] Mersenne 1625, 51.

[46] Carraud pointed out this shift, as he considered the close relationship between mathematics and metaphysics in Mersennian thought. See Carraud 1994, esp. 156–157.

[47] *Questions Inouyes*, Question XIX. In Mersenne 1985, 56. The same argument extends to morality.

[48] Question XXVIII. In Mersenne 1985, 77–78. On Mersenne's relationship with chemistry, see Beaulieu 1993; Clericuzio 2000, 47–50. Mersenne's Alchemist is clearly a chemist and alchemist at once. It was only during the seventeenth centuries, with the development of iatromechanics, that chemistry began to distinguish itself from alchemy. See Califano 2010, I, 65–69.

[49] On Mersenne's interest in Galileo, see Lenoble 1943, 39 ff. and 391 ff.; Beaulieu 1984, 1995, 106–117. On the reception of Copernicanism in Mersenne, see Hine 1973, who argued that though it had garnered Mersenne's favor, he would have considered it no more than a mere mathematical hypothesis. A different interpretation was given by Pierre Costabel, who stressed the correspondence with Jean Rey (1631–1632), where Mersenne focused at length on the defense of the Copernican hypotheses. See Costabel 1994. The issue of the impact of the Galileo *affaire* in seventeenth-century France has been extensively discussed. See Beaulieu 1984 and esp. Lerner 2001; Fabbri 2005; Lewis 2006; Pantin 2011. On Mersenne's general relationships with Italy and Galileo's pupils, see Beaulieu 1986, Beaulieu 1989, 1992.

[50] As is well known, Mersenne translated *Le Mechaniche* and wrote a reworked summary of *Discorsi* (Mersenne 1639). See Lewis 2006, 135 ff. On Mersenne's experience as Galileo's translator and interpretor, see Shea 1977, and Raphael 2008.

[51] Crombie, in particular, focused on this topic, see Crombie, 'Marin Mersenne (1588–1648) and the Seventeenth-Century Problem of Scientific Acceptability'. In Crombie 1990, 402–405.

one of the crucial points of Galilean scientific-philosophical thought,[52] and Mersenne
understands the problematic nature of this aspect. Certain passages of Galileo's
works lead Mersenne to doubt that Galileo actually made all of the empirical observations he describes,[53] prompting him to repeat these experiments to test their actual
validity (the first was the observation of the sphere rolling down an inclined plane[54]).
Experiments came to have a prominent place in Mersenne's scientific perspective[55]
and he showed the public some results of his experimental observations in *Harmonie
universelle* from 1636 to 1637.[56]

However, before moving on to *Harmonie*, it is interesting to look at a very short
treatise about motion, from 1634, in which Mersenne meticulously subjects to
empirical verification the thoughts about the equations of falling bodies that Galileo
had published in his *Dialogue concerning the two chief world systems*. Though the
text of this *Traité des mouvemens*[57] ends with the idea that natural philosophers
must settle for 'explaining the phenomena of nature because the human mind is not
capable of owning the causes and principles,'[58] in the preface, Mersenne explains
his intent to verify Galileo's speculations and the reasons he puts forward are interesting. Mersenne writes that '*true philosophy*' must have '*experience* and *reason* as
its foundation,'[59] meaning that the two factors are part of an inseparable pair.
Furthermore—in thoroughly Galilean spirit—he says that he submitted the data

---

[52] We will return to this topic. See *below*, Chap. 2, §§ 8 ff.

[53] Mersenne 1636a, Livre II (Du Mouvement des Corps), 112, Corollaire I: 'Je doute que le sieur
Galilee ayt fait les experiences des cheutes sur le plan, puis qu'il n'en parle nullement, & que la
proportion qu'il donne contredit souvent l'experience: & desire que plusieurs esprouvent la mesme
chose sur des plans differens avec toutes les precautions dont ils pourront s'aviser, afin qu'il voyent
si leurs experiences respondront aux nostres, & si l'on en pourra tirer assez de lumiere pour faire
un Theoreme en faveur de la vitesse de ces cheutes obliques, dont les vitesses purroient estre mesu-
rees par les differens effets du poids, qui frappera dautant plus fort que le plan sera moins incliné
sur l'horizon, & qu'il approchera davantage de la ligne perpendiculaire.'

[54] See Raphael 2008.

[55] On the experimental element, which is far from devoid of theory in Mersenne, see Crombie's
statement: 'His insistence on the careful specification of experimental procedures, use of controls,
repetition of experiments as well as those calculated from theory, recognition and approximations,
and estimation of experimental errors marked a notable step in experimental method'. 'Marin
Mersenne (1588–1648) and the Seventeenth-Century Problem of Scientific Acceptability.' In
Crombie 1990, 407–408. On Mersenne's relationship with the culture of experimentation, see also
Beaulieu 1995, 265 ff.; Blay 1994; Nardi 1994, and Maury 2003, 211 ff. On the insistence on
empirical confirmation in Galilelian observations, see Lewis 2006, 119 ff. Raphael argues that in
translating Galileo's *Discorsi*, Mersenne identified two different types of 'experiences': universal
and individual (or singular), see Raphael 2008, 32. However, Raphael argues that Mersenne did not
consider the experimental element within the horizon of the scientific method (ibid., 35–36).

[56] The first volume of *Harmonie Universelle* appeared in 1636, precisely when Hobbes was in
Paris.

[57] The text is dated November 23, 1633. It was published the following year. On the importance of
this treatise see Lewis 2006, 122 ff. On evaluating the correctness of the Galilean law of falling
bodies, see Palmerino 1999, 269–328.

[58] Mersenne 1634, 58.

[59] Ibid., *Preface et advertissement au lecteur*, 31 (my italics).

provided by Galileo to empirical evidence, 'so that philosophers who prefer truth to authority can establish some certain fundamental in Physics.'[60]

This interest in experiments translates into a new attention to the relationship between experience and reason. While in *La verité des sciences*, Mersenne maintained that it was up to reason to correct the errors of the senses, in some passages of *Harmonie universelle*, he seems to minimize reason's ability to solve these issues. Though he maintains that 'experience cannot generate a science,' he adds that 'we must not trust only reason too much because it does not always correspond to the truth of appearances, from which it is often far.'[61] Considering the earth's rotation and the force that the earth would have to have to propel rocks as it moves, Mersenne considers it necessary to make use of experience to investigate the nature of these motions: 'As the reasons that we imagine are quite solid often deceive in Physics, as I have shown at several points in this work, I come to experience in order to note the manner that nature acts in these motions, and to draw conclusions from it.'[62] At times, Mersenne even seems to put experience before rational demonstration:

> … I am far from wanting to demonstrate all that I prove by experience, which will be followed by all those who will do it, because understanding must be convinced by evident reason in order to accept a demonstration. I want to say this once and for all, so that we do not believe that I always use the term to demonstrate, or demonstration, in a Mathematical sense. This is easily recognized by those who know the difficulty of demonstrating anything in physics, in which it is very difficult to lay down other maxims more advantageous than well-regulated and well-made experiments ….[63]

These observations lead him to say that, in the realm of physics, it is improper to talk of *principles* and *truth*, because we only truly know what we are able to realize '*de la main, ou de l'esprit*,' which is to say that we only know the foundations of those doctrines that are fully within our power.[64]

[60] Ibid., 32.

[61] Mersenne 1636a, Livre II (*Du mouvement des corps*), 112, Corollaire II. The need that Mersenne saw for empirical verification in physics has been emphasized by Auger 1948, esp. 33–35.

[62] Mersenne 1636a, Livre II, 149–150.

[63] Ibid., Livre III: *Des mouvemens & du son des chordes*, 167 (my trans.): '… je suis bien esloigné de vouloir demonstrer tout ce que je prouve par l'experience, qui sera suivie de tous ceux qui la feront, parce qu'il faut convaincre l'entendement par la raison evidente pour la contraindre d'embrasser une demonstration: ce que je desire que l'on remarque une fois poure toutes, afin que l'on ne croye pas que j'use tousjours de la diction *demonstrer*, ou *demonstration* dans un sens Mathematique; ce que ceux-là concluront aysément qui sçavent la difficulté qui se rencontre à demonstrer aucune chose dans la Physique, dans laquelle il est tres-difficile de poser d'autres maximes plus avantageuses que les experiences bien reglées et bien faites …'

[64] Mersenne 1637, *Nouvelles observations physiques et mathématiques*, 8: 'Il est difficile de rencontrer de principes ou des veritez dans la Physique, dont l'objet appartenant aux choses que Dieu a crées, il ne faut pas s'estonner si nous n'en pouvons trouver les vrayes raisons, & la maniere dont elles agissent, & patissent, puis que nous ne sçavons le vrayes raisons que des choses que nous pouvons faire de la main, ou de l'esprit; & que de toutes les autres choses que Dieu a faites, nous n'en pouvons faire aucune, quelque subtilité & effort que nous y apportions, joint qu'il les a pû autrement faire.' See also Lenoble 1943, 384.

Mersenne's thinking about Galileo's works and, generally, the matter of experiments, led him to develop an epistemological solution that at first glance seems paradoxical: on one hand, he underscores the dimension of experience and empirical observation in the realm of physics, but on the other, he stays true to his theory's probabilistic elements. He reiterates that the true causes and principles in this discipline are not known to those who investigate but only to the creator of nature.[65] This philosophical orientation, rooted in late-medieval scholastic tradition, reinterpreted in modern terms, led one of the leading scholars of Mersennian thought to term his position an '*artificialisme mécaniste*.'[66]

1.2 Certainty and Probability: Mathematical Truth and Physical Hypotheses

In his *Harmonie universelle*, Mersenne argues that the term knowledge is decidedly ill-suited to the realm of physics and that we can only say that we truly have complete knowledge of those disciplines whose foundations we fully understand. Gassendi expressed the same idea,[67] as did Hobbes, who in *De motu, loco et tempore*, forges his distinction between *compositive* and *resolutive* methods based on the dichotomy between the certain knowledge of mathematical science and the probabilistic knowledge of physics.[68]

Regarding the origin and source of all human knowledge, Hobbes argues, starting in the *Elements*,[69] and then in *De motu, loco et tempore,* that it is 'sufficiently accepted and familiar that there is nothing in the human intellect that was not previously in the sense'[70] and his position echoes the words of Mersenne's sceptic in *La verité*.[71] Mersenne, for his part, did not directly express himself on this assumption, but still argued that the senses were reason's 'couriers' and 'messengers.' Nonetheless, he also argued the falseness of the assumption that 'the intellect does

---

[65] Hobbes agreed with this idea, but it was actually common in the earlier philosophical tradition, in particular in the late Paduan Aristotelism in the seventeeth century.

[66] Lenoble 1943, 386. In describing the Mersennian position, Lenoble argues that: 'Cette idée que nos systèmes sont seulement des modèles possibles des choses, n'était certes pas inconnue des Scolastiques. Ce qu'il y a de plus nouveau dans le cas de Mersenne, c'est bien la valeur même attribuée à la science pragmatiste: elle se suffit désormais à elle-même, sans regrets bien vifs d'une science plus parfaite.'

[67] See *Syntagma*. In Gassendi 1658, 95a–95b: 'Philosophamur de iis instar quarum ipsi Authores sumus.' See also: ibid., 122b–123b. On this topic, see Bloch 1971, 150–151; Gregory 2000, 157–189; 174. On Gassendi's epistemology see Fisher 2005, 89 ff. On the role of mathematics in Gassendi's thought, see also Rochot 1957.

[68] See *MLT*, XXX, 10, 352–353.

[69] See *EL*, Part I, chap. II, § 2, 3: 'Originally all conceptions proceed from the action of the thing itself, whereof it is the conception'.

[70] *MLT*, XXX, 3, 349; Eng. Trans. 364.

[71] Mersenne 1625, 149.

not know but that which comes through the external senses,' because 'it knows that air exists and a thousand other things that the senses cannot perceive.'[72] As such, among the things that Mersenne considered inaccessible to sense was air, and he argued that when the senses deceive us, reason intervenes to correct the error, such as when a stick appears broken or crooked when immersed in water. In this case, 'reason opposes itself because the light of dioptrics makes it recognize that it is straight.'[73] In Mersenne's view, optics came to play a fundamental role,[74] (as it did in Descartes[75] and Hobbes[76]). Though Mersenne did not specifically discuss the status of this science, we can consider it a kind of bridge between the aprioristic and axiomatic deductive knowledge of mathematics and the multi-form, varied reality of empirical phenomena. In this context, mathematics is more than the model of certain, formal knowledge that we have mentioned several times, because Mersenne wants to see the mathematical method applied to other disciplines as well.

The extension of the geometric model to every field of knowledge is a clear element both in *De cive*,[77] and in *De motu loco et tempore*.[78] According to Hobbes, however, in the context of physics, only knowledge that is hypothetical and probable can be gained,[79] because the true causes of phenomena—being connected to the motion of the invisible, corpuscular particles that make up reality—are beyond our sensory perception and, consequentially, inaccessible to any empirical verification. Hobbes argues that natural philosophy pertains to indemonstrable things and he explains why natural phenomena escape certain demonstration: they depend on the motion of particles that are 'so subtile as they are invisible' as in the case of '*ayre*' and '*spirits*.'[80]

An interesting analogy emerges when comparing Mersenne's and Hobbes' position: both emphasize the problematic nature of investigating physical phenomena such as *air*, which mostly escape empirical observation, and, accordingly, the best method to use should be founded on demonstrative, deductive, and aprioristic reasoning.[81]

---

[72] Ibid.

[73] Ibid., 222.

[74] Ibid., 148.

[75] The classical opinion that vision is the noblest of the senses is shared not only by Descartes (see *La Dioptrique*, *AT*, VI, 81), but is found often throughout the history of philosophy as well as in Plato (*Timaeus*, 45b–47b) and Aristotle (*Metaphysics*, I, I, 980a)

[76] On the importance of optics in the origins of the Hobbes' mechanism see Giudice 1999, 141–144 and esp. Giudice 2015.

[77] See *De cive*, *Epistle dedicatory*, *OL*, II, 136–137.

[78] See *MLT*, Chap. 23, § 1, 269–270.

[79] See Hobbes to Sir William Cavendish, Earl of Newcastle, from Paris, July 29/Aug. 8 1636, *CH*, I, 33.

[80] Ibid.

[81] This argument, found in Mersenne and Hobbes, seems to confirm Laudan's (odd) hypothesis, according to which the 'hypotheticsm' of certain philosophies of the modern era are the result of a corpuscolar concept of matter, which would be tied to impossibility of verifying these micropar-

Arrigo Pacchi has particularly emphasized the probabilistic elements of Hobbes' natural philosophy. He considered Hobbes a member of the current that Richard Popkin termed *constructive or mitigated scepticism*.[82] Though Pacchi highlighted an important aspect of Hobbes' thought, we should avoid overestimating the extent of Hobbes' *conventionalism*, and we should consider the epistemological status of those hypotheses which, according to Hobbes, develop within the domain of natural philosophy.[83]

Starting in *Elements*, he clearly states that sense 'is by sense to be corrected,'[84] and he considered it indisputable that, as philosophical reasoning has a logical-deductive structure, the information provided by the senses can only be corrected through empirical observation. Probing Hobbes' early scientific thinking lets us consider the problem further and we see additional similarities with some of Mersenne's positions. Hobbes also stressed a fundamental difference between the sciences whose principles are within our power and those concerning the world of 'natural things.'[85] Using terms with 'Duhemian' overtones,[86] Hobbes argues that following the causal chain that links every natural phenomenon to an efficient cause, we could *save the phenomena*, identifying a possible cause for every effect produced (though it was impossible to arrive at any first cause other than attributing the first motion to divine intervention). On the contrary, in sciences in which the starting principles and definitions depend directly on a 'pact and consent among our selves,'[87] we can arrive at first principles that are decided precisely by convention.

He also expressed this idea at the start of *Tractatus opticus II* (late 1640–1642[88]), where we find an exact two-way split between the sciences: mathematical and

---

ticles' behavior. See Laudan 1981, 44. See on the topic, Sophie Roux's critical doubts in Roux 1998, 16 ff.

[82] See Pacchi 1965, 10. As we have noted, Popkin attributed *mitigated scepticism* to Mersenne and Gassendi (Popkin 1979, 148). Lenoble 1943 and Beaulieu 1995, in contrast, considered Mersenne a champion of mechanism and a fervent supporter of modern science. Dear 1988; Garber 2004, 2013, 15–16 argue that Mersenne had, on the whole, kept true to Aristotelism. It is, however, significant that Popkin's position presented both elements at once that appear dichotomous in other interpretations. For a comprehensive overview of Mersenne's relationship with scepticism see Paganini 2008, 129–147.

[83] This topic will be discussed in depth below. See *below*, chap. II, § 8.

[84] Hobbes, *EL*, Part I, chap. II, § 9, 7.

[85] Hobbes to Sir Charles Cavendish, from Paris, [Jan 29.] Feb. 8. 1641, *CH*, I, Letter 31, 83: 'And if a man could make an Hypothesis to salue that contraction of yᵉ sun yet such is the nature of naturall thinges, as a cause may be againe demanded of such Hypothesis, and neuer should one come to an end, wᵗʰout assigning the Immediate hand of God. Whereas in mathematicall sciences wee come at last to a definition wᶜʰ is a beginning or Principle, made true by a pact and consent amongst our selues'.

[86] On the image of scientific theories as mere hypotheses that serve to 'save the phenomena,' see Duhem 2003, esp. 120 ff.

[87] Hobbes to Sir Charles Cavendish, from Paris, 29 Jan. [8 Feb.] 1641, *CH*, I, 83.

[88] Malcolm had originally dated the text to early 1640 (suggesting it as the latest date by which Hobbes left his native land in November 1640). See Malcolm, 'General Introduction', in *CH*, I, lii–lv. More recently, Raylor and Malcolm himself have suggested December 1640 as the earliest

physical.[89] The treatise's incipit is the first clear, precise exposition that Hobbes made regarding the methodological differences between mathematical speculations and physical inquiries. Knowledge gained in the realm of mathematical sciences develops from several first principles that are the vocabulary definitions (*definitiones vocabolorum*) and these *primae veritates* are held true based on an agreement made among us (the intellectual or human community).

In this passage, we see emerging Hobbes' constructivist and conventionalist view of mathematical and geometrical sciences: conclusions we reach in the field of mathematics are absolutely certain and necessarily based on the conventional character of this specific discipline, organized in a deductive, *a priori* logical system.[90] Yet, when describing natural phenomena, we need a different method, or in Hobbes' terms, we must use another kind of principles. In the context of natural philosophy, the reasoning we can produce is exclusively hypothetical (*Hypothesis sive suppositio*).

This hypotheticalist position seems similar to the one his colleague and friend Pierre Gassendi took in the *Syntagma*.[91] When we investigate the nature of a physical event that is within sensory perception, we assume the existence of motions that are the efficient cause of that particular phenomenon. This notwithstanding, it is not completely impossible that heterogeneous motions produce a very similar effect: if we hypothesize that a specific event is determined by a particular motion, and it happens that the hypothesis is wrong in this case, a demonstration is produced that is perfectly correct in formal terms, but that starts from completely inaccurate principles.[92]

But nothing more is asked of the physicist other than to formulate plausible hypotheses (*Amplius igitur a Physico non exigitur, quam ut quos supponit vel fingit motus, sint imaginabiles*) and that, through these hypotheses, we can demonstrate

---

date and April 1643 (with August 1642 most likely) as the latest date that the manuscript was prepared by a Parisian copyist. See Raylor 2005, and Malcolm 2005.

[89] See *TO II*, chap. I, § 1, f. 193 *r*/147.

[90] The division of sciences into certain and probable ones can also be found in *Six Lessons to the Professors of the Mathematics*, EW, VII, 183–184. On this topic, see esp. Giorello 1990. see also Médina 2013a, 86–87. On the concept of mathematics in Hobbes, in general, see Breidert 1979; Grant 1996a, b, and especially Jesseph 1999. See also Sergio 2001, 87–226, and Sergio 2006, 207–254.

[91] *Syntagma*. In Gassendi 1658, I, 286a–286b: 'Si dum aliqua causa invenitur, & tanquam vera, germanaque effertur, ac divenditur, quae verisimilis dumtaxat, ac forte quoque spuria sit, non habeamus, quo respicientes diiudicare illam possimus. Sane & hoc quoque est valde iucundum, nisi assequamur veram causam, verae specie non decipi; & qualiscumque reperta sit, tribuere illi posse legitimum pretium. Hac ratione. Dum causam venati, de alicuius inventae veritate diffidimus, sit nobis occasio invendi, arripiendique aliam viam, iuxta quam fortassis foeliciores simus: nataque adeo exinde est, quae dici Problematica solutio, responsiOve (*sic*) solet, cum vatiae causae proponuntur, ut singulae solum verisimiles, quo aut uberius disquiratur, aliqua ne earum sit vera; aut insinuetur, praeter allatas, requiri quampiam aliam posse.' Gassendi, however—unlike Hobbes—is sceptical about the possibility of establishing a demonstrative science *per causas*. See Gregory 2000, 174–175. See also Fisher 2005, 117, and 2008.

[92] *TO II*, chap. I, § 1, f. 193 *r.*/147.

the observed phenomena through a deductive method (*et per eos concessos neces-sitas demonstretur Phaenomeni*). Nonetheless, Hobbes considered this no small matter (*Neque vero hoc parum est*): if it is possible to give a reason for events, even if they have other causes, they could have been produced in the supposed way: 'as such they are no less useful for human use than if they were known and demonstrated.'[93] Though this result may seem paradoxical, the end of the quoted passage has a Baconian echo regarding the aim of science and the knowledge of physical phenomena, which explains Hobbes' objective: to acquire control of the natural world for the benefit of humanity,[94] an idea that would be stated explicitly in the *Leviathan*.[95] Despite this statement by Hobbes, we should not ignore a funda-mental element in the quoted passage of *Tractatus opticus II*: he explicitly recom-mends that hypotheses be formulated based on events perceived by the senses (*de alicuius eventus sensibus manifesti, quod Phaenomeni appellari solet*), i.e. phenom-ena, whose manifestations and effects can be directly observed.[96]

However, in *La verité*, Mersenne also argued that our limited knowledge of phe-nomena, even if it cannot penetrate the essence of things, was still enough to guide our actions.[97] The example that Hobbes uses in the treatise of the geometric figure of the triangle brings to mind the passage of *La verité* where Mersenne asserts the invariability of mathematical principles, regardless of whether or not a perfect tri-angle exists in reality.[98] Mersenne maintained that physics also pertained to mathe-matically expressible quantities, but as it was connected to motions, it was subject to 'various changes.'[99] The triangle as an example recurs regularly in Hobbes' philo-sophical work: in his *Objections* to Descartes' *Meditations*, Hobbes uses the same image. According to Hobbes, the idea of the triangle is acquired exclusively through abstraction from the sensible objects that have a triangular shape,[100] just as Mersenne

---

[93] Ibid. My translation.

[94] On this topic in Bacon, see Rossi 1974, esp. 3–17; Gaukroger 2001, 6–36 and 132 ff. On the multiple senses of the study of nature in Bacon, see also Fattori 2012, 191–212. For an overview of similarities and differences between Bacon's and Hobbes' models of natural philosophy, see Sergio 2003. See also Bernhardt 1985b, 449–457, Terrel 2015; Milanese 2015; Jacquet 2015; Marquer 2015; Dubos 2015; Ducrocq 2015; Berthier 2015. See also Moreau 1989, 39–47. On the direct relationship between Bacon and Hobbes, an essential text is Bunce 2003.

[95] Hobbes 2012, 74: '*Reason* is the *pace;* Encrease of *Science,* the *way*; and the Benefit of mand-kind, the *end.*'

[96] Horstmann 2001 has precisely described the nature that Hobbes attributed to the *hypothesis* in the context of natural philosophy. In general, on *hypotheses* in modern physics, see Roux 1998.

[97] Mersenne 1625, 14. See also Popkin 1957, 65.

[98] Mersenne 1625, 226–227.

[99] Ibid.

[100] *Objectiones ad Cartesii meditationes*, (henceforth *Objectiones*) *OL*, V, 271–272: 'Si triangulum nullibi gentium existat, non intelligo quomodo naturam aliquam habeat: quod enim nullibi est, non est: neque ergo habet *esse,* seu naturam aliquam. Triangulum in mente oritur ex trinagulo viso, vel ex visis ficto. Cum autem semel rem, unde putamus oriri ideam tranguli, nomine trianguli appel-laverimus, quanquam perit ipsum trangulum, nomen manet. Eodem modo, si cogitatione nostra

had argued in *La verité,* where he had stated that 'the cylindrical figure is found in trees'[101] and the *'entendement'* extrapolates these figures from the phenomenal world through abstraction.

Hobbes—using counterfactual reasoning that refers to *annihilatio mundi*[102] from the medieval tradition—maintains that the triangle, as a universal name, mental entity, or an idea of 'a thing that has three equal angles and two straight lines' would not perish, and neither would its properties, even supposing that there were no longer anything in nature with the shape of the triangle.

In *De motu, loco et tempore* (1642–1643, which we can take as a laboratory for *De corpore*[103]), Hobbes returns to the same image, focusing again on the concept of syllogistic reasoning and knowledge, while emphasizing the need to limit ourselves to what experience proves:

> ... if someone demonstrated some properties of the triangle, it would be necessary, not that the triangle should exist, but only that it should be hypothetically true: if the triangle does exist, then it possesses such a property. For someone to prove that something exists, there is need of the senses, or experience. Even so, the demonstration is not thus established, for someone rigidly demanding the truth from people who says that Socrates lived or existed will tell them to add: 'Unless we have seen [Socrates's] ghost or spirit', or 'Unless we were asleep, we saw Socrates; so Socrates existed, etc.[104]

Nevertheless, both in *Objectiones* and *De motu, loco et tempore,* Hobbes uses the same argument to present the idea of reasoning and philosophy in general as logical calculation.[105] In *De motu,* he presents the entire edifice of philosophical knowledge

---

semel conceperimus angulos trianguli omnes simul aequari duobus rectis, et nomen hoc alterum dederimus triangulo, *habent tres angulos aequales duobus rectis*: etsi nullus angulus existeret in mundo, tamen nomen maneret, et sempiterna erit veritas propositionis istius, *triangulum est habens tres angulos duobus rectis aequales* ... Unde constat *essentiam,* quatenus distinguitur ab *existentia,* nihil aliud esse praeter nominum copulationem per verbum *est.*'

[101] Mersenne 1625, 227.

[102] On the topic of the *annihilatio mundi* in Hobbes, relating to medieval sources, see Zarka 1999, 36–58, and esp. Malherbe 1989, 17–32 and 18–23. For a complete, well-structured discussion, including relating to Gassendi, see Paganini 2004b, 2006a, and b.

[103] As we know, Hobbes decided not to publish his critique of Thomas White's De Mundo and many of the ideas developed in this work can be found later in *De corpore* (1655). See Jacquot 1952a, b. On the reasons for this decision and the importance of *De motu, loco et tempore* in the development of Hobbes' philosophy, see Paganini 2010a, esp. 24 ff.

[104] *MLT,* XXVI, 2, 309; Eng. trans. Hobbes 1976, 305.

[105] See *Objectiones, OL,* V, 257–258: 'Quid jam dicimus, si forte ratiocinatio nihil aliud sit quam copulatio et concatenatio nominum sive appellationum per verbum hoc *est*? Unde colligimus ratione nihil omnino de natura rerum, sed de earum appellationibus: nimirum, utrum copulemus rerum nomina secundum pacta, quae arbitrio nostro fecimus circa ipsarum significationes, vel non. Si hoc sit, sicut esse potest, ratiocinatio dependit a nominibus, nomina ab imaginatione, et imaginatio forte, sicut sentio, ab organorum corporeorum motu: et sic mens nihil aliud erit praeterquam motus in partibus quibusdam corporis organici'.

as 'the same as a faithful, correct and accurate nomenclature of things'[106] and suggests an image of philosophy as a great syllogism, in which concepts are linked together to form unified logical reasoning.[107] For his part, Mersenne had wanted—against sceptical positions—the tool of acquiring knowledge to be based on syllogism, which he associated with the method natural to Euclidean geometry and suggested applying it along with an analytical, synthetic method in the Aristotelian tradition.[108]

In Hobbes, like in Mersenne, syllogism plays a fundamental role, but Hobbes emphasizes that reasoning, meaning a rational process, does not have a direct connection with the existing reality, being nothing more than a calculation of universal names[109]: 'philosophy is the science of general theorems, or of all universals to do with material of any kind, the truth of which can be demonstrated by natural reason.'[110] In other words, correct philosophical reasoning is only a logical calculation, a syllogism, devoid of any sort of paralogism. Indeed: '"reason" is nothing but the faculty of syllogising, reasoning being merely a continuous linking of propositions, or their gathering under one head, or, to put it more briefly, "the calculation of names."'[111]

In the first chapter of *De motu, loco et tempore*, Hobbes associates this syllogistic conception of philosophy with a strictly geometric method—just as Mersenne had done before him in *La verité des sciences*—and he insists on the concept of the

---

[106]*MLT*, XIV, 1, 201–202: Philosophia vera, plane idem est quod vera, propria & accurata rerum nomenclatura, consistit enim in cognitione differentiarum; differentias autem rerum nosse is solus videtur qui singulis rebus suas appellationes proprias attribuere didicerit; praeterea ratiocinatio recta, quae philosophorum opus est, nihil aliud est quam recta verarum propositionum in syllogismum combinatio, vera autem propositio constat ex recta copulatione nominum, nimirum subiecti & praedicati secundum proprias & adaequatas earum significationes; ex quo colligitur philosophiam veram esse non posse, quae fundamentum non habeat in adaequata rerum nomenclatura.'

[107]According to Hobbes, the truth or falseness of philosophical claims depend only on propositions, which in turn depend only on the meaning of the names. See *MLT*, XXX, 17, 357: 'veritas autem & falsitas, idem sunt quod vera & falsa propositio'. On the formal and syllogistic concept of reason in Hobbes, see Zarka 1989 and esp. Minerbi Belgrado 1993, 63 ff.

[108]See Mersenne 1625, 199 and 203.

[109]The names themselves are determined by human will, and there is no connection between the name and the thing it describes. According to Arrigo Pacchi, in his nominalism, Hobbes 'goes beyond Abelardian conceptualism, and even positions more radical than Occamism: indeed, for Occam, the universal has no mental reflection, being an *intentio animae* ... Hobbes, in contrast, removes this final concrete quality from the universal and sweeps away any kind of predetermined relationship between individual things. They no longer have any need for the constraints of Aristotelian hierarchy of genus and species, or of any hierarchy; any connection between the names of things is thus purely arbitrary'. See the footnote by Arrigo Pacchi in Hobbes 1968, 36. On Hobbes' nominalism and conceptualism, see Dal Pra 1962, 411–433. Zarka (1999, 85–93) discusses at length the comparison between Ockham and Hobbes. See also Jesseph 2013, 380–383.

[110]*MLT*, I, 1, 105; Eng. trans. Hobbes 1976, 23.

[111]*MLT*, XXX, 22, 358; Eng. trans. Hobbes 1976, 377.

demonstration part of geometry[112] that could be potentially extended to every branch of knowledge.[113] Hobbes goes on to explicitly state that the method of mathematics can be applied to all other realms of philosophical knowledge, and if this has not happened thus far, this shortcoming can only be blamed on mistakes made by the 'professors of the other sciences.' Although 'some have spoken more plausibly than others have,'[114] it seems that 'none of them has taught anything that was not open to question.'[115]

In the next paragraph, Hobbes distinguishes *logic* from *history*, *rhetoric* and *poetics*. Only logical science has the role of teaching, that is 'to demonstrate the truth of some universal assertion'[116] and, as a result, it becomes clear that philosophy 'should therefore be treated logically:'

> Now since philosophy, i.e. every science, must be treated in such a way that we shall know from necessary deductions that the conclusions we draw are true, it should therefore be treated logically. Indeed, the aim of its students is not to impress [others], but to know with certainty; and hence philosophy is not concerned with rhetoric. Again, its aim is to know the necessity of consequences and the truth of universal propositions: therefore philosophy is not to do with history, and much less with poetry, for the latter relates singular events, and moreover it sets aside truth.[117]

But regarding the processes of knowledge acquisition, Mersenne suggested the joint application of the *analytical* and *synthetic* methods, of *compositio* and *resolutio* (from the Paduan Aristotelian tradition,[118] which Mersenne had found in Biancani), in which he identified '*deus* (sic) *manieres pour apprendre les sciences:*'

> ... because analysis leads us so admirably from the summit of each science to the first principles, and the very simple elements, and the way of composition,

---

[112] On the particular status that geometry has in Hobbes' philosophical system, see Sacksteder 1980, 1992. Hanson 1991 underscores the extension of the geometric method to all realms of knowledge as a uniquely Hobbes' conventionalism, by highlighting also the underlying relationship with scepticism, inherent to the problem and Mersenne's interest in Hobbes' philosophy. Ibid., 636 ff.

[113] *MLT*, I, 1, 106: 'Alia pars philosophiae considerat spatii ad spatium, temporis ad tempus, figurae ad figuram, numeri ad numerum relationes; atque haec pars constituit Geometriam et Arithmeticam, quae solent ambo comprehendi uno nomine Mathematicae, cuius appellationis causa est quod scriptores Geometriae agnoscerentur docuisse, id est manifestum fecisse discipulis et lectoribus suis doctrinam quam tradiderunt esse veram, omnemque eis dubitationem ademisse, qui ideo non modo audiisse sed certo didicisse aliquid dicerentur, ideoque ἀπό τοῦ μανθάνειν, id est discere, vocabant Geometriam et Arithmeticam scientias mathematicas.'

[114] Ibid., 106.

[115] Ibid., Eng. trans. Hobbes 1976, 24, modified.

[116] Ibid., *MLT*, I, 2, 106.

[117] *MLT*, I, 3, 107; Eng. trans. Hobbes 1976, 26.

[118] These concepts were much used in the school of Padua and by its leading sixteenth-century exponents, such as Agostino Nifo and Jacopo Zabarella. See Randall 1961, 30–31, and Crombie 1953, 228–230. On Renaissance Aristotelianism, see Schmitt 1985, and Bianchi 2003. On Hobbes' relationship with humanistic and Renaissance thought, in addition to the mentioned studies by Skinner, see Schuhmann 1985, 1990; Paganini 1999, 2006c, 2007, 2010b.

τάξις ἤ ὁδὸς συνθετιχὴ, leads us so perfectly, and so certainly, from the first principles of sciences to their perfection ....[119]

Aldo Gargani has explored this problem in Hobbesian thought, emphasizing the similarities with the methods then in vogue in the School of Padua[120]; the particular connotation that this terminology took on in Mersenne suggests, if not direct parentage, at least strong parallels with Hobbes' treatment of the problem. He writes in *De motu, loco et tempore*:

> If the progression indeed proceeds from the imagination of the cause to the imagination of the effect, and thus towards the goal (which is always the final effect), the mind's discourse is called *composition*, σύνθεσις; if, on the other hand, it proceeds from effect to cause and then to premises, it is called *resolution*, ἀνάλυσις. Both of these are called *reminiscence* (Lat.: *reminiscentia*) ... If every time we imagine an end, the imagination followed the same order of means, proceeding from the cause to the effect, this very reminiscence of means suited to an end would be called *art*, and, conversely, if it proceeded from the effect to the cause, it would be called *science of causes*.[121]

The *compositive method*—the method that leads from cause to effect—is intended for *art*, meant here as the conventional and constructive knowledge that can produce artifacts (not necessary products of technique). The *resolutive method*, on the other hand, is applied to the *sciences of causes*, and it sets itself the objective of arriving at the causal chain, investigating the origin of a given effect. The question is a development of Hobbes' thought, and in *De corpore*, he would change his approach, envisioning a joint application of *composition* and *resolution*.[122] This issue is also closely linked to Hobbes' epistemological position, but in order to fully understand Hobbes' position, we must examine Galileo's thought and the contribution he made to the origins of Hobbes' philosophy.

---

[119] Mersenne 1625, 203 (my trans.): '... car l'Analyse nous conduit si admirablement depuis le sommet de chaque science jusques aus premiers principes, & les tres-simples élemens, & la voye de composition, τάξις ἤ ὁδὸς συνθετιχὴ, nous meine si parfaictement, & si asseurément depuis les premiers principes des sciences jusques à leur perfection ...' (my translation).

[120] In particular, Gargani gave Agostino Nifo the credit for having formulated an interpretation of the two methods, especially fruitful for developing the scientific-philosophical disciplines, which was also particular to Hobbes' intellectual paradigm. Gargani 1971, 33–38; esp. 34: 'According to Nifo, scientific research starts from cognition of the objects of the sensory experience, on the basis of which the resolutive method arrives at the discovery of cause, a process named *demonstratio signi*. This is the start of the conceptual elaboration (*intellectus negotiatio*) phase, involving an operation of composition and division of the induced or discovered causal term or "discovery" of the resolutive technique.'

[121] *MLT*, XXX, 10, 352–353; Eng. trans. Hobbes 1976, 368–369, subtantially modified.

[122] *De corpore*, VI, 10, *OL*, I, 70: 'Interea manifestum est quod in causarum investigatione partim methodo analytica partim synthetica opus est. Analytica, ad effectus circumstantias sigillatim concipiendas, synthetica ad ea quae singulae per se efficiunt in unum componenda.' In *De corpore* itself, Hobbes does, in fact, seem to make a distinction between the two methods (chap. III, § 9 and chap. VI, § 13), resolving the problem of the quoted passage. On the combination of *compositio* and *resolutio* in Hobbes' philosophy, see Jesseph 2006, 123–124; See also Médina 2013a, b, 105.

In *De motu, loco et tempore*, Hobbes repeated what he had written in the *Elements*, that the method of inquiry could have no other starting point than sensory perception[123]: through the perception of changes that we perceive in the external world we can investigate natural phenomena. The starting point for any inquiry into the physical world is a reality in motion that acts on our sense organs. Hobbes specifies, however, that not all events can be experienced by our senses and that, there is in fact a myriad of objects that do not immediately fall within our sensory perception. Significantly, in Chapter 7—taking up a passage of Galileo's *Dialogue*—he pursues an interesting point, explaining why 'The variety of things is unknown to us.'[124] After repeating that bodies' participles act on us through countless motions that are unknown to us, he emphasizes that the slighter the interaction with specific phenomena with our senses—and our opportunity to develop empirical observations—the less chance we will have to achieve a certain knowledge of these specific phenomena.[125] The topic was already addressed in *Tractatus opticus II*[126] and, as we will see, Hobbes' scientific inquiry relies on a certain interpretation of Galileo's thought, of which Hobbes considered himself an admirer and a follower in some senses.

However, underlying the relationship between the empirical origin of knowledge and the hypothetical interpretation of science are many hidden issues in which we again see agreement with some of Mersenne's ideas. One of these themes is among Hobbes' philosophy's most critical, problematic aspects: the troubled interaction between Hobbes' radical materialism (at its height in his *Objectiones* to Descartes, with the corporization of the mind[127]) and his phenomenalistic approach, according to which the existence of substance can be inferred only through rational

---

[123] *MLT*, VII, 1, 145–146: 'Methodus autem inquirendi unica videtur esse posse, nempe ea quae incipit a varietate phantasmatum sive imaginum quae a rebus ipsis in sensoria agentibus efficiuntur, sine quibus imaginibus inquirere de re qualibet aeque potest lapis atque homo. Quoniam igitur ea immutata esse dicimus quae eodem modo apparent ac prius, et mutata quae aliter; mutation rerum in eo consistit quod sensoriis immutatis ipsae tamen eandem non efficiant speciem, seu imaginem in animo, hoc est consistit in motu aliquo partium obiecti adventitio.'

[124] *MLT*, VII, 4, 147–148: 'Cum ergo corpus sive materia prima mutari potest, & per partes moveri modis innumerabilibus, et per motus huiusmodi efficere innumerabilia phantasmata in animis sentientium, hoc est specierum innumerabiles varietates, & impossibile sit scire quos motus habeant totius mundi singulae particulae, sequitur etiam impossibile esse ut sciamus quot sunt rerum varietates, & proinde an sint in coelo corpora nostris analoga, necne. Potest fieri ut sint; potest fieri ut omnes chimaerae & monstra imaginationis humanae habeant in coelis res ipsis analogas, potest etiam fieri ut non sit ibi ullum grave aut leve, ullus homo, aut animal aut arbor, quippe quae horum nihil scire possumus, propterea quod non operentur in sensus nostros a tanta distantia.' Cf. Galilei, *Dialogo sopra i due massimi sistemi*, *OG*, VII, 86.

[125] Hobbes returns to a quite similar discussion of lunar motions whose causes are particularly difficult to investigate. See *MLT*, XXIV, 1, 289–290; See also *De homine*, II, 6, *OL*, II, 16. See Bitpol-Hespériès 2005, 166, who quotes a passage as an example of the absolute predominance of rational demonstration over experimental observation.

[126] See Giudice 2016, 92 ff.

[127] *Objectiones*, Objectio IV, *OL*, V, 258: '…mens nihil aliud erit praeterquam motus in partibus quibusdam corporis organici'.

conjecture.[128] As Paganini noted,[129] this dichotomy is resolved in the mature formulation of *De corpore*, in which Hobbes develops a correlation between the concept of *phantasma* (in chap. 25)[130] and a particular theory of *accident* (in chap. 8).[131] He removed any ontological reality from sensible qualities,[132] considering them not inherent to objects but to the *phantasmata* of the sentient being. This idea should be considered in light of Hobbes' theory of accident, according to which '*accidens esse concipiendi corporis modum*',[133] which led to equating *accident* and *phantasm*.[134] Paganini also emphasized the influence of modern sceptical tradition in Hobbes' development of phenomenalism and, in particular, the contribution of the French philosophers that Popkin considered exponents of constructive scepticism: Mersenne and Gassendi.[135] In identifying the accident with that manifested by the substance, Hobbes gave priority to the level of the 'apparent truths,' i.e. phenomena, and this development had a precedent in Mersenne's ideas. As Popkin was the first to note,[136] Mersenne had argued in *La verité* that, though we do not know the inner nature of things, phenomenal knowledge is, nonetheless, absolutely enough to develop science and human knowledge.[137] Mersenne's response to the sceptic's

---

[128] On the status of science and the relationships between nominalism and empiria, see Malherbe 1989. See also Sergio 2001, 209–224, which reviews and discusses the theories that have emerged over the last three decades of studying the issue.

[129] See Paganini 2003, 29 ff. See also Paganini 2004a. I find Paganini's argument convincing, as he says that it is much more probable that Hobbes discovered sceptical themes through the translation of Montaigne's *Essais* in John Florio's version (as well as Mersenne and Gassendi's writings, which problematize some of Aenesidemus' tropes), rather than by reading the works of the Spanish writer Francisco Sanchez (as Lupoli maintains. See Lupoli 2004, 2006, 74 ff.

[130] See *De corpore*, XXV, 9, *OL*, I, 325–327.

[131] Ibid., VIII, 2, 91–92.

[132] On the process of reducing palpable qualities to mere perceptions, see Leijenhorst 2002, 84–89, developing interesting ideas both about Hobbesian natural philosophy's debt to late scholasticism (on the concepts of space and time, body and accident), and analyzing how these concepts were absorbed and fit into the framework of a mechanical philosophy.

[133] *De corpore*, VIII, 2, *OL*, I, 92.

[134] On the concept of *phantasma* in Hobbes, see Zarka 1992, who he emphasizes the Aristotelian origin of this concept (Ibid., 24). On this topic, see also Pécharman 1992, esp. pp. 51–52. See also Milanese 2013, 50–53.

[135] See Paganini 2003, 31–32. Paganini gives a lengthy discussion of Hobbes' relationship with both ancient and modern scepticism. See also Paganini 2004a, 2008, 171–227, and Paganini 2015. Popkin had already briefly discussed the topic (see Popkin 1982, 133–48, republished with a lengthy appendix in Popkin 1992, 9–26 and 27–49). Richard Tuck had emphasized Mersenne and Hobbes' shared proposition of providing the certain foundations of knowledge with an 'anti-sceptic' function (see Tuck 1988). I find it difficult to support Sorell's position on the subject, as he argues that Hobbes' thought is essentially far from sceptical topics (see Sorell 1993, 121–135.

[136] See Popkin 1957, 65 ff.

[137] Mersenne 1625, 13–15: '...tout ce que vous apportez contre l'Aristote monstre seulement que nous ne sçavons pas la dernieres differences des individus, & des especes, & que l'entendement ne penetre point la substance que par les accidens: ce qui est veritable, car nous nous servons des effets pour nous élever a Dieu, & aus autres substances invisibles, comme si les effets étoient des cristaus à travers lesquels nous apperçeussions ce qui est dedans: or ce peu de science suffit pour

claim that Aenesidemus' ten tropes show that we do not know 'the essence and nature of things' was that this is not at all necessary to 'establish some truths.'[138]

Gassendi expressed the same idea in similar words to those of Hobbes: he argues that, although the condition of our knowledge does not allow us to see the profound nature of things, nevertheless we can know the phenomenal nature of natural reality, through the effects that we can experience and, based on these, we can also identify a possible cause of these given phenomena.[139]

However, faced with the assumed impossibility of having certain, incontrovertible knowledge due to the heterogeneity of sensory perception, Mersenne responded by referring to concepts of *quantity* and *proportion*, that let different sizes and figures be compared and related, providing a valid, uniform criterion.[140] Significantly, distinctive traits of Mersenne's epistemology identified by Alistair Crombie include attention to the quantitative element—a particular trait of the mathematization of reality—physical probabilism, and the dialectic between the speculative element and experimental observation.[141]

The idea that merely phenomenal knowledge was enough for science and practical, working knowledge naturally evokes the position stated by Hobbes in many of his works, especially in the above-mentioned passage in *Tractatus opticus II*. Hobbes' speculation, however, compared to Mersenne's, has a greater theoretical development of the first, fundamental principles of philosophy.

For example, on the nature of the *accident*—as we will see in more detail when we discuss the relationships between Hobbes and Galileo[142]– Hobbes makes a distinction between accidents that can be born and perish independently from the body's existence, and accidents without which the body could not even be conceived (and not even exist), such as *extension* and *figure*.[143] The fundamental concept of Hobbesian natural philosophy applies to this second group of accidents: the concept of *cause*, which has enormous importance in Hobbes' philosophical system,

---

nous servir de guide en nos actions ... c'est donc assez pour avoir la science de quelque chose, de sçavoir ses effets, ses operations, & son usage, par lesquels nous la distinguons de tout autre individu, ou d'avec les autres especes: nous ne voulons pas nous attribuer une science plus grande, ny plus particuliere que celle-là'.

[138] Ibid., 150.

[139] *Syntagma*. In Gassendi 1658, I, 207b: 'Ea nempe nostrae perspicaciae, cognitionisque conditio est, ut, cum pervidere naturas rerum intimas non possimus, aliquos effectus possimus; contentos nos esse oporteat, si hariolati quidpiam circa illas ex quibusdam effectibus, nostras qualescumque de ipsis notiones adnitamur aliis effectibus accomodare, cum eorum causas poscimur, seu quomodo a suis naturis originem habeant, rogamur.' On this topic see also Fisher 2005, 19 ff. and 346 ff.

[140] See Mersenne 1625, 151.

[141] See 'Marin Mersenne (1588–1648) and the Seventeenth-Century Problem of Scientific Acceptability,' in Crombie 1990, 405.

[142] See *below*, chap. 2, § 2.

[143] *De corpore*, VIII, 3, *OL*, I, 92–93. On accidents inseparable from the concept of body, which, we will see are fundamental to Hobbesian philosophy, see Schuhmann 1992, which identifies these accidents in *size, extension*, and *local motion* (ibid., 75). See also Malherbe 1989, 22.

serving as a bridge between the universe of perception and phenomena and the material reality that populates the universe.[144]

This particular connotation of Hobbes' phenomenalism, as well as the concepts of *extension*, *figure*, and *cause* are key to understanding how differing realms of knowledge were presented and evolved in Hobbes. He focuses on the quantitative element of science, and, significantly, pays particular attention to sciences commonly termed *mixed mathematics*.[145] We have already noted how important the idea of mixed mathematics was in Mersenne's thinking, but this concept is worth further consideration as it is key to Hobbes' perspective as well, and Mersenne's ideas on the subject can aptly be related to those that Hobbes later developed.

Both thinkers gave special attention to optics. Hobbes focused on its mathematical aspects, allowing him to conceive it mainly as a rational science based on deductive reasoning on the model of pure mathematics.[146] This particular view of optics is the foundation of a 'mathematized physics,' forming a kind of bridge between the deductive, aprioristic certainty of mathematical disciplines and the problematic nature of physics, always shaped by the impossibility of knowing everything that has to do with motion at a microparticle level. Central to Hobbes' position is always the idea that rational inquiry is applied to the physical world through the geometric notion of quantity. Mersenne also expressed this idea, arguing—in a dedicatory epistle to *Questions théologiques*—that men could not penetrate 'the external surface of bodies,' but that this was in no way an insurmountable obstacle to founding a solid physical science, because through the concept of *quantity*—by using *lines* and *figures*—we are capable of mathematically measuring natural reality.[147]

Following the principles of modern science, Mersenne removes any scientific validity from the concept of *quality*: because human knowledge is limited to the body's external surfaces, any scientific, rational inquiry must be limited to the *numbers*, the *lines* and the *figures*, i.e. the conceptual elements that let the physical reality be mathematized and make up the fundamental feature of Galileo's scientific

---

[144] The concept of cause is the common denominator linking matter, the reality of bodies in motion and the phenomenal world of *phantasms* that define our perceptions and, as a result, our concepts as well. On the idea of cause in Hobbes, see Leijenhorst 2005, who underscores that this is the bridge between the subjective world of our *phantasmata* and the objective world of bodies in motion (ibid., 90). It is used with the same meaning in Malherbe 1984, 84 ff. and Paganini 2008, 213–217. Zarka prefered a more distinctly formal sense, an interpretive principle more than as a law governing the phenomena of reality (see Zarka 1999, 73 ff.).

[145] See *De homine*, X, 5, *OL*, II, 93.

[146] Some scholars of Hobbes' thought have highlighted the importance that optics has in his system as a bridge between pure mathematics and physical disciplines. See Tuck 1988; Stroud 1983, 39 ff.; Médina 1997; Giudice 1999, 141.

[147] *Questions theologiques, physiques, morales, et matematiques.* In Mersenne 1985, Epître dédicatoire, 202: '…il semble que la capacité des hommes est bornée par l'écorce, et par la surface des choses corporelles, et qu'ils ne peuvent penetrer plus avant que la quantité, avec une entiere satisfaction. C'est pourquoy les anciens n'ont pu donner aucune demonstration de ce qui appartient aux qualités, et se sont restreints aux nombres, aux lignes, et aux figures, si l'on excepte la pesanteur, dont Archimede a parlé dans ses Isorropiques'. See also Lenoble 1943, 353 ff.

thought.[148] It is in this perspective that Mersenne interprets the work and intellectual figure of this *'philosophe très excellent:'* Galileo Galilei.[149]

Hobbes shares Mersenne's assumption that our concept of reality is limited to the surface of bodies that take up a geometrically quantifiable space. His epistemological position—which does differ from Mersenne's in many respects[150]—emerges clearly, only considering the dual relationship of his scientific thought. Though he inherited the same doubting attitude about scepticism as Mersenne (and Gassendi, to an extent), he also came to grapple with Galileo's ideas in particular, which would become the foundation of his philosophy.[151]

1.3 Astronomical Problems, Animal Analogies, and Mechanical Explanations

In *Questions théologiques, physiques et mathématiques,* Mersenne expresses his opposition to the abilities commonly attributed to talismans to attract the influence of the stars.[152] According to Mersenne, this idea is merely a popular superstition. He believes that we can rule out that stars have any *influence* at all on the earth and the

---

[148] See Lewis 2006, 118.

[149] In several passages of Mersenne's works, he expressly calls Galileo a *'philosophe'* and not a mere mathematician. See, for example the *advertissement* placed at the end of the first book of *Harmonie Universelle,* which introduces the topic of the second book, titled: *Des mouvements de toytes sortes de corps:* Mersenne 1636a, Livre I, 84. Lenoble had also already discussed the topic. See Lenoble 1943, 357. I believe that Mersenne, broadly speaking, followed the principles of 'Galilean' philosophy, though his epistemological perspective differs from Galileo's, and in some respects is much closer to that of Gassendi, as Popkin has suggested. I do not think that Mersenne considered Galileo a mere mathematician or just an acute observer of reality, as Garber suggests (Garber 2004) and Massimo Bucciantini, following Garber. See Bucciantini 2007, 44 (conversely, I agree with Bucciantini that this was, more or less, the idea that Descartes had of Galileo).

[150] Hobbes' materialism is marked by his metaphysical perspective, which greatly reduces the scope of his so-called conventionalism (as we will discuss further below, by comparing Hobbes' position with that of Galileo). On the topic, see Malherbe 1989, 24, diverging from the more Ockhamist interpretations of Hobbesian philosophy suggested by Bernhardt and Zarka (See Bernhardt 1985a, and Zarka 1985). See also Milanese 2013, 53 ff.

[151] My interpretation of Hobbes' perspective diverges from that of Malet, who emphasizes its distance from Galileo's methodology and epistemology. See Malet 2001, 319 ff.

[152] Mersenne, *Questions théologiques, physiques et mathématiques,* in Mersenne 1985, 279: 'Ce qui est si ridicule, et si inepte qu'il n'y a plus que les vieilles, et les trop credules qui ne s'en moquent comme d'une pure fable: car outre que l'experience fait voir que ces graveures, et ces figures n'ont nulle force, ou aptitude pour determiner, et pour attirer les vertus des Astres, la raison y repugne entierement, qui mesme persuade que les astres n'ont pas la force, ni les influences qu'on leur attribuë, car chaque astre n'a point d'autre force sur nous que celle qu'il exerce avec sa lumiere, et sa chaleur; de sorte que si l'on disposoit autant de chandelles autour de la terre, comme il y a d'estoiles au Ciel, dont elle fust aussi illuminée, et échauffée, comme elle est par lesdites estoiles, nous sentirions les mesmes influences.' In Mersenne 1634, the *question* III, (see Mersenne 1985, 565–588), Mersenne refutes the principles of astrology and judicial astrology.

human beings who live here, other than in terms of *light* and *heat*. This is an essential observation because it rejects any explanation that is not expressed mechanically: only motion can make two bodies interact that are far from each other, such as stars and planets, and Mersenne actually connects the phenomenon of light as well to the motion of air.[153] Following this general orientation, he continues his attack on the ideas of *sympathy* and *antipathy* as well as against that of '*vertus occultes*.' Indeed, 'These qualities are only hidden to the ignorant, because learned men who know the origin of the actions that the common people call *sympathy*, or *antipathy* do not use these terms at all and demonstrate that what is called *hidden* is clear to them.'[154] As such, if learned men cannot identify the cause of a given phenomenon, they simply candidly admit their lack of competence on the subject, without resorting to hidden virtues or other explanations that have no rational basis. Mersenne continues his discussion, observing that chemists, if pressed, often have to admit their ignorance of numerous physical phenomena whose causes they claim to know. He concludes with an assumption that we have already discussed at length: we have certain and perfect knowledge only of those disciplines of which we fully understand the first principles.[155]

Beyond the difficulties inevitably linked to the study of physical phenomena, which we have seen that Mersenne's and Hobbes' epistemologies share, there is no question that the two thinkers are also akin in their distaste for ideas from the magical-hermetic tradition of *sympathy* and *antipathy*.

In *Cogitata* (1644) Mersenne would present Hobbes as the mechanist philosopher par excellence,[156] and in Chapter 26 of *De motu, loco et tempore*, Hobbes openly states his opposition to the concepts of *influence*, *sympathy* and *antipathy*, with an argument that seems to follow the same lines as Mersenne's: we cannot conceive any action produced by one body upon another other than in terms of *local motion*. The stars can act on our planet only through light and heat. Any notion of sympathy, antipathy, or a hidden quality must necessarily be excluded from any explanation worthy of the status of scientific hypothesis.[157]

---

[153] Mersenne 1636a, Livre II, 99. The idea that the phenomenon of visual perception is created only by a moving column of air has very deep similarities with Hobbes' optical theory, which we will discuss further below.

[154] Ibid., 299–300. In Chapter 24 of *De motu, loco et tempore*, Hobbes adds a paragraph (21) with a title that would seem to contradict Mersenne: '*Why not all influence do constitute light*', in which he argues that the earth's rotation motion can influence the motion of light though without producing heat (cf. *MLT*, XXIV, 21, 303–304). However, this aspect of Hobbes' thoughts should be evaluated in light of his and Mersenne's astronomical perspectives, that we will discuss below.

[155] Mersenne 1636a, Livre II, 300: '…l'on est contraint d'avoüer que l'homme n'est pas capable de sçavoir la raison d'autre chose que de ce qu'il peut faire, ny d'autres sciences, que de celles, dont il fait luy-mesme les principes, comme l'on peut demonstrer en considerant les Mathematiques.'

[156] *Ballistica et acontismologia*, in Mersenne 1644, *Praefatio utilis ad lectorem* (p. not num., first page of the preface).

[157] *MLT*, XXXVI, 2, 397: '…an fortuita dependeant ab astris, pertinet ut aliquid dicatur de modo quo procedere possit operatio ad res longinquas; omnis quidem actio, ut dictum saepe est, est motus, nimirum motus localis, quies enim nihil mutat, id est nihil efficit. Quando vero ex motu

The two thinkers' arguments have quite significant parallels, and both undoubtedly had a precedent in Galileo's *Dialogue* where he criticized Kepler for explaining the phenomenon of the tides by resorting to the 'moon's dominion over the
waters, to occult properties, and to such puerilities.'[158] According to Galileo, any
sort of interaction between bodies that were not in the mechanical physics framework was inconceivable. This idea, a founding principle of modern science and
philosophy, would be inherited by Mersenne to an extent, and, more so, by Hobbes.[159]

Hobbes, however, had considerable difficulty in explaining astronomical phenomena, and in Chapter 24 of *De motu, loco et tempore*, where he addresses the
phenomena of the tides and the interaction between lunar and terrestrial motions—
which Hobbes felt Galileo had not successfully explained[160]—he seemed to resort
to an element completely outside of his basic mechanical orientation. He maintains
that the earth 'can exert some power over the moon not only by means of its daily
action, but also by a kind of influence-as if a magnetic one.'[161] Hobbes sought to
explain the reasons for certain physical phenomena, such as magnetism and the
tides, which seemed to escape any mechanical explanation. In the final paragraph of
the same chapter, he identifies the cause of certain lunar motions in the earth's rotation. He assumed that the earth was endowed with an expansion and contraction
motion like that which had been identified in the sun, both in his latin optical treatises[162] and in *De motu*.[163] As a result, 'so the Sun, by means of its light, together
with its spin about its own axis, not only draws the earth circularly in the latter's
annual motion but also causes parts of the earth to be turned Sun-ward one after

---

effectum aliquem produci oculis videmus, ibi operationem & motum, eandem rem esse facile
omnes consentimus; sed quando effectum vident, motum autem non vident, ibi motum esse negant
plerique, et operationem, qua effectus producti sunt, secundum differentias quas homines in seipsis
sentiunt diverse appellant, modo calorem, modo frigus, modo humorem, modo siccitatem, modo
lumen; et quando haerent, sympathiam aut antipathiam, aut occultam qualitatem, aut denique
influentiam, sed nunquam motum; quasi qualitates naturae, et potentiae corporum eodem modo
infunderentur in corpora, quo aqua, aut alia res fluida infunditur, vel influit in vasculum. Sciendum
igitur est motum a corpore ad corpus propagari in quantacunque distantia per continuam proximi
et contigui corporis protrusionem, ita ut producendo effectui non sit opus corpus aliquod, sive
partem astri aliquam euclare ad terram.'

[158] Galilei, *Dialogo sopra i due massimi sistemi del mondo*, *OG*, VII, 486. Eng. trans. Galilei 1967,
462: 'Ma...più mi meraviglio del Keplero che di altri, il quale, d'ingegno libero ed acuto, e che
aveva in mano i moti attribuiti alla Terra, abbia poi dato orecchio ed assenso a predominii della
Luna sopra l'acqua, ed a proprietà occulte, e simili fanciullezze.' On the difference between
Galileo's and Kepler's epistemologies, see Bucciantini 2003.

[159] We will later discuss this subject at length. See *below*, chap. 2, § 4.

[160] Hobbes noted that Galileo had not taken into account the lunar motions coinciding with the
tides, and, therefore, had developed a flawed explanation of the phenomenon that did not involve
the moon. See *MLT*, XVI, 2, 211.

[161] Ibid., XXIV, 1, 289; Eng. trans. Hobbes 1976, 278.

[162] *TO I, OL*, V, 218–220; *TO II*, chap. I, §§ 4–8, ff. 193v–197r/148–150.

[163] *MLT*, IX, 2, 161–162.

another in the daily [axial] motion. Likewise the earth draws the Moon circularly by means of its influence in conjunction with [its] daily motion.'[164]

He was not the first to attribute systolic and diastolic motion to the earth: Mersenne's *Questions Inouyes* preceded him, in which Mersenne explained the reasons for the phenomena of the tides by hypothesizing that the earth had a motion that 'refers to that of the heart, which is called Systole and Diastole.'[165] Hobbes was faced with the problem of having to explain the differences between the Sun's action and that of the earth: first and foremost, the production of light. In order to explain this problem, he made use of the internal isomorphic structure of the body of the sun, completely different from the earth in this respect. He developed a cumbersome explanation with the aim of describing some physical events that are difficult to link to mechanism: the earth is endowed with the same dilation and contraction motion as luminous bodies (such as the sun and the stars), but, unlike those bodies, which have a homogeneous constitution, the heterogeneity of the parts of the earth produces certain phenomena, such as the properties of some metals.[166]

The problem that Hobbes encountered in applying mechanism to all physical realities is particularly evident in astronomical problems. He gives an especially odd explanation for the earth's rotation.

> ... when one part of the Earth is exposed to the Sun's rays at very close range, another part, one benefiting less from its warmth, strives to press forward into the same position, and the part that was there first is now sated and moves away. This is the cause of diurnal motion, I think, which is not unlike the reason why living creatures are moved in accordance with the action of the factors that they like, i.e. those whose action aids and strengthens the creatures' inner and congenial motion. For we draw up to the fire, and the parts of the body that are growing cold we turn towards it, one after another, even if we are doing something else. I do not, however, think that the Earth is animate; yet all consistent bodies, or those possessing coherent parts, have this in common: they maintain habitual motion inasfar as other motions, gravity especially, do not hinder [them].[167]

The discussion has clear marks of vitalism and finalism, and Cees Leijenhorst has argued that it has an echo of Tommaso Campanella's *pansensism*.[168] Hobbes

---

[164] Ibid., XXIV, 21, 303–304; Eng. trans. Hobbes 1976, 297.

[165] *Questions Inouyes*, question XXXVII. In Mersenne 1985, 102.

[166] *MLT*, XXIV, 21, 303–304: 'Poterit quis interrogare quare terra non sit per se lucida ut sol, si quidem enim lux (quae solis est influentia) fiat per solum motum expansivum, terra autem influentiam in lunam exerceat per similem expansionem, erit quoque terrae virtus illa expansiva lux. Dicendum igitur [est] influentiam quidem eorum quorum natura per omnes particulas minimas, et ut ita dicam, mathematice homogenea est, lucem esse eamque fortem aut debilem pro velocitatis gradu quo se dilatant. In corporibus vero heterogeneis, quale est terra, et quodlibet aliud corpus cuius partes motus specificos habent et proprios sibi; et pugnantes cum motibus particularum adiacentium, etsi unaquaeque pars, ac proinde totum, motum habeat illum quem diximus expansionis sive dilatationis, [dico] non tamen illum motum esse lucem, sed influentiam a luce differentem, et quia corpora essentialiter inter se differunt per motum internum partium et innumerabiles sunt differentiae corporum specificae, innumerabiles quoque esse differentias influentiarum.'

[167] Ibid., XIX, 7, 246; Eng. Trans. Hobbes 1976, 221, modified.

[168] Leijenhorst gave the theory of the diurnal rotation of the earth that appears in *De motu, loco et tempore* as evidence of the influence of Campanella's *pansensism* and *panpsychism*, on Hobbes'

himself is quick to maintain that he is not equating the earth with an animal, but it is, regardless, quite difficult to try to fit such an explanation in the framework of physical mechanism. Nevertheless, as Paganini has pointed out, in Pierre Gassendi's *Syntagma* there is a very similar argument comparing the behavior of animals and inanimate beings.[169] In this case as well, it is quite likely that Hobbes had direct contact with Gassendi's ideas.

However, investigating Hobbes' position more in depth and comparing the quoted passage with others from *De motu, loco et tempore*, we can understand this apparent idiosyncrasy; before we examine Hobbes' position, we should note that a very similar argument had been made in Mersenne's *Questions*. Mersenne made the following observation, in addressing diurnal motion, using rigorous mathematical calculations:

> Since the earth needs the sun, it must seek it, as we seek fire, which we need. Therefore, if we do not want the cities and the countryside to turn, when we go to the top of the towers to contemplate them, we must not expect the sun and the stars to turn to consider the earth.[170]

There are clear parallels between the two thinkers: both present the image of the earth that turns to the Sun to benefit equally from the light and warmth that it emits. Yet, as Horstmann has noted, a similar argument about the motion of the earth was also in Kepler's *Astronomia nova*,[171] which, in *De corpore*, Hobbes would tie together with the theory of elliptical orbits of the planets' revolving motions, in contrast to Copernicus and Galileo, who considered those motions circular.[172] However, the fact that this particular argument, with an almost pansensist orientation, is in Mersenne and Gassendi gives us a clue to help understand why this reasoning is in Hobbes, though so seemingly far from his philosophical princi-

philosophy. See Leijenhorst 1997, 120. See also Sergio 2006, 207–208; Sergio 2007, 311; Sergio 2016, 86. On Campanella's pansensism, see especially the first chapters of the second book of *De sensu rerum et magia*: Campanella 2007, 33–64. See also Ernst 2010, 108 ff.

[169] Paganini noted the mechanical explanation of sensation and animal motions in Hobbes, highlighting the similarities with Gassendi's arguments. See Paganini 1990, 369 ff. Cf. *Syntagma*. In Gassendi 1658, II, 328a–b; see also ibid., 345b. Gassendi, like Hobbes, believes that animals are endowed with the faculty of reason. See, for example, his *Objectiones* to Descartes' *Meditations* (See *Objectiones Quintae, AT*, VII, 271. See also *Exercitationes paradoxicae adversos Aristoteleos*, Praefatio. In Gassendi 1658, III, 102; ibid., 202b–203a). As Gregory has noted, this *topos* is common in scholastic anthropocentrism, as well as in libertine literature, such as in La Mothe Le Vayer, and Cyrano de Bergerac. See Gregory 2000, 51, and 102–104. See also Bloch 1971, 368 ff.

[170] Mersenne, *Questions théologiques, physiques et mathématiques*. In Mersenne 1985, 341–342 (my trans.): 'Puisque la terre a besoin du Soleil, elle doit l'aller chercher, comme nous cherchons le feu, dont nous avons besoin: car si nous ne desirons pas que les villes, et les campagnes se tournent, quand nous montons au haut des tours pour les contempler, aussi, ne devons nous pas desiderer que le Soleil et les estoiles se tournent pour envisager la terre.'

[171] Horstmann saw interesting similarities between Kepler and Hobbes' writings, from *Tractatus opticus II*, to *Decameron physiologicum*. See Horstmann 1998, 142 ff.

[172] Ibid., 138.

ples.[173] Hobbes returns to the topic several times. In Chapter 18 of *De motu*, he makes the same earth-animal analogy: animals, when exposed to a source of heat: 'of their own accord alter their position so that the vital motion may be kept at a constant temperature most suitable to its nature'[174] and this same reasoning is applied to the earth.[175] As Paganini has noted,[176] he sees a correlation between the earth and animals, not because he considered the earth endowed with sensitivity and intentionality, but because he considered every animal action, including voluntary ones, as a kind of reflection conditioned by external stimulus. Hobbes explicates this in Chapter 25 of *De corpore*[177] where desire and aversion are explained in strictly mechanical terms. In other words, an animal's motion shifts to turn towards the heat source on the surface of its skin that has not yet had the benefit of its warmth. This is conceived exclusively as a direct reaction to the action of the warm air, moved by the heat-producing body, which comes into contact with the receptors on the animal's skin.[178] But, even considering these observations, the theory that Mersenne and Hobbes suggested still seems quite specious, and in Hobbes' later works, not surprisingly, there are no more comparisons between the earth's motions and that of animals.

Hobbes did, however, clarify the assumptions of his argument, when discussing the eccentricity of the orbital motion in Chapter 25 of *De motu, loco et tempore*.

---

[173] In fact, an argument very similar to that in Mersenne and Hobbes is also seen in William Gilbert's *De Magnete*, where, when discussing the earth's rotation, he seems to suggest that the earth's motion is seeking the light from the sun and the benefits of that light. See Gilbert 1600, 224. See also Westfall 1971, 27.

[174] *MLT*, XVIII, 15, 238; Eng. Trans. Hobbes 1976, 211.

[175] Ibid.: 'Quod si voluntaria dicatur conversio esse animalium, nihil obstat hoc quin terrae conveniat quanquam enim in terra motus ille voluntarius non sit, tamen in animalibus voluntas illa, sive appetitus caloris motus est, videtur igitur terra ideo convertere se ad solem, hoc est, moveri motu diurno, propterea quod partes a sole aversae ad conservandam naturam, sive motum suum essentialem ad solem accedant, partes vero soli proximae calore solis saturatae in eodem loco non amplius se bene habeant, & proinde ut motus illorum naturalis liberius exerceatur rursus se recipient, atque hoc moto fit conversio ea qua definitur dies, ideoque diurna appellatur.'

[176] See Paganini 1990, 2010a.

[177] *De corpore, OL*, I, 315–334. On the topic of animal reactions to a heat source, Hobbes' explanation is much like Descartes' in *L'homme*, though Descartes certainly did not apply this explanation to human beings like Hobbes did. (See Descartes, *AT*, XI; 187 ff. See also ibid., *AT*, XI, 119–120. Bitpol-Hespériès drew a comparison between Hobbes' *De homine* and Descartes' *L'homme*, identifying similarities in content and focusing on the fact that both discussed the phenomenon of vision. See Bitpol-Hespériès 2005. The analogy of the animal-machine is also found in the *Discours de la méthode*. See Descartes, *Discours de la méthode*, *AT*, VI, 46 ff. See also *Traité de l'homme, AT*, XI, 119–120. On the *animal machine* theory in Descartes, see Morris 2000; Gaukroger 2002, 196 ff.

[178] In *De corpore*, XXV, 5, *OL*, I, 320, Hobbes writes: 'Etsi autem sensio, uti diximus, omnis fiat per reactionem, ut tamen quicquid reagit sensiat necessario non est', referring to a form of pansensism very similar to that of Telesio and Campanella: 'Scio fuisse philosohos quosdam, eosdemque viros doctos, qui corpora omnia sensu praedita esse sustinuerunt; nec video, si natura sensionis in reactione collocaretur, quo modo refutari possint.'

Here he admitted that he could not offer a comprehensive explanation of eccentricity. One of the possible solutions he proposed was linked to the constitution of hard bodies. First, we should note that according to Hobbes 'any solid body, i.e. a hard one, whose parts cannot be torn apart or separated one from another save by [the action of] a great motion from outside, will owe this property to internal motion of its parts, namely to that motion of the matter which determines the body's form.'[179] This same motion is also attributed to the earth which, if stopped in its 'natural' motion, reacts like a spinning top that strikes a surface.[180] This specific motion of internal parts is also the origin of diurnal motion,[181] but to understand this reasoning, we need to refer to Hobbes' theory of hard bodies, which first emerged in his correspondence with Descartes in 1641.[182] According to Hobbes, the hardness of bodies is determined by a whirling motion that affects the internal particles of which these bodies are composed.[183] His argument was based on the assumption that hard bodies—and the earth is such a body—owe their hardness to the motions of the internal parts, and that any other alteration of this internal motion effected by other bodies (the Sun, in this case) would produce an opposite motion that tends to reestablish the original balance.

Aside from the specific argument, what emerges here is Hobbes' particular propensity for developing mechanical explanations, a tendency certainly not far from Mersenne's intellectual perspective, as we have seen. Though this methodological tendency is much more marked in Hobbes, there is a clear convergence of the two thinkers towards a generally mechanical orientation of physics, both in the astronomical problems that we have analyzed and in investigations of the nature of light and sound found throughout *Harmonie universelle*.

---

[179] *MLT*, XXIV, 11, 299; Eng. trans. Hobbes 1976, 291, modified.

[180] *MLT*, XXIV, 11, 299: 'Itaque et terra habet motum talem materiae internum per quem est terra. Rursum omne corpus durum, si pars eius aliqua ab alio corpore incurrente impellatur, hanc proprietatem habet ut & se plus vel minus, prout durities sua & vis incurrentis maior vel minor est, restituat, & liberando se; si aliter non possit, resilit, ut constitutionem suam specificam conservet, non ut agens sensibile quod cognoscit quid sui sit conservativum, sed ut turbo rotatus velociter & impingens contra parietem statim resilit non ex appetitu conservandi motum circularem, sed ex vi illius ipsius motus qui materiam ad figuram certam determinverat quae cum impactu in parietem consistere non potuit. Itaque et terra eam quoque habet proprietatem, ut si agens aliquod in ipsam ita agat, ut ea actione motus ille eius internus, per quem est terra, impediatur, statim se recipiat.'

[181] Ibid., XXIV, 13, 300.

[182] See Baldin 2017.

[183] See Hobbes to Mersenne for Descartes, February 7, 1641, *AT*, III, 302. Descartes replied by arguing that these motions were the cause of the softness of bodies. He also maintained that the hardness of a given object was due to the internal cohesion of its parts. (*AT*, III, 321–322).

1.4   Light and Optics Between Geometry and Physics

In *Harmonie universelle*, Mersenne argued that speculations on the nature of sound did not bear results exclusively in the narrow field of acoustics, because 'we can represent everything that exists in the world, and, as a result, in all sciences, by means of sound; given that all things consist of *weight, number*, and *measure*, [and] sound represents these three properties.'[184] Although biblical references[185]—which had a profound theological meaning in Mersenne's work[186]—were widely used in the Middle Ages (such as in Augustine and Roger Bacon[187]), nevertheless, it is particularly significant analyzed in the context of modern science and philosophy, where it is used to express the geometricization of reality.[188]

According to Mersenne, every physical body has its own sound and this sound coincides with motion, and, as such, inquiry on motion allows the discovery of the fundamental characteristics of the entity that produced it. The idea that every physical object has a specific sound was already in the *Traité de l'harmonie universelle*[189] and in *Questions*.[190] The most interesting aspect is the parallel that Mersenne draws here between optics and music.[191] From the argument that Mersenne developed, we can infer that the principles and media through which the action of sound is

---

[184] Mersenne 1636a (*De la nature & des proprietez du Son*), 43. The topic was already addressed in *Traité de l'harmonie universelle*. See Mersenne 2003, 52. Cozzoli ignored *Traité* in his study of Mersenne's optics (see Cozzoli 2007, 2010) and he mentions only briefly *Harmonie Universelle* (ibid., 14–15).

[185] See *Sapientia*, XI, 20

[186] See Fabbri 2003, 158–170.

[187] See Alessio 1985, 74 ff. On the expression's recurrence in the Middle Ages, see Parodi 1984, and Bianchi and Randi 1990.

[188] Not incidentally, this topic is also found in Galileo Galilei. See Galluzzi 1979a, b, 252; Bucciantini 2003, 300.

[189] See Mersenne 2003, 41

[190] *Questions physiques et mathématiques*, Question 35. In Mersenne 1985, 405: 'Il faut seulement remarquer qu'il n'y a quasi nul corps dans toute la nature qui n'ayt un son particulier … l'on peut dire que toutes les impressions que les objects font sur nos sens, ne sont autre chose qu'une espece de sons puisque elles consistent dans un mouvement, par lequel les corps nous communiquent leurs proprietez, et nous enseignent ce qu'ils peuvent, et ce qu'ils sont, et toute que sorte de mouvement faict un son, ou plustost que le son, et le mouvement sont une mesme chose.'

[191] Ibid., 406: 'Si l'on cognoissoit la vistesse de la lumiere, et du mouvement qu'elle fait dans l'air, et dans l'œil, et le mouvement, ou l'impression que les autres objects impriment sur nous, l'on pourroit determiner, et expliquer leurs raisons, et leurs Analogies par le moyen des sons; d'où l'on infereroit leurs vertus, et leurs proprietez; et parce que les raisons sont mieux cognuës, et plus aysées à concevoir, à veriffier, et à expliquer dans les sons, que dans les autres objects, l'on en tireroit de la lumiere pour toutes les autres sciences'. A basic form of the analogy was already seen in the *Traité* from 1627, but here, Mersenne maintained that investigation was easier in the field of optics, because sound, unlike light, is invisible. See Mersenne 2003, 20. On Mersenne's musical and acoustic theories, see Bailhache 1994, and esp Fabbri 2003.

expressed are the same as in the field of light and vision.[192] These ideas are returned to and developed at length both in *Harmonie universelle*, where Mersenne compares light and sound on several occasions and in *Harmonicorum libri* (1636), which are something of a Latin summary of the previous book, written for those who 'do not know French.'[193]

First, he observes that the principle difference between light and sound is in light's instantaneous propagation in contrast to sound, which takes time to spread in the surrounding space.[194] It is, however, the 'subtle' nature of sound that lets it cross opaque bodies.[195] Both propagate, however, through air in concentric circles, 'like a drop of oil poured on a sheet of paper or on fabric, or like the rings produced in water when a stone is thrown.'[196]

The problem of comparing sound and light takes up the entire Proposition 25 of the first chapter of Mersenne's *Harmonie*, in which he argues that light lets us perceive color due to its different incidences on the reflection plane, and, as a result, 'colors are nothing other than the different incidence and reflection of rays, as sound is nothing other than different motions of air.'[197] According to Mersenne, sound was nothing other than a motion of air[198] and in places, Mersenne seems to extend this reasoning to the phenomenon of light as well: 'If we consider the nature of light, with great attention, we may find that it is nothing other than a motion of air that brings with it the image of its first mover, i.e. the luminous body.'[199] Here, Mersenne seems to favor a *theory* of light that only includes transmitting the motion of the media, compared to the traditional perspective that referred to *species* (though he does write of a transported 'image'). He later developed several ideas that highlighted the problematic nature of this position. First, regarding the propagation of

---

[192] Brandt had already advanced the theory that the ideas about optics and music in *Harmonie universelle* could have been very interesting for Hobbes. See Brandt 1928, 158–160.

[193] Mersenne 1636b, Epistle dedicatory (p. not num.). Interestingly, some copies of the work are dedicated to Charles Cavendish. See Jacquot 1952a, b, 16.

[194] Mersenne 1636a, Livre I, (*De la nature & des proprietez du son*), 15: '...la lumiere s'estend dans toutes la sphere de son activité dans un instant, ou si elle a besoin de quelque temps, il est si court que nous ne pouvons pas le remarquer: mais le Son ne peut pas remplir la sphere de son activité que dans un espace de temps, qui est d'autant plus long que le lieu où se fait le Son est plus esloigné de l'oreille.' The *Harmonicorum libri* explain how light and sound are different and similar in the *Propositiones* V and VI (2 and 3).

[195] Mersenne 1636a, Livre I, 18. The issue is addressed again on page 24 where Mersenne seeks to provide an explanation of the propagation of sound through solid bodies and finds the it in the air vibrations in the pores of solid bodies.

[196] Ibid., 20. See also Mersenne 1636b, 2–3 (See also Mersenne 2003, 69).

[197] Mersenne 1636a, Livre I, 45.

[198] See Mersenne 1636b, 1–2. See also Auger 1948, 40.

[199] Mersenne 1636a, Livre I, 45 (my italics). See also ibid., I, 16. In his *Traité de l'harmonie universelle*, Mersenne considered the existence of species as an established fact. See Mersenne 2003, 47. On Mersenne's theory of light, see Auger 1948, 38; Beaulieu 1982a, b, 311–316.

light and sound, he argues that though both spread in a circular form, the luminous
or acoustic source necessarily emits rays.[200] Specifically regarding light, he added:

> The movement of light is, it seems, subtler than that of sound, and penetrates further into
> the substance of air, such that it fills with a certain liquor similar to a very subtle and a very
> light oil, and it moves so that it affects the eye and the optic nerve. This [the eye] begins to
> discover all external objects, as soon as the moved air has entered its pores to transmit a
> similar movement to the internal air of the membrane which is called arachnoid.[201]

He develops a similar explanation to describe the phenomenon of acoustic perception, though it is not completely clear what Mersenne is referring to when he
writes 'liquor similar to a very subtle and very light oil' (*liqueur semblable à de
l'huile tres-subtile & tres-claire*).[202]

Investigating the phenomena of the echo and sound reflection, he assumes that
light can be conceived as an 'accident,' which would let us resolve the problems tied
to reflection. If a light ray coincides exactly with the column of air—or if it were
made up of particles—it would be quite difficult to conceive of the reflection and the
intersection of light rays in a single point:

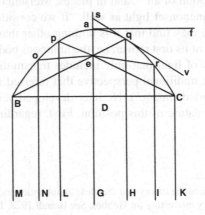

---

[200] Mersenne 1636a, Livre I, 46.

[201] Ibid., 47–48 (my trans.): 'Le mouvement de la lumiere est ce semble plus subtile que celuy des
Sons, & penetre plus avant dans la substance de l'air, qu'il remplit d'une certaine liqueur semblable à de l'huile tres-subtile & tres-claire, qui se meut de telle sorte qu'elle affecte l'œil & le nerf
optique, qui commence à descouvrir tous les objets exterieurs, si tost que l'air esmeu s'est introduit
dans ses pores pour imprimer un semblable mouvement à l'air interieur de la membrane qu'on
appelle *aranée*.'

[202] Here, as elsewhere, we can see the influence of the image of light as 'very spirited, tenuous, and
fast substance', that Galileo suggested in his *Copernican letters* (see Galilei, Letter to Mons. Pietro
Dini, March 23, 1615, *OG*, V, 301–303), and Descartes' conception of subtle matter, which he
describes as 'matiere fort subtile et fort fluide, qui s'estende sans interruption depuis les Astres
iusques à nous.' Descartes, *La dioptrique*, *AT*, VI, 87. On the Descartes' concept of light, see Roux
1997, 49–66.

And it is very difficult to imagine how all the light which passes through the plane BC (*see figure*) (which is supposed to be as wide as the sky) can be collected in a point, since there is no point in this surface which is not covered and filled. Consequently, this light is continuous, without any pores and without any void, and that this gathering at the point *e* cannot be made without penetration of an infinity of rays which condense themselves to infinity. Nevertheless, it is, it seems to me, even more difficult to understand how all the solid of the air that strikes the mirror *a*CB, is reflected at the point *e*. Therefore, it can be said that light is an accident which is not determined in any place, that it cannot occupy and cover a larger place, or a lesser one. But the air is a body, whose different parts cannot naturally be penetrated; and although it has an infinity of small empty spaces, nevertheless it cannot be reduced to a point like the light.[203]

Throughout *Harmonie*, many similarities and comparisons are made between optics and music, though Mersenne focuses particularly on sound, and presents fewer ideas on light than on sound. We can, nonetheless, still suppose that some of Mersenne's briefly delineated ideas had an effect on Hobbes' thinking if we consider some Hobbes' optical texts.

Alan Shapiro, one of the first scholars to look in depth at Hobbes' optics, had assumed that he was influenced by Mersenne to develop a new theory of light, based exclusively on the idea of the transmission of motion in the medium.[204] Indeed, until recently, Hobbes was almost unanimously considered the author of the *Short tract on first principles*,[205] a text which includes the theory of *species* (envisioning the passage of particles from the light source to the sentient being's receptors), and Shapiro suggested he was 'converted' to a mediumistic theory of light during the years of his third and final Grand Tour.[206] Many clues now lead us to believe that the *Short tract* is actually by Robert Payne, and therefore we cannot be certain that Hobbes had already conceived the origins of a new optical theory before 1634.[207]

---

[203] Mersenne 1636a, Livre I, 49: 'Or encore qu'il soit tres-difficile de s'imaginer comment toute la lumiere qui passe par le plan BC, (quoy qu'on la suppose aussi large que le Ciel) peut estre rassemblée dans un point, attendu qu'il n'y a nul point dans ladite surface qui n'en soit couvert & rempli, & consequemment que ladite lumiere est continuë sans aucuns pores & sans aucune vuide, & que ce rassemblement au point *e* ne se peut faire sans penetration d'une infinité de rayons qui se condensent jusques à l'infini, neantmoins il est ce me semble encore plus difficile de comprendre comment tout le solide de l'air qui va frapper la glace *a*CB, se reflechit au point *e*; car l'on peut dire que la lumiere est un accident qui n'est pas tellement determiné aux lieux, qu'il ne puisse occuper & couvrir tantost un plus grand lieu, & tantost un moindre: mais l'air est un corps, dont les differentes parties ne peuvent naturellement se penetrer: & bien qu'il eust une infinité de petits espaces vuides, neantmoins il ne peut estre reduit à un point comme la lumiere.'

[204] See Shapiro 1973, 166–167. See also Beaulieu 1990, 84; Malcolm 2002, 123–125; Giudice 2015, 151.

[205] See below, *Appendix*.

[206] Hobbes' first mention of a theory of light that considers only the transmission of motion in the medium is in the letter to William Cavendish, Earl of Newcastle, mentioned above several times, from October 16/26, 1636.

[207] Hobbes says on several occasions that he started considering the nature of light around 1630. See, for example, Mersenne's letter for Descartes from March 1641 (Hobbes to Marin Mersenne, March [20/] 30, 1641, *CH*, I, 102–103), and the mention in the *First Draught*. See Hobbes, *FD*, not num. (f. 3 *r*) / 76–7. On the authorship of the *Short Tract* see the appendix, below.

However, in the *Appendix* to the Latin edition of *Leviathan* (1668), Hobbes seems to suggest that in his early years he supported the theories of *species*,[208] and though we do not have the facts to settle the question, it is significant that in Hobbes' correspondence we find a profound interest in optics and the theories of vision in the years 1634–1636. The above-mentioned letter to William Cavendish in Newcastle, from October 1636, has some first mentions of a complex theory of vision, explaining the phenomenon only by referring to the motion of the medium pushed by the action of the light source.[209] These ideas about vision return in *Elements of law,* of which the second chapter of the first part is entirely devoted to this question.[210] Considering the Latin treatises on optics, we can see several correlations between Hobbes' and Mersenne's thinking, and they prove deeper than mere parallels.

First, in *Tractatus opticus I* (which was published in Mersenne's *Universae geometriae mixtaeque synopsis* from 1644), Hobbes wrote that his theory did not consider the passage of any particle of the light source to the sentient being.[211] In rejecting the theory of *species*, he offered his own alternative solution, which involved no particles entering the eye: the propagation of light occurs only through the medium, as the source generates motion in the surrounding air, which in turn transmits the adjacent air to the objects.[212] Hobbes assumes that the light source is subject to a contraction and expansion motion that he terms systolic and diastolic, drawing a physiological analogy. Indeed, that which the sentient being perceives as

---

[208] Hobbes, *Appendix ad Leviathan*, in Hobbes 2012, 1185 (*OL*, III, 537): 'Memini tamen quod *Corpus* putarem aliquando id solum esse, quod Tactui meo vel Visui obstaret. Itaque speciem quoque corporis in speculo, aut somno, aut tenebris apparentem, quanquam miratus, corpus tamen esse arbitrabar. Sed consideranti postea Species illas evanescere, ut quarum existentia dependeret non a seipsis, sed a natura animata, non amplius mihi visae sunt reales, sed Phantasmata & effectus rerum in organa sensuum, agentium; & proinde esse *incorporeas*.' Karl Schuhmann quotes this passage in support of his theory that the *Short Tract* was written by Hobbes, see Schuhmann 1995b, 20. In Hobbes 1988a, b, 38, we read indeed that: 'Species are substances'.

[209] Hobbes to William Cavendish, Earl of Newcastle, *CH*, I, 37–38. Jean Bernhardt and Yves Charles Zarka have noted that some of Hobbes' theories (such as that light is only motion, that its propagation is instant, and the rejection of any concept of *species*) were already found in Descartes' *Dioptrique* (See Bernhardt 1979, and Zarka, 1988, 86–87), which Hobbes only received in October 1637, as shown by the letter he sent to Sir Kenelm Digby on October 4/14 1637 (See Sir Kenelm Digby to Hobbes, October 4/14, 1637, *CH*, I, 51), which means we can say that Hobbes formulated his theory independently. See also Marquer 2005, 15–16. On the controversy between Descartes and Hobbes about *Meditations*, see Curley 1995, 97–109; Guenancia, 1990. See also Marion 2005. The controversy continued later as well, as Gianluca Mori has well demonstrated, see Mori 2010a, 2010b, 166–172. On Hobbes' materialism in *Objections*, see Terrel 2016, 13–31.

[210] *EL*, Part I, chap. II, 3–7.

[211] *TO I, OL*, V, 217: '…in visione, neque objectum, neque pars ejus quæcunque transit a loco suo ad oculum.'

[212] Ibid., 217–218: 'Ut motus possit motum generare ad quamlibet distantiam, non est necessarium ut corpus illud a quo motus generatur, transeat per totum illud spatium per quod motus propagatur; sufficit enim ut parum, imo insensibiliter motum, protrudat id quod proxime adstat; nam id quod adstat, pulsum suo loco, pellit quoque quod est proximum sibi, atque eo modo motus propagabitur quantum libueris…'

'*scintillationem:*' 'is nothing but … *systole* and *diastole*.'[213] He explains how this phenomenon happens and how visual perception forms:

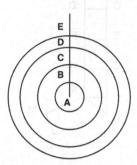

For, at the very moment when the movement starts from B to C (*see figure*), the movement must start from C to D, and from D to E, and from E further on. That is why, if the eye is placed at any distance from the Sun, let us suppose in E, at the very moment when the Sun begins to expand at B, at the same instant, the eye is struck in E. Therefore, the movement will be propagated to the retina and from there to the brain through the optic nerve due to the *conatus* of the retina; and this happens at the same time that the movement starts in B.[214]

This model of light's circular propagation brings to mind Mersenne's speculations, such as the idea that 'light propagates instantly at any distance.'[215] We should note, however, that Hobbes' theory is much more elaborate and that he extensively discusses the nature of the light ray. However, just as Mersenne had argued in *Harmonicorum libri*,[216] Hobbes also argues that though light propagates circularly, it is made up of rays that expand in the surrounding space, starting from the light source. He maintains that, in terms of the ray's entity and physical constitution, it should be considered a solid space[217]—because motion could not otherwise propagate itself—but he also saw the need to clarify that it does not consist exactly in physical space nor in air:

---

[213] Ibid., 218 (my italics).

[214] Ibid., 219–220 (my trans.): 'Nam quo instante incipit motus a B versus C, necesse est ut incipiat motus a C versus D, et a D versus E, et ab E prorsum. Quare si statuatur oculos in qualibet distantia a sole, puta in E: quo istante incipit sol dilatare se in B, eodem ferietur oculus in E. Unde propagabitur motus ad retinam, et inde per conatum retinae nervum optimum usque ad cerebrum: et hoc fit eodem instante, quo motus incipit in B.' See also Prins 1996, 133 ff.

[215] *TO I, OL*, V, 221.

[216] In Mersenne 1636b, 3, Mersenne writes that, though both light and sound are propagated 'in orbem'. Nevertheless, 'soni non solum linea recta, seu directis radiis, sed etiam circularibus, ellipticis, parabolicis, & aliis, quibusvis lineis feruntur ad aurem; lux vero solummodo rectis lineis fertur ad oculum, sive directis, sive reflexis, sive refractis.'

[217] *TO I, OL*, V, 222: 'Quoniam enim radius est via per quam motus projicitur a lucido, neque potest esse motus nisi corporis: sequitur radium locom esse corporis, et proinde habere tres dimensiones. Est ergo radius spatium solidum.'

I call *ray* the way through which the movement propagates from the light source through
the medium. For example: suppose that the light source is in AB (*see figure*), from which it
is moved to the part of the medium which lies between AB and CD, and it is pushed to EF,
and from the part of the medium that was between CD and EF, advancing further to GH, it
pushed that part which was between EF and GH, going further to IK, and so on in succession, whether directly or not, for example, in the direction of LM. The space which was
contained between the lines AIOL and BKM, is what I call ray, or the way through which
the motion is transmitted from the light source through the medium.[218]

In *Tractatus Opticus II*, Hobbes repeats the idea that the ray does not at all coincide with air but, rather, with the motion that is propagated through the air (without
changing the fact that no motion can be conceived without a medium that consists
of a physical body).[219] Therefore, the ray is often inaccurately called '*radium*;' a
more correct definition would be '*radiatio*.'[220]

From what we can glean from the above passages, Hobbes seems to have encountered the same problem that Mersenne raised: though the light phenomenon may
coincide with the motion of air, completely identifying light with the air proved
quite problematic, especially if we consider phenomena, such as refraction and
reflection, to which he gave ample attention. Closely linked to the definition of the
ray is that of 'line of light' (*linea lucis*)[221] and, using these two terms (along with

---

[218] Ibid., 221–222 (my trans.): '*Radium* appello, viam per quam motus a lucido per medium propagatur. Exempli gratia: sit lucidum AB, a quo moto ad CD pars medii quae interjacet inter AB et CD,
protrudatur ad EF: et a parte medii quae erat inter CD et EF, promota ulterius ad GH, propellatur
pars illa quae erat inter EF et GH, ulterius ad IK, et sic deinceps, sive directe sive non, puta versus
LM. Spatium jam quod continetur inter lineas AIOL, et BKM, est id quod voco radium, sive viam
per quam motus a lucido per medium propagatur.'

[219] *TO II*, chap. II, § 2, f. 204v/p. 160: 'Cum vero directa haec motus a lucido propagatio, non sit
ipsum Corpus per quod motus propagatur (nam differentia magna est jnter ipsum aerem et motum
in aere) neque aliud corpus praeter ipsum, non potest radius lucis dici corpus, ut radius rotae lig-
neae lignum, sed tantum via motus propagatio.'

[220] Ibid.

[221] *TO I*, *OL*, V, 222–223: 'Lineam unde radii latera incipiunt: exempli gratia, lineam AB, unde
incipiunt latera AI et BK: appello lineam *lucis simpliciter.* Linearum autem quae a linea lucis con-

that of *propagated line of light*), Hobbes developed the concept of *wave front*.[222] The very concept of ray (a key element for understanding the relationship that forms in Hobbes' system between geometric optics and physical optics[223]) implies the idea of an infinitesimal portion of the wave front. Considering it an infinitesimal element and ignoring its physical dimension, Hobbes manages a transition from physical rays to mathematical ones, though 'we must consider the width of the ray, inferior to any given size.'[224]

Yet, this transition of the ray, from a physical size to a geometric entity, is not without its problems, as seen in the second optical treatise:

> Since motion cannot be imagined except in a body, and since each body has three dimensions, length, breadth, and depth, the path of motion must also have these dimensions. Thus, the ray is not a length without width, but a solid whose length ends with the surface of the luminous or radiating body. Although it may sometimes be considered not as a surface but as a point, and yet, with reasoning, [because] the magnitude of the object or luminous body is not considered; but we do not say that each mathematical point, or line, or surface does not have dimensions in reality, but [we only say] that these [dimensions] are not considered in the argument.[225]

First, Hobbes repeats that no motion can be propagated without bodies, and then that the points, lines, and surfaces are not without dimension, but this dimension is not considered in the mathematical demonstration.

Nonetheless, if we consider the ray as '*radiatio*' and not as solid space, this lets further ideas be developed:

---

tinua protrusione derivantur, quales sunt CD, EF etc., unamquamque appello lineam lucis *eousque propagatam*'.

[222] See Shapiro 1973, 150–151. See also Bernhardt 1977, 9–10; Médina 1997, 40–41; Giudice 1999, 69 ff.

[223] For a discussion of this topic in *First draught,* see Stroud 1983, 39–40; Giudice 1999, 70; Médina 2016, 49 ff.

[224] *TO I, OL,* V, 228. See also Shapiro 1973, 160–161.

[225] *TO II,* chap. II, § 2, f. 204*v*/160 (my trans.): 'Rursum quoniam motus intelligi non potest nisi in corpore, habeatque omne Corpus, tres dimensiones, Longitudinem, Latitudinem, et crassitiem, necesse est ut etiam via motus constet dimensionibus iisdem, Non est ergo radius longitudo sine latitudine, sed solidum, cuius longitudo terminatur superficie corporis lucidi sive radiantis; quamquam possit interdum illa considerari non ut superficies, sed ut punctum, nimirum cum ratiocinatione, obiecti sive lucidi magnitudo non consideratur; neque dicitur aliquid punctum vel linea, vel superficie mathematica propterea quod dimensionibus careat, sed quia in argumentum non assumuntur.'

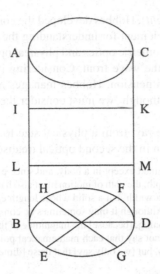

In the radiation, however, we consider the length, and the width, but not the depth. This last indeed can be found from an established width in all the positions. For example, if we consider a radiation cylindrical, the length will be AB or CD (*see figure*), the width will be BD, or EF, or GH, and what can be demonstrated about one width will be demonstrated for all. We understand from one width how all the radiation advances, since we must consider more dimensions than lenghth and width if it was not a contemplation of the radiation. On the other hand, the radiation which I consider here is that which comes from any minimal and imperceptible part of the light. However, because both dimensions must be considered I will give to each radiation a certain width which is enough to mark it in notes, or in letters, by which each dimension can be distinguished and enumerated more comfortably; because the width, once the demonstration is finished, can be reduce by anyone imagining the slimness of the line. Radiation is therefore called length which comes from A.[226]

As Shapiro was the first to note,[227] the conceptual shift that Hobbes made, considering light rays both as physical entities and as geometric lines, made it possible for him to apply refraction to curved surfaces. In such surfaces, each single point on the ray front has a different angle of incidence with the surface.[228] In *Tractatus opti-*

---

[226] Ibid. (my trans.): 'In radiatione vero considerabimus longitudinem, et latitudinem non autem Crassitiem; haec enim intelligitur ex sumpta latitudine per omnes positiones; Exempli causa, si fiat radiatio cylindrice, secundum longitudinem AB vel CD, latitudo erit BD, vel EF, vel GH, cum autem quod demonstratur de una latitudine, demonstretur de omnibus, intelligetur ex una latitudine quomodo tota radiatio solida progredietur, ut non opus sit in radiationis contemplatione, plures dimensiones considerare quam longitudinem et latitudinem. Rursum quamquam radiationem hic considero eam quae fit a qualibet minima et imperceptibile parte lucidi, tamen quia et sic ambae dimensiones aliquando contemplandae sunt, dabo omni irradiationi latitudinem conspicuam, quantum sufficit adscriptioni notarum, sive literarum, quibus commodius omnis dimensio distingui et nominari possit, quam latitudinem, finita demonstratione, ad exilitatem linearem revocare imaginatione sua unusquisque potest; radiationem ergo vocabimus longitudinem quae est ab A.

[227] See Shapiro 1973, 161.

[228] It is not necessary to analyze the refraction phenomenon in detail here, as there is an extensive thorough literature on it (see Médina 1997, 40–41; Giudice 1999, 70 ff.; Horstmann 2000. We can

cus I, he addresses the question of the incident ray obliquely in a non-uniform medium. He argued that the refraction develops as if this ray hit the point of contact with a flat surface contingent to that curve.[229] To demonstrate it, we must assume that the length of the front of the ray is 'less than any given size' (*minorem quavis magnitudinem data*).[230] In his second optical treatise (and later in the *First draught*), Hobbes explained his thinking and clarified that the rays are not perfectly parallel because they are conical. However, because of their infinitesimal constitution, they can also not be considered rectangular as the difference is imperceptible.[231]

If we consider the ray's incidence on a flat surface, we encounter a problem: in his first optical treatise, Hobbes argued that we can consider the incident ray perpendicularly on a flat surface, such as a mathematical line, but this is impossible if the ray intersects the plane obliquely[232]:

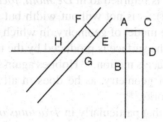

We can therefore consider the ray ABCD (*see figure*) without width, that is to say, as a mathematical line. But in the oblique incidence, where the distance is greater in the action on the plane from F to H than from E to G, it is impossible to consider EFGH as a mathematical line; because if EF is considered as a mathematical point, it must nevertheless be considered that it acts further in one side than in the other. Therefore, it must be considered as having sides, not as a point. If we consider the oblique incident oblique line as mathematical, we must consider EF as a point, and what is absurd, as a non-point.[233]

---

simply remark that in considering refraction, Hobbes argued that when a ray of light propagates uniformly, whether it be dense or rare, it behaves like a cylinder, but when one part of the ray propagates less easily than the other, as in refraction, it behaves like a 'truncated cone'. Giudice 1999, 72. See also Giudice 2015, 156.

[229] *TO I, OL*, V, 245.

[230] Ibid., 228.

[231] See *TO II*, chap. II, § 2, f. 204v/160 The topic is discussed in more detail in *First draught*, see *FD*, fol. 13 ff./122 ff. See also Stroud 1983, 39 ff.

[232] *TO I, OL*, V, 225.

[233] Ibid., 226 (my trans.): 'Possumus ergo considerare radium ABCD sine latitudine, hoc est ut linea mathematica. Sed incidentia obliqua, ubi operatio ab F ad planum in H in majori est distantia quam ab E in G, non potest considerari EFGH ut linea mathematica: quia sic consideratur EF ut punctum mathematicum, quod tamen consideratur uno termino operari longius quam altero, hoc est, consideratur ut habens terminos, hoc est, non ut punctum. Itaque si consideraremus lineam oblique incidentem ut mathematicam, consideraremus EF ut punctum et non punctum, quod est absurdum.'

The analysis of the ray's incidence on a curved surface is only possible if we consider rays as mathematical lines, though in the case of a flat surface, mathematical lines can be used only in considering the perpendicular incidence of the rays. If, however, we consider the oblique incidence, we must use the physical dimension. In *Tractatus opticus II*, Hobbes maintains, however, that the term *ray* can be used incorrectly and lead philosophers into two errors.[234] As we know, the first error is to consider the ray a body. Hobbes did not consider light to consist truly in the medium through which the motion is propagated but in the motion itself.[235] The second error is considering the ray a simple mathematical line that only has length and lacks width.

In the second optical treatise, these ideas on the nature of the ray are closely linked to the nature of geometric entities. Hobbes emphasizes that they are not at all lacking dimension but that their dimension is not taken into consideration in the demonstration.[236] The topic is returned to in *De motu, loco et tempore*,[237] where he states that 'γραμμήν (line) … is not without width but is considered as such.'[238] He presents his constructive model of geometry, in which the line is generated by a point's motion, and just as the surface is produced by the motion of a line, the body is drawn, in turn, by the surface's motion.[239] Hobbes again suggests here a 'physical' and 'materialist' concept of geometry, as he does in all of his subsequent philosophical and mathematic works.[240]

We have seen, however, that particularly in *Tractatus opticus II* (and later in the *First draught*), a fundamental link emerges between Hobbes' optical inquiries and his developing the concept of geometrical entities and geometry in general.[241] Hobbes had developed his position based on thinking about the nature of the light ray, a problem that Mersenne had already grappled with in *Harmonie universelle*.

The relationship between geometric optics and physical optics is linked to the relationship that Hobbes sees, in general, between physics and geometry. He presents a clear dichotomy in the realm of the sciences, in his philosophical and scientific works. However, this distinction does not mark a clean, impassable split between the mathematical sciences (that can obtain the level of philosophy) and physics, which must settle in the limits of opinion. Of interest on this topic is a passage of Chapter 10 of *De homine*, about the classification of the disciplines.[242] Here he says, 'One cannot proceed in reasoning about natural things that are brought

---

[234] *TO II*, chap. II, § 1, f. 204*r*/159–160.

[235] Ibid., chap. II, § 3, f. 205*r*/160.

[236] Ibid.

[237] See *MLT*, II, 8, 114.

[238] Ibid. My translation.

[239] Ibid., 114–115.

[240] See *De corpore*, VIII, 12, *OL*, I, 98–99; *Six Lessons*, *EW*, VII, 200–201. See Jesseph 1999, 79 ff. On the concept of the point in Hobbes' mathematics see also Grant 1996a, b, 112 ff.

[241] See Médina 2015, 82.

[242] See *De homine*, X, 5, *OL*, II, 93. As Médina has demonstrated, Hobbes' definition of mixed mathematics follows that of Aristotle (See Aristotle, *Posterior analytics*, I, 13, 79a 7–10; *Physics*, II, 194a 10). See Médina 2015, 56.

about by motion from the effects to the causes without a knowledge of those things that follow from that kind of motion,' and 'one cannot proceed to the consequences of motions without a knowledge of quantity,' which is geometry.

As a result, he considers the demonstration to always be necessary *a priori*, which is a mathematical-geometric demonstration because the variety of motions cannot be fully understood without knowing the science that measures sizes and figures. In other words, Hobbes argues that physics must necessarily be the foundation of the *geometricization of reality*, i.e. the process of translating nature into a mathematical language of which Galileo is the 'patron saint.' Significantly, in Hobbes, the foundations and objects of geometry and physics partly coincide: only bodies are the object of philosophy[243] and the principles of what Hobbes calls the 'first philosophy' are in the motion of these bodies.[244]

This is why Hobbes' thoughts on the foundations of sciences—which has often been termed 'post-sceptic'[245]—though influenced by the problemetizing of Pyrrhonian themes, as seen in Mersenne (and Gassendi), are also worth investigating in the light of Galileo's thought. Though Galileo is the founder of physical sciences, this is, of course, not because he was the first to investigate natural phenomena, but rather he was the first to conceive the correct scientific method, giving physics a mathematical structure. Mathematicizing and translating the Aristotelian physics that had run aground on the concept of *quality*, in quantitative terms[246] is the most important undertaking in natural philosophy, which is why Galileo was rightly called: 'the greatest philosopher not only of our century, but of all time.'[247]

---

[243] *De corpore*, I, 8, *OL*, I, 9.

[244] *De corpore*, VI, 5, *OL*, I, 62: 'Causae autem universalium (eorum quorum causae aliquae omnino sunt) manifestae sunt per se sive *naturae* (ut dicunt) nota ... causa enim eorum omnium universalis una, est motus; nam et figurarum omnium varietas ex varietate oritur motuum quibus construuntur.' The words that François du Verdus wrote on the subject to Hobbes in December 1655 are of interest: '...puis que les Corps agissent selon [leurs grandeurs et leurs figures *altered to* leur grandeur et leur figure] et par leurs mouvemens qui sont les Objects de la Géometrie et des Mécaniques, il est impossible d'estre Philosophe a moins que d'estre Géométre Et aussi tot j'ai veu ces Gens-là (*which he had termed* 'Pédans' *and* 'perroquets') s'éfaroucher croyans qu'une Philosophie Geometrique c'est a dire la vraye Philosophie [tendit *deleted* > tende] a renverser les verités les plus importantes. Mesmes quand je luy ay allegué que du vray on ne peut rien inferer que de vray [Et que mesmes je leur ay fait *deleted* > jusqu'a leur faire] voir dans l'Écriture Sainte la Nécéssité qu'il y a d'estre Geométre pour estre Philosophe où il est dit Que Dieu á fait toutes choses *En poids En nombre En mesure*. Car (dis-je a ces Messieurs là) *la Mesure* veut dire l'étenduë et la grandeur du Corps, *le nombre* veut dire en Effect le nombre de ses Angles, qui fait la figure; Et *le Poids* c'est l'effort a tendre vers quelque part.' François du Verdus to Hobbes, from Bordeaux, December [13/] 23, 1655, *CH*, I, 217.

[245] See Tuck 1988. See also Malcolm 2002, 184–189 and esp. Paganini 2003, 29 ff. See also Paganini 2004a, and Paganini 2015.

[246] See Leijenhorst 2002, 84 ff.

[247] *MLT*, X, 9, 178; Eng. Trans. Hobbes 1976, 123 modified.

References

Works of Thomas Hobbes:

Hobbes, Thomas. 1968. *Elementi di legge naturale e politica*, ed. Arrigo Pacchi. Florence: La Nuova Italia.
———. 1976. *Thomas White's De Mundo Examined*, ed. Harold Whitmore Jones. Bradford: Bradford University Press.
———. 1988a. (probably by Robert Payne), *Court traité des premiers principes*, ed. Jean Bernhardt. Paris: Presses Universitaires de France.
———. 1988b. *Of Passions* (British Library: Ms. Harl. 6083), ed. Anna Minerbi Belgrado. *Rivista di Storia della Filosofia* (4): 729–738.
———. 2012. *Leviathan*, ed. Noel Malcolm, 3 Vols. Oxford, Clarendon Press.

Other Works:

Bacon, Roger. 1897–1900. *Opus Maius*, ed. John Henry Bridges, 2 Vols. Oxford: Clarendon Press. (Reprint: Frankfut am Main: Minerva, 1964).
Campanella, Tommaso. 2007. *Del senso delle cose e della magia*, ed. Germana Ernst. Rome-Bari: Laterza.
Clavius, Christoph. 1591. *Euclidis elementorum libri XV*. Köln: Ciotti.
Gassendi, Pierre. 1658. *Opera Omnia*, 6 Vols. Lyon: Anisson, & Devenet, 1658 (Stuttgart-Bad Cannstatt: Fromman Holzboog, 1994).
Gilbert, William. 1600. *De Magnete*. London: Peter Short.
Mersenne, Marin. 1623. *Quaestiones Celeberrimae in Genesim...* Paris: Cramoisy.
———. 1625. *La Verité des Sciences. Contre les Sceptiques ou Pyrrhonines*. Paris: Toussainct du Bray. (Reprint: Stuttgart-Bad Cannstatt: Friedrich Frommann Verlag, 1969).
———. 1634. *Les Preludes de l'harmonie universelle ou questions curieuses*. Paris: Henry Guenon (Reprint in Mersenne. 1985: 515–660).
———. 1634. *Traité des Mouvemens, et de la chute des corps pesans, & de la proportion de leurs differentes vitesses. Dans lequel l'on verra plusierurs experiences tres-exactes*. Paris: Jacques Villery (Reprint in *Corpus*, 2/1986: 25–58).
———. 1636a. *Harmonie Universelle*, 1st Vol. Paris: Sebastien Cramoisy.
———. 1636b. *Harmonicorum libri, in qvibus agitur de sonorvm natvra, cavsis, & effectibus...* Paris: Gvillelmi Bavdry.
———. 1637. *Harmonie Universelle*, 2nd Vol. Paris: Pierre Ballard.
———. 1639. *Les nouvelles pensees de Galileée, Mathematicien et Ingenieur du Duc de Florence*. Paris: Pierre Rocolet.
———. 1985. *Questions Inouyes*, (Corpus des œuvres de philosophie en langue française) (Paris: Fayard, 1985).
———. 2003. *Traité de l'Harmonie Universelle*, ed. Claudio Buccolini. Paris: Fayard (or. Ed. 1627).

Secondary Sources:

Alessio, Franco. 1985. *Introduzione a Ruggero Bacone*. Rome/Bari: Laterza.
Aquinas, Thomas. 1964. *Quaestiones Disputatae*. 2 Vols. Turin-Rome: Marietti.

Armogathe, Jean-Robert. 1992. Le groupe de Mersenne et la vie académique parisienne. *Dix-septième siècle* 175: 131–139.

Auger, Léon. 1948. Le R. P. Mersenne et la physique. *Revue d'histoire des sciences et de leurs applications* 2 (1): 33–52.

Bailhache, Patrice. 1994. L'harmonie universelle: la musique entre les mathématiques, la physique, la métaphysique et la religion. In *Études sur Marin Mersenne*, Special issue of *Les Études Philosophiques* 68 (Iss. 1–2), ed. Jean-Robert Armogathe and Michel Blay, 13–24.

Baldin, Gregorio. 2016b. 'La reflexion de l'arc' et le *conatus*: aux origines de la physique de Hobbes. *Philosophical Enquiries. Revue des philosophies anglophones* 7: 15–42.

———. 2017. Archi, spiriti e *conatus*. Hobbes e Descartes sui principi della fisica. *Historia Philosophica* 15: 147–165.

Beaulieu, Armand. 1982a. Découverte d'un livre de Mersenne. *Revue d'histoire des sciences* 25: 55–56.

———. 1982b. Lumière et matière chez Mersenne. *Dix-Septième Siècle* 136 (3): 311–316.

———. 1984. Les réactions des savants français au début du XVIIe siècle devant l'heliocentrisme de Galilée. In *Novità celesti e crisi del sapere*, ed. Paolo Galluzzi, 373–381. Florence: Giunti-Barbera.

———. 1986. Mersenne et l'Italie. In *La France et l'Italie au temps de Mazarin*, ed. Jean Serroy, 69–77. Grenoble: Presses Universitaires de Grenoble.

———. 1989. Torricelli et Mersenne. In *L'œuvre de Torricelli: science galiléenne et nouvelle géométrie*, ed. François De Gandt, 39–51. Paris: Les Belles Lettres.

———. 1990. Les relations de Hobbes et de Mersenne. In *Thomas Hobbes. Philosophie première, théorie de la science et politique*, ed. Yves-Charles Zarka and Jean Bernhardt, 81–90. Paris: Presses Universitaires de France.

———. 1992. Le group de Mersenne. Ce que l'Italie lui a donné—Ce qu'il a donné à l'Italie. In *Geometria e atomismo nella scuola galileiana (Atti del convegno)*, ed. Massimo Bucciantini and Maurizio Torrini, 17–34. Florence: Olschki.

———. 1993. L'attitude nuancée de Mersenne vers la chimie. In *Alchimie et philosophie à la Renaissance*, ed. Jean-Claude Margolin and Sylvain Matton, 395–403. Paris: Vrin.

———. 1995. *Mersenne. Le grand minime*. Bruxelles: Fondation Nicolas-Claude de Peiresc.

Bernhardt, Jean. 1977. Hobbes et le mouvement de la lumière. *Revue d'histoire des sciences* 30: 3–24.

———. 1979. La polémique de Hobbes contre la *Dioptrique* de Descartes dans le *Tractatus Opticus II* (1644). *Revue Internationale de Philosophie* 33: 432–442.

———. 1985a. Nominalisme et mécanisme chez Hobbes. *Archives de philosophie* 48: 235–249.

———. 1985b. Sur le passage de F. Bacon à Th. Hobbes. *Les Études Philosophiques* (4): 449–457.

Berthier, Jauffrey. 2015. Prudence juridique et prudence politique chez Bacon et Hobbes (I): Le philosophe, le juriste et l'homme politique chez Francis Bacon. *Philosophical Enquiries: Revue des philosophies anglophones* 4: 93–126. http://www.philosophicalenquiries.com/numero4article5Berthier.html.

Bianchi, Luca. 2003. *Studi sull'aristotelismo del Rinascimento*. Padua: Il Poligrafo.

Bianchi, Luca, and Eugenio Randi. 1990. *Le verità dissonanti*. Rome-Bari: Laterza.

Bitpol-Hespériès, Annie. 2005. *L'Homme* de Descartes et le *De Homine* de Hobbes. In *Hobbes, Descartes et la métaphysique*, ed. Dominique Weber, 155–186. Paris: Vrin.

Blay, Michel. 1994. Mersenne expérimentateur: les études sur les mouvements des fluides jusqu'en 1644. In *Études sur Marin Mersenne*, Special issue of *Les Études Philosophiques* 68 (Iss. 1–2), ed. Jean-Robert Armogathe and Michel Blay, 69–86.

Bloch, Olivier René. 1971. *La philosophie de Gassendi: Nominalisme, matérialisme et métaphysique*. The Hague: Martinus Nijhoff.

Brandt, Frithiof. 1928. *Thomas Hobbes' Mechanical Conception of Nature*. Copenhagen/London: Levin & Mungsgaard-Librairie Hachette (or. Ed. 1921).

Breidert, Wolfgang. 1979. Les mathématiques et la méthode mathématique chez Hobbes. *Revue internationale de philosophie* 129: 414–431.

Bucciantini, Massimo. 2003. *Galileo e Keplero. Filosofia, cosmologia e teologia nell'Età della Controriforma*. Turin: Einaudi.

———. 2007. Descartes, Mersenne e la filosofia invisibile di Galileo. *Giornale critico della filosofia italiana* 86 (1): 38–52.

Buccolini, Claudio. 1997. Il ruolo del sillogismo nelle dimostrazioni geometriche della *Vérité des Sciences* di Marin Mersenne. *Nouvelles de la République des Lettres* 1: 7–36.

———. 2002. Mersenne traduttore di Bacon? *Nouvelles de la République des Lettres* (2): 7–19.

———. 2014. Mersenne et la philosophie baconienne en France à l'époque de Descartes. In *Bacon et Descartes. Genèse de la modernité philosophique*, ed. Élodie Cassan, 115–134. Lyon: ENS Éditions.

———. 2015. Il 'mos geometricus' fra usi teologici ed esiti materialistici: le obiezioni di Mersenne contro la metafisica cartesiana. In *La ragione e le sue vie. Saperi e procedure di prova in età moderna*, ed. Carlo Borghero and Claudio Buccolini, 59–81. Florence: Le Lettere.

Bunce, Robin. 2003. Thomas Hobbes' Relationship with Francis Bacon – An Introduction. *Hobbes Studies* 16: 41–83.

Califano, Salvatore. 2010. *Storia della chimica*, 2 Vols. Turin: Bollati Boringhieri.

Carraud, Vincent. 1994. Mathématique et métaphysique: les sciences du possible. In *Études sur Marin Mersenne*, Special issue of *Les Études Philosophiques* 68 (Iss. 1-2), ed. Jean-Robert Armogathe and Michel Blay, 145–159.

Clericuzio, Antonio. 2000. *Elements, Principles and Corpuscles. A Study of Atomism and Chemistry in Seventeenth Century*. Dordrecht: Kluwer.

Costabel, Pierre. 1994. Mersenne et la cosmologie. In *Quatrième centenaire de la naissance de Marin Mersenne (Actes du colloque)*, ed. Jean-Marie Constant and Anne Fillon, 47–55. Le Mans: Université du Maine.

Cozzoli, Daniele. 2007. Light, Motion, Space and Time in Mersenne's Optics. In *Notions of Space and Time. Early Modern Concepts and Fundamental Theories (Begriffe von Raum and Zeit. Frühneuzeitliche Konzepte und Fundamentale Theorien)*, ed. Frank Linhard and Peter Eisenhardt, 29–43. Frankfurt am Mein: Klostermann.

———. 2010. The Development of Mersenne's Optics. *Perspectives on Science* 18 (1): 9–25.

Crapulli, Giovanni. 1969. *Mathesis Universalis. Genesi di un'idea nel XVI secolo*. Rome: Edizioni dell'Ateneo & Bizzarri.

Crombie, Alistair C. 1953. *Augustine to Galileo. The History of Science A.D. 400–1650*. Cambridge, MA: Harvard University Press.

———. 1990. *Science, Optics and Music in Medieval and Early Modern Thought*. London/Ronceverte: The Hambledon press.

Curley, Edwin M. 1995. Hobbes Versus Descartes. In *Descartes and his contemporaries. Meditations, Objections, and Replies*, ed. Roger Ariew and Marjorie Grene, 97–109. Chicago: The University of Chicago Press.

Dal Pra, Mario. 1962. Note sulla logica di Hobbes. *Rivista critica di storia della filosofia* 17: 411–433.

Dear, Peter. 1988. *Mersenne and the Learning of the Schools*. Ithaca/London: Cornell University Press.

Dubos, Nicolas. 2015. Le passage de Bacon à Hobbes: remarques sur la question de l'histoire. *Philosophical Enquiries: Revue des philosophies anglophones* 4: 59–76. http://www.philosophicalenquiries.com/numero4article6Dubos.html.

Ducrocq, Myriam-Isabelle. 2015. Le silence, le secret, la transparence: de l'art du gouvernement à la science civile chez Francis Bacon et Thomas Hobbes. *Philosophical Enquiries: Revue des philosophies anglophones* 4: 77–91. http://www.philosophicalenquiries.com/numero4article-7Ducrocq.html.

Duhem, Pierre. 2003. *Sauver les apparences. Sur la notion de théorie physique de Platon à Galilée*. Paris: Vrin (or. Ed. 1908)

Ernst, Germana. 2010. *Tommaso Campanella. Il libro e il corpo della natura*. Rome/Bari: Laterza.

Fabbri, Natacha. 2003. *Cosmologia e armonia in Kepler e Mersenne. Contrappunto a due voci sul tema dell'*Harmonice Mundi. Florence: Olschki.

———. 2005. Echi della condanna di Galilei in Marin Mersenne. In *I primi Lincei e il Sant'Uffizio. Questioni di scienza e di fede. (Atti del convegno, Roma, 12 e 13 giugno 2003)*, 449–480. Rome: Bardi Editore.

Fattori, Marta. 2000. Fortin de la Hoguette tra Francis Bacon e Marin Mersenne: intorno all'edizione francese del *De Augmentis Scientiarum* (1624). In *Linguaggio e filosofia nel Seicento europeo*, ed. Marta Fattori, 385–411. Florence: Olschki.

———. 2012. *Études sur Francis Bacon*. Paris: Presses Universitaires de France.

Fisher, Saul. 2005. *Pierre Gassendi's Philosophy and Science*. Leiden/Boston: Brill.

———. 2008. Gassendi et l'hypothèse dans la méthode scientifique. In *Gassendi et la modernité*, ed. Sylvie Taussig, 399–425. Turnhout: Brepols.

Fumaroli, Marc. 2015. *La République des Lettres*. Paris: Gallimard.

Galilei, Galileo. 1967. *Dialogue Concerning the Two Chief Wold System*, Eng. Trans. Stillman Drake. Los Angeles-Berkeley: University of California Press.

Galluzzi, Paolo. 1973. Il 'Platonismo' del tardo Cinquecento e la filosofia di Galileo. In *Ricerche sulla cultura dell'Italia moderna*, ed. Paola Zambelli, 39–79. Rome/Bari: Laterza.

———. 1979a. *Momento. Studi galileiani*. Rome: Edizioni dell'Ateneo & Bizzarri.

———. 1979b. Il tema dell''ordine' in Galileo. In *Ordo. Atti del II colloquio internazionale del Lessico Intellettuale Europeo*, ed. Marta Fattori and Massimo L. Bianchi, 235–277. Rome: Ateneo & Bizzarri.

Garber, Daniel. 2004. On the Frontlines of Scientific Revolution: How Mersenne Learned to Love Galileo. *Perspectives on Science* 12 (2): 135–163.

———. 2013. Remarks on the Pre-History of Mechanical Philosophy. In *The Mechanization of Natural Philosophy*, ed. Daniel Garber and Sophie Roux, 3–26. Dordrecht: Springer.

Gargani, Aldo G. 1971. *Hobbes e la scienza*. Einaudi: Turin.

Gaukroger, Stephen. 2001. *Francis Bacon and the Transformation of Early-Modern Philosophy*. Cambridge: Cambridge University Press.

———. 2002. *Descartes' System of Natural Philosophy*. Cambridge: Cambridge University Press.

Giorello, Giulio. 1990. Pratica geometrica e immagine della matematica in Thomas Hobbes. In *Hobbes Oggi*, ed. Bernard Willms et al., 215–255. Milan: Franco Angeli.

Giudice, Franco. 1999. *Luce e visione. Thomas Hobbes e la scienza dell'ottica*. Florence: Olschki.

———. 2015. The Most Curious of Science: Hobbes's Optics. In *The Oxford Handbook of Hobbes*, ed. Aloysius P. Martinich and Kinch Hoekstra, 147–165. Oxford: Oxford University Press.

———. 2016. Optics in Hobbes's Natural Philosophy. *Hobbes Studies* 29 (1): 86–102.

Grant, Edward. 1996a. *The Foundation of Modern Science in the Middle Ages*. Cambridge: Cambridge University Press.

Grant, Hardy. 1996b. Hobbes and Mathematics. In *The Cambridge Companion to Hobbes*, ed. Tom Sorell, 108–128. Cambridge: Cambridge University Press.

Gregory, Tullio. 2000. *Genèse de la raison classique de Charron à Descartes*. Paris: Presses Universitaires de France.

Guenancia, Pierre. 1990. Hobbes—Descartes: le nom et la chose. In *Thomas Hobbes. Philosophie première, théorie de la science et politique*, ed. Yves-Charles Zarka and Jean Bernhardt, 67–79. Paris: Presses Universitaires de France.

Hanson, Donald W. 1991. Reconsidering Hobbes's Conventionalism. *The Review of Politics* 53 (4): 627–651.

Hine, William L. 1973. Mersenne and Copernicanism. *Isis* 64: 18–32.

Horstmann, Frank. 1998. Ein Baustein zur Kepler-Rezeption: Thomas Hobbes'. *Physica Coelestis Studia Leibnitiana* 30 (2): 135–160.

———. 2000. Hobbes und Sinusgesetz der Refraktion. *Annals of Science* 57: 415–440.

———. 2001. Hobbes on Hypotheses in Natural Philosophy. *The Monist* 84 (4): 487–501.

Jacquet, Chantal. 2015. Des mots aux choses: le problème du langage chez Bacon et Hobbes. *Philosophical Enquiries: Revue des philosophies anglophones* 4: 27–39. http://www.philosophicalenquiries.com/numero4article3Jaquet.html.

Jacquot, Jean. 1952a. Notes on an Unpublished Work of Thomas Hobbes. *Notes and Records of the Royal Society of London* 9: 188–195.

———. 1952b. Sir Charles Cavendish and His Learned Friends: A Contribution to the History of Scientific Relation between England and the Continent in the Earlier Part of the 17th Century, *Annals of Science* 8 (1): 13–27; and. 8: (2): 175–191.

Jesseph, Douglas M. 1999. *Squaring the Circle. The War Between Hobbes and Wallis*. Chicago/ London: University of Chicago Press.

———. 2006. Hobbesian Mechanics. *Oxford Studies in Early Modern Philosophy* 3: 119–152.

———. 2013. Logic and Demonstrative Knowledge. In *The Oxford Handbook of British Philosophy in the Seventeenth Century*, ed. Peter Anstey, 373–392. Oxford: Oxford University Press.

Jullien, Vincent. 2006. Gassendi, Roberval à l'académie de Mersenne. Lieux et occasions de contact entre ces deux auteurs. *Dix-septième siècle* 233 (4): 601–613.

Laudan, Larry. 1981. *Science and Hypothesis. Historical Essays on Scientific Methodology.* Dordrecht: Springer.

Leijenhorst, Cees. 1997. Motion, Monks and Golden Mountains: Campanella, and Hobbes on Perception and Cognition. *Bruniana & Campanelliana* 3 (1): 93–121.

———. 2002. *The Mechanisation of Aristotelianism. The Late Aristotelian Setting of Thomas Hobbes' Natural Philosophy*. Leiden/Boston/Köln: Brill.

———. 2005. La causalité chez Hobbes et Descartes. In *Hobbes, Descartes et la métaphysique*, ed. Dominique Weber, 79–119. Paris: Vrin.

Lenoble, Robert. 1943. *Mersenne ou la naissance du mécanisme*. Paris: Vrin.

Lerner, Michel-Pierre. 2001. La réception de la condamnation de Galilée en France. In *Largo campo di filosofare. Eurosymposium Galileo 2001*, ed. José Montesinos and Carlos Solis, 513–547. La Orotava: Fundaciòn Canaria Orotava de Historia de la Ciencia.

Lewis, John. 2006. *Galileo in France. French Reactions to the Theories and Trial of Galileo.* New York: Peter Lang.

Lupoli, Agostino. 2004. Hobbes e Sanchez. In *Nuove prospettive critiche sul Leviatano di Hobbes*, ed. Luc Foisneau and George Wright, 263–301. Milan: Franco Angeli.

———. 2006. *Nei limiti della materia. Hobbes e Boyle: materialismo epistemologico, filosofia corpuscolare e Dio corporeo*. Milan: Baldini, Castoldi, Dalai.

Malcolm, Noel. 2002. *Aspects of Hobbes*. Oxford: Clarendon Press.

———. 2005. Hobbes, the Latin Optical Manuscript, and the Parisian Scribe. *English Manuscript Studies 1100–1700* 12: 210–232.

Malet, Antoni. 2001. The Power of Images: Mathematics and Metaphysics in Hobbes's Optics. *Studies in History and Philosophy of Science* 32 (2): 303–333.

Malherbe, Michel. 1984. *Thomas Hobbes ou l'oeuvre de la raison*. Paris: Vrin.

———. 1989. Hobbes et la fondation de la philosophie première. In *Thomas Hobbes: de la métaphysique à la politique*, ed. Martin Bertman and Michel Malherbe, 17–32. Paris: Vrin.

Marion, Jean-Luc. 1994. Le concept de métaphysique selon Mersenne. In *Études sur Marin Mersenne*, Special issue of *Les Études Philosophiques* 68 (Iss. 1–2), ed. Jean-Robert Armogathe and Michel Blay, 129–143.

———. 2005. Hobbes et Descartes: l'étant comme corps. In *Hobbes, Descartes et la métaphysique*, ed. Dominique Weber, 59–77. Paris: Vrin.

Marquer, Éric. 2005. Ce que sa polémique avec Descartes a modifié dans la pensée de Hobbes. In *Hobbes, Descartes et la métaphysique*, ed. Dominique Weber, 15–32. Paris: Vrin.

———. 2015. Analyse du langage et science politique selon Bacon et Hobbes. *Philosophical Enquiries: Revue des philosophies anglophones* 4: 41–57. http://www.philosophicalenquiries. com/numero4article4Marquer.html.

Maury, Jean-Pierre. 2003. *À l'origine de la recherche scientifique: Mersenne*. Paris: Vuibert.

Médina, José. 1997. Nature de la lumière et science de l'optique chez Hobbes. In *Le siècle de la lumière 1600–1715*, ed. Christian Biet and Vincent Jullien, 33–48. Paris: ENS Éditions Fonteany-St. Cloud.

———. 2013a. Mathématique et philosophie chez Thomas Hobbes: une pensée du mouvement en mouvement. In *Lectures de Hobbes*, ed. Jauffrey Berthier, Nicolas Dubos, Arnaud Milanese, and Jean Terrel, 85–132. Paris: Ellipses.

———. 2013b. Physiologie mécaniste et mouvement cardiaque: Hobbes, Harvey et Descartes. In *Lectures de Hobbes*, ed. Jauffrey Berthier, Nicolas Dubos, Arnaud Milanese, and Jean Terrel, 133–162. Paris: Ellipses.

———. 2015. La philosophie de l'optique du *De Homine*. In *Thomas Hobbes, De l'Homme*, ed. Jean Terrel et al., 31–82. Paris: Vrin.

———. 2016. Hobbes's Geometrical Optics. *Hobbes Studies* 29 (1): 39–65.

Milanese, Arnaud. 2013. Philosophie première et philosophie de la nature. In *Lectures de Hobbes*, ed. Jauffrey Berthier, Nicolas Dubos, Arnaud Milanese, and Jean Terrel, 35–62. Paris: Ellipses.

———. 2015. Sur le passage de Bacon à Hobbes: un système et ses tensions. *Philosophical Enquiries: Revue des philosophies anglophones* 4: 7–26. http://www.philosophicalenquiries. com/numero4article2Milanese.html.

Minerbi Belgrado, Anna. 1993. *Linguaggio e mondo in Hobbes*. Rome: Editori Riuniti.

Moreau, Pierre-François. 1989. *Hobbes. Philosophie, Science, Religion*. Paris: Presses Universitaires de France.

Mori, Gianluca. 2010a. Hobbes, Cartesio e le idee: un dibattito segreto. *Rivista di storia della filosofia* (2): 229–246.

———. 2010b. *Cartesio*. Rome: Carocci.

Morris, Katherine. 2000. Bêtes-machines. In *Descartes' Natural Philosophy*, ed. Stephen Gaukroger, John Schuster, and John Sutton, 401–419. London/New York: Routledge.

Nardi, Antonio. 1994. Théorème de Torricelli ou Théorème de Mersenne. In *Études sur Marin Mersenne*, Special issue of *Les Études Philosophiques* 68 (Iss. 1–2), ed. Jean-Robert Armogathe and Michel Blay, 87–118.

Pacchi, Arrigo. 1964. Il '*De motu, loco et tempore*' e un inedito hobbesiano. *Rivista critica di storia della filosofia* 19: 159–168.

———. 1965. *Convenzione e ipotesi nella filosofia naturale di Thomas Hobbes*. Florence: La Nuova Italia.

Paganini, Gianni. 1990. Hobbes, Gassendi e la psicologia del meccanicismo. In *Hobbes Oggi*, ed. Bernard Willms et al., 351–445. Milan: Franco Angeli.

———. 1999. Thomas Hobbes e Lorenzo Valla. Critica umanistica e filosofia moderna. *Rinascimento* 39: 515–568.

———. 2003. Hobbes Among Ancient and Modern Sceptics: Phenomena and Bodies. In *The Return of Scepticism from Hobbes and Descartes to Bayle*, ed. Gianni Paganini, 3–35. Dordrecht/Boston/Leiden: Kluwer.

———. 2004a. Hobbes e lo scetticismo continentale. In *Nuove prospettive critiche sul Leviatano di Hobbes*, ed. Luc Foisneau and George Wright, 303–328. Milan: Franco Angeli.

———. 2004b. Hobbes, Gassendi und die Hypothese der Weltvernichtung. In *Konstellationsforschung*, ed. Martin Mulsow and Marcelo Stamm, 258–339. Frankfurt am Mein: Surkhamp.

———. 2006a. Hobbes, Gassendi e l'ipotesi annichilatoria. *Giornale critico della filosofia italiana* 85 (1): 55–81.

———. 2006b. Le lieu du néant. Gassendi et l'hypothèse de l'annihilatio mundi. *Dix-septième siècle* 58: 587–600.

———. 2006c. Alle origini del 'mortal God': Hobbes, Lipsius e il *Corpus Hermeticum*. *Rivista di Storia della Filosofia* (3): 509–532.

———. 2007. Hobbes's Critique of Essence and Its Source. In *The Cambridge Companion to Hobbes's Leviathan*, ed. Patricia Springborg, 337–357. Cambridge: Cambridge University Press.

————. 2008. *Skepsis. Le débat des modernes sur le scepticisme*. Paris: Vrin.

————. 2010a. Introduzione. In *Moto, luogo e tempo*, ed. Thomas Hobbes, 9–104. Turin: Utet.

————. 2010b. Hobbes e l'umanesimo. In *L'umanesimo scientifico dal Rinascimento all'Illuminismo*, ed. Lorenzo Bianchi and Gianni Paganini, 135–158. Naples: Liguori.

————. 2015. Hobbes and the French Skeptics. In *Skepticism and Political Thought in Seventeenth and Eighteenth Centuries*, ed. John Christian Laursen and Gianni Paganini, 55–82. n. p: University of Toronto Press.

Palmerino, Carla Rita. 1999. Infinite Degrees of Speed: Marin Mersenne and the Debate Over Galileo's Law of Free Fall. *Early Science and Medicine* 4 (4): 269–328.

Pantin, Isabelle. 2011. Premières répercussions de l'affaire Galilée en France chez les philosophes et les libertins. In *Il caso Galileo. Una rilettura storica, filosofica, teologica. Convegno internazionale di studi. Firenze, 26–30 maggio 2009*, ed. Massimo Bucciantini, Michele Camerota, and Franco Giudice, 237–257. Florence: Olschki.

Parodi, Massimo. 1984. Misura, analogia e peso. Un'analogia del XII secolo. In *La storia della filosofia come sapere critico. Studi offerti a Mario Dal Pra*, ed. Nicola Badaloni et al., 52–71. Milan: Franco Angeli.

Pécharman, Martine. 1992. Le vocabulaire de l'être dans la philosophie première: *ens, esse, essentia*. In *Hobbes et son vocabulaire*, ed. Yves-Charles Zarka, 31–59. Paris: Vrin.

Popkin, Richard H. 1957. Father Mersenne's War Against Pyrrhonism. *The Modern Schoolman* 34: 61–78.

————. 1979. *The History of Scepticism from Erasmus to Spinoza*. Berkeley/Los Angeles/London: University of California Press (or. Ed. 1960).

————. 1982. Hobbes and Scepticism. In *History of Philosophy in the Making. A Symposium of Essays to Honor Professor James D. Collins on His 65th Birthday*, ed. Linus J. Thro, 133–148. Washington, DC: University Press of America. Re-published in 1992, with an appendix in *The Third Force in 17th Century Thought*, ed. by Richard H. Popkin, 9–26 and 27–49. Leiden/New York/København/Köln: Brill, 1992.

————. 1992. Hobbes and Scepticism. In *The Third Force in 17th Century Thought*, ed. Richard H. Popkin, 9–49. Leiden-New York-København-Köln: Brill.

Prins, Jan. 1996. Hobbes on Light and Vision. In *The Cambridge Companion to Hobbes*, ed. Tom Sorell, 129–156. Cambridge: Cambridge University Press.

Randall, John H. 1961. *The School of Padua and the Emergence of Modern Science*. Padua: Antenore.

Raphael, Renée. 2008. Galileo's *Discorsi* and Mersenne's *Nouvelles Pensées*: Mersenne as a Reader of Galilean 'Experience'. *Nuncius* 23: 7–36.

Raylor, Timothy. 2005. The Date and Script of Hobbes's Latin Optical Manuscript. *English Manuscript Studies 1100–1700* 12: 201–209.

Robertson Croom, George. 1886. *Hobbes*. Edinburgh/London: William Blackwood and Sons. (Reprint: Thoemmes Press, 1993).

Rochot, Bernard. 1957. Gassendi et les mathématiques. *Revue d'histoire des sciences* 10 (1): 69–78.

Rossi, Paolo. 1974. *Francesco Bacone. Dalla magia alla scienza*. Turin: Einaudi (or. Ed. 1957).

Roux, Sophie. 1997. La nature de la lumière selon Descartes. In *Le siècle de la lumière 1600–1715*, ed. Christian Biet and Vincent Jullien, 49–66. Paris: ENS Éditions Fonteany-St. Cloud.

————. 1998. Le scepticisme et les hypothèses de la physique. *Revue de Synthèse* 4e s (2–3): 211–255.

Sacksteder, William. 1980. The Art of the Geometricians. *Journal of the History of philosophy* 18: 131–146.

————. 1992. Three Diverse Sciences in Hobbes: First Philosophy, Geometry, and Physics. *The Review of Metaphysics* 45 (4): 739–772.

Schmitt, Charles B. 1985. *Problemi dell'aristotelismo rinascimentale*. Naples: Bibliopolis.

Schuhmann, Karl. 1985. Rapidità del pensiero e ascensione al cielo: alcuni motivi ermetici in Hobbes. *Rivista di Storia della Filosofia* 2: 203–227.

————. 1990. Hobbes and Reinassance Philosophy. In *Hobbes Oggi*, ed. Bernard Willms et alii, 331–349. Milan: Franco Angeli.

————. 1992. Le vocabulaire de l'espace. In *Hobbes et son vocabulaire*, ed. Yves-Charles Zarka, 61–82. Paris: Vrin.

————. 1995a. Hobbes dans les publications de Mersenne en 1644. *Archives de Philosophie* 58 (2): 2–7.

————. 1995b. Le *Short Tract*. Première oeuvre philosophique de Hobbes. *Hobbes Studies* 8: 3–36.

Sergio, Emilio. 2001. *Contro il Leviatano. Hobbes e le controversie scientifiche 1650–1665*. Soveria Mannelli: Rubbettino.

————. 2003. Bacon, Hobbes e l'idea di '*Philosophia naturalis*': due modelli di riforma della scienza. *Nuncius* 18 (2): 515–547.

————. 2006. *Verità matematiche e forme della natura da Galileo a Newton*. Rome: Aracne.

————. 2007. Campanella e Galileo in un 'english play' del circolo di Newcastle: 'Wit's triumvirate, or the philosopher' (1633–1635). *Giornale critico della filosofia italiana* 86 (2): 298–315.

————. 2016. Hobbes lecteur de Campanella: autour des sources cachées du matérialisme hobbesien. In *Hobbes et le matérialisme*, ed. Jauffrey Berthier and Arnaud Milanese, 75–89. Paris: Éditions matérilogiques.

Shapiro, Alan E. 1973. Kinematics Optics: A Study of the Wave Theory of Light in the Seventeenth Century. *Archive for History of Exact Sciences* 11: 134–266.

Shea, William R. 1977. Marin Mersenne: Galileo's 'traduttore-traditore'. *Annali dell'Istituto e Museo di Storia della Scienza di Firenze* A. 2 (1): 55–70.

Sorell, Tom. 1993. Hobbes Without Doubt. *History of Philosophy Quarterly* 10: 121–135.

Stroud, Elaine C. 1983. *Thomas Hobbes' A Minute or First Draught of the Optiques: A Critical Edition*. PhD Dissertation. Madison: University of Winsconsin-Madison.

Taton, René. 1994. Le P. Marin Mersenne et la communauté scientifique parisienne au XVIIᵉ siècle. In *Quatrième centenaire de la naissance de Marin Mersenne (Actes du colloque)*, ed. Jean-Marie Constant and Anne Fillon, 13–25. Le Mans: Université du Maine.

Terrel, Jean. 2015. Hobbes et Bacon: une relation decisive. *Philosophical Enquiries. Revue des philosophies anglophones* 4: 1–6. http://www.philosophicalenquiries.com/numero4article-1Terrel.html.

————. 2016. Le matérialisme de Hobbes dans les *Troisièmes Objections*. In *Hobbes et le matérialisme*, ed. Jauffrey Berthier and Arnaud Milanese, 13–31. Paris: Éditions Matériologiques.

Tönnies, Ferdinand. 1925. *Thomas Hobbes. Leben und Lehre*. Stuttgart: Fromman Verlag (Faksimile-Neudruck: Stuttgart-Bad Cannstatt: Fromman Holzboog, 1971).

Tuck, Richard. 1988. Optics and Sceptics: The Philosophical Foundation of Hobbes's political Thought. In *Conscience and Casuistry in Early Modern Europe*, ed. Edmund Leites, 235–263. Cambridge: Cambridge University Press.

Westfall, Richard S. 1971. *The Construction of Modern Science. Mechanisms and Mechanics*. Cambridge: Cambridge University Press.

Zarka, Yves-Charles. 1985. Empirisme, nominalisme et matérialisme. *Archives de philosophie* 48: 177–233.

————. 1988. La matière et la représentation: Hobbes lecteur de *La Dioptrique* de Descartes. In *Problématique et réception du Discours de la méthode*, ed. Henri Méchoulan, 81–98. Paris: Vrin.

————. 1989. Aspects sémantiques, syntaxiques et pragmatiques dans la pensée de Hobbes. In *Thomas Hobbes: de la métaphysique à la politique*, ed. Martin Bertman and Michel Malherbe, 33–46. Paris: Vrin.

————. 1992. Le vocabulaire de l'apparaître: le champ sémantique de la notion de *phantasma*. In *Hobbes et son vocabulaire*, ed. Yves-Charles Zarka, 13–29. Paris: Vrin.

————. 1999. *La décision métaphysique de Hobbes*. Paris: Vrin (1st Ed.: 1987).

Chapter 2
Hobbes: Principles of Galilean Philosophy

2.1 Real Qualities and Pure Names

Hobbes praises Galileo in the dedicatory epistle of *De corpore*, presenting the Italian scientist as a sort of Prometheus of physics for having uncovered the mysteries of the nature of motion.[1] However, because of the central role played by the notion of movement in Hobbesian philosophy, we have to imagine that Galileo's influence was not limited to the field of physics, but that it represented the very foundations of the mechanical image of the world developed by Hobbes.[2]

We know that Hobbes met Galileo in person in the fall of 1635 thanks to a passage from Galileo's correspondence. On December 1 of that year he wrote to Fulgenzio Micanzio that he had received a visit from some 'persons from beyond the alps in the last few days, among them an English Lord who tells me that my unfortunate *Dialogue* is to be translated into that language, something that can only be considered to my disadvantage.'[3] The remark regarding the English version of the *Dialogue Concerning the Two Chief World Systems* seems to refer to the translation by Joseph Webbe, commissioned by Cavendish.[4] We can therefore assume that

---

[1] See *De corpore*, *Epistola dedicatoria*, *OL*, I, n. pag.

[2] See Jesseph 2004, who specifies that the principles of Hobbes' philosophy are rooted in the concepts elaborated by Galileo and, in particular, that Hobbes drew 'the fundamental idea that the world is a mechanical system in which everything can be understood in terms of mathematically-specifiable laws of motion' from the Italian scientist and philosopher (ibid., 191). For a detailed examination of this matter see Paganini 2010, 24 ff.; Paganini 2015. Interesting observations on the relationship between Galileo and Hobbes can also be found in Brandt 1928, esp. 77–84; Pacchi 1965, esp. 74–75; Gargani 1971, 50 ff., and 170–173. As regards the debt owed by Hobbesian physics to Galileo, see Leijenhorst 2004, which we will discuss in greater depth in the next chapter.

[3] Galileo to Fulgenzio Micanzio, 1 December 1635, *OG*, XVI, 355.

[4] After saying that he was in London to look for a copy of the *Dialogue*, which he was unable to find, Hobbes comments: 'I heare say it is called in, in Italy, as a booke that will do more hurt to their Religion then all the bookes have done of Luther and Calvin, such opposition they thinke is

---

© Springer Nature Switzerland AG 2020
G. Baldin, *Hobbes and Galileo: Method, Matter and the Science of Motion*,
International Archives of the History of Ideas Archives internationales
d'histoire des idées 230, https://doi.org/10.1007/978-3-030-41414-6_2

the 'English Lord' mentioned by Galileo is William Cavendish, who went on to become the Third Earl of Devonshire..[5]

Above and beyond this brief meeting, Hobbes was profoundly influenced by Galileo's writings, as suggested by an oft-cited passage from his autobiography in Latin verse, steeped in Galilean echoes.[6] The author not only describes the natural world as a *book*, but maintains that every '*phantasia*' (a term that, as we know, has the twofold meaning of sensory and mental representation in Hobbes' philosophical lexicon) is simply the product of matter 'agitated' by *motion*.[7] It is no coincidence that, in a letter sent from Paris in October 1636, Hobbes not only refers explicitly to Galileo's cogitations in *Dialogue Concerning the Two Chief World Systems*,[8] but also suggests that colors are nothing more than modifications produced by motion in the brain of the sentient being.[9]

This passage is often rightly cited as an indicator of the genesis of Hobbesian mechanistic philosophy.[10] Nevertheless, it is also very important to emphasize that Hobbes' statement—that colors simply consist of motion in the receptive organs of the sentient being—also suggests an important Galilean legacy. In fact, it is Hobbes' first explicit reference to the distinction made by Galileo in *The Assayer* between the different qualities of bodies: the qualities inherent to the perceived object and

---

betweene their Religion, and naturall reason, I doubt not but the Translation of it will here be publiquely embraced, and therefore wish extreamely that Dr Webbe would hasten it'. Hobbes to William Cavendish, Earl of Newcastle, from London, 26 January [/5 February] 1634, *CH*, I, 19. The text of the translation, in two volumes, which is complete but was never published, is stored at the British Library: MS Harley 6230.

[5] See Giudice 2011; Bucciantini, Camerota and Giudice 2012, 133–159. Richard Blackbourne, in his biography of Hobbes, writes that Hobbes actually had the opportunity to spend a lot of time with Galileo in Pisa and that the two of them developed a firm friendship (see Blackbourne, *Vitae Hobbianae Auctarium, OL*, I, xxviii). However, the passage contains several inaccuracies: firstly, Galileo was not in Pisa at the time, but under arrest at his villa in Arcetri. What is more, the presumed encounter between the pair must have been quite brief, since by 26 December 1635, Hobbes and Cavendish were already in Rome, staying at the Jesuit English College. See Schuhmann 1998, 47.

[6] See *Vita Carmine Expressa, OL*, I, lxxxix. See also above, 'Introduction.'

[7] See Paganini 2010, 28; Jesseph 2004, 202.

[8] Hobbes to William Cavendish, Earl of Newcastle, from Paris, 16 [/26] October 1636, *CH*, I, 38. See Galilei, *Dialogo sopra i due massimi sistemi, OG*, VII, 102. The first to identify the reference to Hobbes in the first day of the *Dialogue* was Brandt. See Brandt 1928, 145.

[9] Hobbes to William Cavendish, Earl of Newcastle, from Paris, 16 [/26] October 1636, *CH*, I, p. 38. Hobbes' 'Galilean' position must also have been shared by Sir William Cavendish, Earl of Newcastle, as can be deduced from an earlier letter from Hobbes to William Cavendish, where he wrote that he shared the Lord's opinion 'That the variety of thinges is but the variety of locall motion in ye spirits or invisible partes of bodies. And That such motion is heate.' Hobbes to William Cavendish, Earl of Newcastle, from Paris, 29 July/8 August 1636, *CH*, I, 33.

[10] Regarding the genesis of Hobbesian mechanism in the 1630s and Hobbes' first analyses of the phenomenon of sensation, see Brandt 1928, 143 ff.; Pacchi 1965, 15 ff.; Gargani 1971, 209 ff.; Giudice 1999, 134 ff.

those that instead reside solely in the organs of the sentient being, later catalogued by Robert Boyle as the 'primary' and 'secondary qualities' of sensible objects.[11]

The accuracy of this observation seems to be confirmed by Hobbes himself, in a passage from the *Tractatus Opticus II*, where he writes that, when reflecting upon visual phenomena, he could not 'escape from the evidence' of considering sensible qualities as the mere result of modifications produced by motion in the organs of the sentient being:

> When I was reflecting on the nature of the diaphanous, I considered firstly all the knowledge which we could acquire of the sensible qualitities, that comes from the action of the objects on the sensory organs. However, since the sense is nothing but motion in some parts of the brain, I could not escape from this conclusion, that the difference in sensible qualities comes from the difference of motions.[12]

The subject had already been developed extensively in *Elements of Law* where, upon discussing the senses, Hobbes warned against the pitfalls of considering sensible qualities as really existing in the external world. In fact, 'because the image in vision consisting in colour and shape is the knowledge we have of the qualities of the object of that sense; it is no hard matter for a man to fall into this opinion, that the same colour and shape are the very qualities themselves.'[13] On the contrary,

---

[11] See Boyle, *The Origine of Formes and Qualities*. In Boyle 1999–2000, V, 284–285. On this aspect of Boyle's thought see Anstey 2000, 15–112. Kargon claimed that the mechanist philosopher did not 'create' the distinction between primary and secondary qualities from scratch, but had instead relegated some of the Aristotelian qualities to the level of secondary qualities. See Kargon 1966. As we will see further on in the chapter and in Chapter 6, the translation of Aristotelian language and concepts in order to adapt them to the requirements of mechanism are aspects peculiar both to Galileo and Hobbes' thinking. Regarding the importance of the distinction between 'primary' and 'secondary qualities' in reference to the translation of the natural reality into quantifiable terms that can be expressed mathematically, see Crombie, 'The primary and secondary qualities in Galileo Galilei's natural philosophy', (1969). In Crombie 1990, 323–343. On the importance of the distinction for the genesis of modern science and the central role played by Galileo's *The Assayer* in the development of the concept of the primary or essential qualities of bodies, see Gaukroger 2006, 326 ff. Sorell and Leijenhorst observe that the Lockean distinction between primary and secondary qualities is not present in Hobbesian thinking since, according to Hobbes, reality is composed purely of bodies in motion and every quality is simply a *phantasia* in the sentient being. See Sorell 1986, 79 and Leijenhorst 2007, esp. 93. In a certain sense (as we will see later on), their intuition is correct, as Hobbes reduces the 'primary qualities' to just one: *corporality*. Nevertheless, I feel it is less credible to state that Hobbes would support a position similar to that of Berkeley: 'that apparent form and extension vary with the perceiver just like secondary colors such a color, which means that size and other primary qualities cannot be in the alleged external substance either'. (Ibid.).

[12] *TO II*, chap. I, § 23, fol. 203r/159 (my trans.): 'Ego cum de natura diaphani cogitarem, consideravi imprimis omnem notitiam quam habere possumus qualitatum sensibilium, derivari ab obiectorum in organa sensum actione; sensum autem esse motum in partibus cerebri; non potui effugere conclusionem hanc, diversitatem qualitatem sensibilium provenire a diversitate motuum.' See also ibid. chap. IV, § 16, fol. 249r/209: 'Cum autem id quod propter apparentiam externam appellatur Color, aliud non sit realiter praeter motum internum in partibus a fundo oculi ad Cor (in quo origo est omnis sensionis) propagatum, sequitur diversitatem colorum apparentium aliud non esse praeter diversitatem illorum motuum.'

[13] *EL*, Part I, chap. II, § 4, 3. See, in this regard, the words of the author of the *Short Tract*. Hobbes 1988a, 44: 'Light, Colour, Heate, and other proper obiects of sense, when they are perceiv'd by

Hobbes establishes four principles that associate the sensation solely with motion in the medium and in the organs of the sentient being:

(1) That the subject wherein colour and image are inherent, is not the object or thing seen.
(2) That there is nothing without us really which we call an image or colour.
(3) That the said image or colour is but an apparition unto us of that motion, agitation, or alteration, which the object worketh in the brain or spirits, or some internal substance of the head.
(4) That as in conception by vision, so also in the conceptions that arise from other senses, the subject of their inherence is not the object, but the sentient.[14]

As he continues, Hobbes goes back to the idea expressed in his letter to Newcastle, that color is nothing other than perturbed light,[15] and applies the same reasoning to sound and the other senses. So, 'likewise the heat we feel from the fire is manifestly in us, and is quite different from the heat which is in the fire: for our heat is pleasure or pain, according as it is extreme or moderate; *but in the coal there is no such thing.*'[16] Hobbes concludes that:

> Whatsoever accidents or qualities our senses make us think there be in the world, they are not there, but are *seemings* and *apparitions* only. The things that really are in the world without us, are those *motions* by which these seemings are caused.[17]

An almost identical argument, with the same example of *heat* and coal, also features in the *Tractatus Opticus I.*[18] This recurrence is anything but coincidental, because the Hobbesian treatise contains a number of significant analogies with the passage from *The Assayer*, in which Galileo set out the principle of the subjective nature of the so-called secondary qualities. In fact, the Italian scientist divided the qualities into two categories: those directly inherent to the perceived object and the qualities that, instead, are manifestations that develop solely within the sensory organs of the sentient being and are therefore determined by the physical conformations of the object, or rather by its primary qualities.

---

sense, are nothing but the severall Actions of Externall things upon the Animal spirits, by severall Organs. and when they are not actually perceiv'd, then they be powers of the Agents to produce such actions. For if Light and heate were qualityes actually inherent in the species, and not severall manners of action, seing the species enter, by all the organs, to the spirits, heat should be seene, and Light felt. contrary to Experience.' This passage provides further confirmation of the interest shown in Galileo by those who frequented the circle of Newcastle.

[14] *EL*, Part I, chap. II, § 4, 4. On the importance of *The Assayer* in terms of the development of the mechanical concept of vision proposed by Hobbes, see Brandt 1928, 80; Pacchi 1965, 74 ff. Médina 1997, 41–42

[15] See *EL*, Part I, chap. II, § 8, 6. See *De corpore*, XXV, 10, *OL*, I, 329; Ibid. XXVII, 13, 374.

[16] *EL*, Part I, chap. II., § 9, 7 (my italics).

[17] Ibid., § 10, 7 (my italics).

[18] *TO I*, *OL*, V, 217: 'Item dum carbo ignitus calefacit hominem, etsi neque carbo neque homo suo loco exeat, neque ideo moveatur, est tamen aliquid materiae sive corporis subtilis in carbone, quod movetur, et motum ciet in medio usque ad hominem; et est in homine stantei mmoto, motus aliquis in partibus internis inde generatus. Motus autem hic in partibus hominis internis est calor; et sic moveri, calefieri, hoc est pati; et motus ille qui est in partibus carbonis igniti, est actio ejus, sive calefactio; et sic moveri, calefacere'.

Galileo's argument was based on the statement in the *Libra Astronomica* by his critic Lotario Sarsi (i.e. the Jesuit Orazio Grassi), who, on the basis of a passage in *De Caelo*,[19] maintained—in agreement with Aristotle—that the air surrounding projectiles hurled at great speed was transformed into fire.[20] In responding to his interlocutor, Galileo shows off all his rhetorical ability. He started by setting out his significant doubts that, regarding what we call *heat*, 'people in general have a concept of this that is very remote from the truth. For they believe that heat is a real phenomenon or property, or qualitie, which actually resides in the material by which we feel ourselves warmed.'[21] As he continues, Galileo sets out the famous theory taken up again by Hobbes in his *Elements*[22]: people are inclined to believe that heat, smell, and any other action of objects external to the sentient being are qualities inherent to the bodies, while instead they consist solely of subjective manifestations of the interaction between the perceived object and the receptive organs; an interaction expressed simply in terms of matter and motion. Indeed, with regard to every 'material or corporeal substance,' Galileo stated that he was forced to think of it as '*bounded*, and as having this or that *shape*; as being *large or small* in relation to other things, and in some specific *place* at any given *time*; as being *in motion* or at rest; as *touching or not touching some other body*; and as being *one in number*, or *few*, or *many*. From these conditions I cannot separate such a substance by any stretch of my imagination.'[23] On the contrary, continued Galileo, 'that it must be white or red, bitter or sweet, noisy or silent, and of sweet or foul odor, my mind does not feel compelled to bring in as necessary accompaniments. Without the senses as our guides, reason or imagination unaided would probably never arrive at qualities like these.'[24] The conclusion that Galileo drew from this is of fundamental importance, because it is one of the cornerstones of modern science:

> Hence I think that tastes, odors, colors, and so on are no more than mere names so far as the object in which we place them is concerned, and that they reside only in the sensorial body. Hence if the living creature were removed, all these qualities would be wiped away and annihilated. But since we have imposed upon them special names, distinct from those of the

---

[19] See Aristotle, *De Caelo*, II, 289a, 24 ff. Eng. Trans. Aristotle 1986, 179–181: 'Compare the case of flying missiles. These are themselves set on fire so that leaden balls are melted, and if the missiles themselves catch fire, the air which surrounds them must be affected likewise. These then become heated themselves by reason of their flight through the air, which owing to the impact upon it is made fire by the movement'.

[20] See Sarsi 1619 in *OG*, VI, 162.

[21] Galilei, *Il saggiatore*, *OG*, VI, 347; Eng. trans. in Drake 1957, 274.

[22] Anna Minerbi Belgrado underscored the presence of the distinction between 'primary' and 'secondary' qualities in the *Elements of law*. See Minerbi Belgrado 1993, 11–13. Malcolm also has interesting things to say about this matter: see: Malcolm, 'Robert Payne, the Hobbes Manuscripts, and the "Short Tract",' In Malcolm 2002, 122 ff.; Malcolm, 'Hobbes and Spinoza'. Ibid., 27–52; esp. 29. See also Anstey 2013, esp. 248–250.

[23] Galilei, *Il saggiatore*, *OG*, VI, 347–348, my italics; Eng. trans. Drake 1957, 274.

[24] Ibid.; Eng. trans. ibid.

other and real qualities mentioned previously, we wish to believe that they really exist as actually different from those.[25]

Galileo maintained that only certain qualities could be considered 'essential' to the bodies; others are 'mere names' and have a purely subjective dimension.[26] What is more, he specified which elements these are: '*shapes, numbers,* and slow or rapid *movements,*' so that 'if ears, tongues, and noses were removed, shapes and numbers and motions would remain, but not odors or tastes or sounds. The latter, I believe, are nothing more than names when separated from living beings, just as tickling and titillation are nothing but names in the absence of such things as noses and armpits.'[27]

Although it has been observed that the radical dichotomy developed by Galileo—which differentiates between primary and secondary qualities—recalls the Aristotelian distinction between 'proper' and 'common sensibles,'[28] nevertheless, in the epistemological dimension of Galileo's, thought, it takes on a radically different scope because it is founded on his conception of matter, which is a fundamental element of mechanism.

There is an evident conceptual analogy between the Hobbesian and the Galilean argument and it is reasonable to believe that Hobbes was well aware of this page from *The Assayer* when he wrote his *Elements of Law,* although the only work by Galileo to be cited in the text is his 'first dialogue concerning local motions,'[29] that is the first day of his *Discourses and Mathematical Demonstrations Relating to Two New Sciences.*[30]

---

[25] Ibid.; Eng. trans. ibid., modified.

[26] Galileo's expression is reused almost verbatim by Hobbes, in *TO II*, where he states that *species* are '*nihil nisi verba*'. See *TO II*, chap. I, § 10, fol. 197v/151. See Médina 2015, 131 and note.

[27] Galilei, *Il saggiatore*, *OG*, VI, 350; Eng. trans. Drake 1957, 276–277.

[28] See Aristotle, *De Anima*, II, 6, 418a 8–22; Thomas Aquinas, *Summa Theologiae*, Pars I, quaestio 78, art. 3. On the relationship between the Aristotelian-scholastic distinction and Galileo: Crombie, 'The primary and secondary qualities in Galileo Galilei's natural philosophy'; Shea 1970, esp. 23–24. See also the interesting observations by Redondi 2009, 66 ff., on Ockham and the distinction between *intentiones primae* and *secundae*.

[29] *EL*, part. I, chap. VIII, § 1, 33.

[30] In *EL*, part I, Chapt. 8, § 2, 32–33, Hobbes writes: 'Sounds that differ in any height, please by inequality and equality alternate, that is to say, the higher note striketh twice, for one stroke of the other, whereby they strike together every second time; as is well proved by Galileo, in the first dialogue concerning local motions, where he also sheweth, that two sounds differing a fifth, delight the ear by an equality of striking after two inequalities.' Cf. Galilei, *Discorsi e dimostrazioni matematiche intorno a due nuove scienze*, *OG*, VIII, 147.

2.2 'Essential' Accidents

The reduction of bodies' secondary qualities to the action of motion on the sense organs of the perceiver, as seen in *Elements*, also appears in the nearly coeval *Tractatus Opticus I*.[31] But we should emphasize that there was an internal development of the Hobbesian philosophical system starting with the concepts developed in *De motu, loco et tempore*. In keeping with his general orientation, in *Elements of Law*, Hobbes notes that the only factors considered in his philosophy to explain natural phenomena are *matter* and *motion,* with no further determinations about the *qualities* that Galileo considered inherent to bodies, i.e. the *size,* the *figure,* and the *number.* The question is returned to in *De motu, loco et tempore,* where Hobbes develops interesting ideas about the concept of *matter* and establishes a basic dichotomy of reality, distinguishing between *body* and *accident.*

> There being two classes of things, one of which Aristotle called τὸ ὄv, i.e. *ens,* and the other τὸ εἶναι, that is *esse,* we cannot conceive of a passage from the first, i.e. *ens,* to non-*ens.* Hence, as every philosopher believes, the entities (*entia*) do not, by reason of a natural or ordinary virtue, decay absolutely ... Hence there can be created and there can perish completely not the entities (*entia*) themselves but the acts, forms and accidents which distinguished them from other beings.[32]

According to Hobbes, the *body* is that which can be conceived as 'that which has dimensions or which occupies an imaginary space'[33] and this idea is inextricably linked to that of *matter,* about which Hobbes is heir to the thinking of Democritus and Lucretius, as emerges even more clearly in *De corpore.* He considers *body* and *matter* as 'names of the same thing, but interpreted in different ways. When considered *simpliciter,* an object that exists is termed *body,* but when considered as capable of assuming a new form or a new figure is called *matter.*'[34] He maintains that there is one matter that, in his words, is the 'matter of everything that is.'[35]

The *matter* has the same ontological status as *universals*: it is a pure name; however, it is highly significant that Hobbes identifies the *body* with *matter,* which can be conceived as the mere corporeity or materiality of the body. It is not, therefore,

---

[31] See *TO I, OL,* V, 220–221.

[32] *MLT,* XXXV, 1, 387–388; Eng. Trans. Hobbes 1976, 422, largely modified (my italics).

[33] Ibid., XXVII, 1, 312. On the issues of identifying the body with the *ens* or, that which occupies an imagined space, and on the conceptual distinction between *ens and essentia,* see Pécharman 1992, 35 ff. On the real space/imagined space distinction and the relative relationships of these two concepts with the fundamental idea of Hobbesian philosophy, i.e. that of *body,* see Bernhardt 1990, and Schuhmann 1992.

[34] *MLT,* XXVII,1312, modified

[35] Ibid., VII, 3, 146–147: 'Ut autem quid sit materia prima facilius intelligamus, sciendum est primo *corpus & materiam,* sive *corporeum & materiale,* hoc est quicquid spatio quod imaginamur subest, idem esse, vocari autem corpus simpliciter, materiam autem quando comparatur cu meo quod ex ipsa factum est, ut lignum simpliciter *corpus* dicitur, idem autem quatenus ex eo fit scamnum, *materia* scamni appellatur. Deinde consideranda est *materia* sub paucis mutationibus, ut ex iis intelligatur quid sit materia omnium.'

a reality existing in and of itself, nor any entity that inhabits the physical world, but it is, rather, the result of a process of abstraction by human faculties that, isolating the reality of the body as such, only conceives its materiality.

Yet, this matter, which has no determination of its own, is defined by certain attributes or particular characteristics: the *accidents*, the second type of elements that Aristotle mentioned, distinguishing between entity (*ens*) and being (*esse*).[36] This latter kind of conceptual reality is fundamental because it is what allows us to conceive of the singular entities: the bodies, with their determinations[37]:

> The result is that *esse* is nothing but an *accident of the body,* by which is determined and characterized the way of conceiving it. Therefore 'to be moved,' 'to be at rest,' 'to be white' and the like we call *accidents* of bodies, and we believe them to be present in the bodies, because they are the different ways of perceiving it.[38]

However, '*accidents, essences,* and *forms,* (*corporeity* excluded) are produced and perish every day'[39] and they are, therefore, the transient properties of the one matter.[40] In *De motu, loco et tempore*, Hobbes seems to take further the position Galileo developed in *The Assayer*: through a dichotomous solution that entails dividing reality into two types of distinct entities, he seems to restrain the entirety of the primary qualities of bodies as conceived by Galileo into a single quality: *corporeity.*[41]

However, starting from an idea found in a passage of *De motu,* according to which the accident is a way of conceiving the body,[42] Hobbes comes to reflect further on the question. In a draft of the *De corpore* from about 1644, he develops a more complex solution to the problem, suggesting an internal differentiation of accidents, on the foundation of Aristotle's definition of accident, that he had already mentioned in *De motu, loco et tempore*.[43] Here, he returned to the fundamental distinction between body and accident, referring to the former as 'whatsoever not depending our cogitation, is coincident or coextended with any part of space,' and

---

[36] See Pécharman 1992, 41 ff.

[37] Hobbes' concept of accident has been extensively discussed and conflicting interpretations of it have been given, some of which are more 'phenomenalist' and others more 'realist' (we will return to this topic in the following paragraph). See Malherbe 1984, 76 ff. and Malherbe 1989; Leijenhorst 2002, 155–162; Lupoli 2006, 101–138; Paganini 2008, 221 ff.; Milanese 2013, 53–54.

[38] *MLT*, XXVII, 1, 313; Eng. Trans. Hobbes 1976, 312, modified.

[39] Ibid., 314.

[40] Hobbes suggests several times in the work that this matter, incapable of generation and incorruptible, must necessarily be eternal. See, for example, *MLT*, V, 3, 130, where Hobbes also argues that 'Materia tamen eius, in qua consistit natura corporis, non interit: sicut vini cotyla infusa oceano, quanquam desinat esse vinum, non tamen desinit esse corpus.'

[41] On this, see Paganini 2010, 64. On the 'corporization' of the Aristotelian concept of substance in Hobbes, see Spragens 1973, 77 ff.

[42] See, *MLT*, XXVII, 1, 313.

[43] Tom Sorell emphasizes the dramatic difference between the Hobbesian position and the Aristotelian concept of substance and accident, see Sorell 1986, 51–52.

defining the accident as 'not a part of natural things' but as 'the manner (*modus*) of conceiving a body or according to which a body is conceived.'[44]

His argument focused in particular on an internal division of the latter category:

> Aristotle's definition (*Accidens inest in subjecto non tamquam pars sic tamen ut sine subjecti interitu abesse potest*) is right save that some accidents may not be from the body without the destruction thereof, for a body cannot be conceived without *extension* and *figure*. Other accidents which are not common to all bodies but proper to some as *rest, motion, colour, hardness* etc. do continually perish, others succeeding, so as the body never perishes.[45]

Accidents are not truly contained in bodies[46]; however, the mutation, or change, of a body is determined by the existence or absence of certain accidents.[47] Hobbes also reiterated that one matter persists despite any modification, or even destruction to which a body is subject: according to Aristotelian philosophical tradition, this is called *first matter*.[48]

Yet, as Schuhmann noted,[49] there are some accidents that cannot be removed through a process of abstraction, without simultaneously conceiving the body's destruction. The accidents that Hobbes identified as being completely inseparable from the body are *extension* and *figure*; it bears no repeating that size and figure were two of the qualities that Galileo gave in *The Assayer* as essential qualities of bodies.[50]

---

[44] *De Principiis* (National Library of Wales, Ms. 5297), Appendix II, *MLT*, 452.

[45] Ibid.

[46] Hobbes clarifies that the expression '*accidentia in corpore inesse*' could be misleading because it suggests the presence of quality inside the object: 'When we say *accidentia in corpore inesse*, it must to be understood as if something were contained in the body: for example, as if redness were in blood as blood is in a bloody cloth i.e. ut pars in toto, for so an accident were also a body; but as *magnitude, rest, motion* etc. in that which is *magnum, quiescens, motum*, so every other accident is in his subject.' Ibid., 453. This passage is transferred almost whole into the final version of Chapter 8 of *De corpore*, *OL*, I, 92–93.

[47] *De Principiis* (National Library of Wales, Ms. 5297), Appendix II, *MLT*, 457: 'The production or destruction of any accident is the cause that the subject is said to be changed, but only of the form that it is said to be generated or corrupted (destroyed)'.

[48] Ibid.: 'The common matter of all things which the philosophers call materia prima is not a distinct body from all other bodies nor one of them but a name only, signifying a body to be considered without considering any form or any accident except only *magnitude* or *extension* and aptitude to receive *form and accident*. So as materia prima is corpus universale i.e. a body considered universally whereof it cannot be said that there is no form or no accident, but in which it may be said no form or accident besides quantity and aptitude to receive form or accident is considered i.e. brought into argument or account.' Cf. Aristotle, *De generatione et corruptione*, I, 5320b; *Metaphysics*, E, 1029b.

[49] See Schuhmann 1992, 75–76, who stresses that these are accidents of *size, extension* and *local motion*, without, however, mentioning the 'Galilean' roots of these concepts.

[50] For this reason, I consider Sorell and Leijenhorst's suggestion only partially correct, and it should be interpreted in the light of these additional considerations, with respect to the 'essential' accidents of bodies.

The relationship that Hobbes establishes between the concept of *body* and the two accidents of *extension* and *figure* is fundamental to his concept of philosophy.[51] To develop the question fully, we must return to the relationship that Hobbes had established between these two elements in *De motu, loco et tempore*. Though in Chapter 1 he had given a highly formal, rational definition of philosophy as the 'science of general theorems, or of all the universals in any matter, the truth of which can be discovered by natural reason,'[52] in Chapter 27—where he distinguished between *imaginable* entities and those not in the realm of philosophical speculation (thereby excluding from the field of inquiry God and other elements termed incomprehensible)—he argued that philosophy applies to the study of *imaginable entities* that are, as such, understandable.

> In this sense, therefore, *ens* is everything that occupies space, or which can be measured as to *length*, *breadth*, and *depth*. From this definition it appears that *entity* (*ens*), and *body* are the same thing ... *body* is that which has dimensions, or which occupy an imaginary space.[53]

The perfect coincidence of *entity* and *body* is the foundation of Hobbes' materialist philosophy of which bodies are, precisely, the objects of inquiry. It is, however, important to emphasize that the definition of the body as such that can take up an *imaginary space*.[54] The condition of the object that occupies an imagined space is what makes the body itself thinkable, because this is what allows it to be distinguished from the first matter, and therefore serves as a *principium individuationis*.[55]

In the final version of *De corpore*, in Chapter 8, Hobbes considered it worth developing the issue more comprehensively than in his previous works. He specifically discusses *body* and *accident*, though restating that the phrase: 'accident is in a body... it is not to be understood, as if any thing were contained in that body.'[56] Referencing the Aristotelian adage according to which the accident is in the subject 'not as any part thereof, but so as that it may be away, the subject still remains,'[57] Hobbes wrote that this definition is right 'saving that there are certain accidents which can never perish except the body perish also; for no body can be conceived without *extension*, or without *figure*.'[58]

Hobbes went on to make a highly significant clarification:

> And as for the opinion that some may have, that all other accidents are not in their bodies (lat. *inesse*) in the same manner that *extension*, *motion*, *rest*, or *figure*, are in the same; for example, that colour, heat, odour, virtue, vice, and the like, are otherwise in them, and, as we say *inherent;* I desire they would suspend their judgement for the present, and expect a

---

[51] See Terrel 1994, 72 ff.

[52] *MLT*, I, 1, 105; Eng. Trans. Hobbes 1976, 23, modified.

[53] Ibid., XXVII, 1, 312, Eng. Trans. Hobbes 1976, 311 modified (my italics).

[54] See Malherbe 1989, 21–23

[55] See Minerbi Belgrado 1993, 63.

[56] *De corpore*, VIII, 3, *OL*, I, 92; Eng. Trans. *EW*, I, 104.

[57] Ibid. Hobbes cites, though not word for word, Aristotle, *Metaphysics*, E, 2, 1026b.

[58] *De corpore*, VIII, 3, *OL*, I, 92–93; Eng. Trans. *EW*, I, 104 (my italics).

little, till it be found out by ratiocination, whether these are very accidents are not also certain motions either of the mind of the perceiver, or of the bodies themselves which are perceived; for in the search of this, a great part of natural philosophy consists.[59]

Hobbes seems to suggest here that *figure* and *extension*, though they are essential accidents of bodies—in the sense that if they were to be removed, the body itself would perish—do not in any way have an ontologically special *status* that distinguishes them from other accidents.[60] Every accident is a modification perceptible with the senses of the internal motions of the body, which are conveyed directly, or through a vehicle, to the perceiving subject. However, in the next paragraph Hobbes makes a further clarification about *extension*:

> The *extension* of a body is the same thing with the *magnitude* of it, or that which some call *real space*. But this *magnitude* does not depend upon our cogitation, as imaginary space doth; for this is an effect of our imagination, but *magnitude* is the cause of it; this is an accident of the mind, that of a body existing out of the mind.[61]

When defining *space* and *place*, Hobbes also writes that the latter 'is nothing out of the mind, nor *magnitude* any thing within it. And lastly, *place* is feigned extension, but *magnitude* true extension; and a placed body is not extension, but a thing extended.'[62]

Hobbes ties every accident to the motion caused by the action of the object on the perceiver; still, the particular accident of the extension, which is nothing other than how we perceive the size of the body, corresponds exactly to a real property of the body, that of being extended, i.e. limited in its 'magnitude.'

The concept of *accident* is, as such, essential as it is the key element of Hobbesian philosophy. According to the definition given in Chapter 1 of *De corpore*, the 'subject of Philosophy' is 'every body of which we can conceive any generation, and which we may, by any consideration thereof compare with other bodies ... that is to say, every body of whose generation or properties we can have any knowledge.'[63] It is clear—from what has been described above—that the idea of body necessarily entails the existence of an extended size[64] of a particular magnitude, that serves as a *principium individuationis* and lets the body be identified in its individuality.

In *De motu, loco et tempore*, Hobbes had already situated the *principium individuationis* in matter and in form,[65] though giving these two terms a different

---

[59] Ibid., 93; Eng. Trans. 104–105.
[60] On the concept of accident in Hobbes, see Milanese 2011, esp. 209 ff., offering an alternative interpretation from 'phenomenical.' See also Terrel 1994, 73–75.
[61] *De corpore*, VIII, 4, *OL*, I, 93; Eng. Trans. *EW*, I, 105.
[62] Ibid., VIII, 5, 94; Eng. Trans. *EW*, I, 105.
[63] Ibid., I, 8, *OL*, I, 9; Eng. Trans. *EW*, I, 10.
[64] This aspect has been particularly highlighted by Schuhmann 1992, 72 ff.
[65] *MLT*, XII, 2, 189–190: 'Iam si quaeratur de aliqua re nominatim, cuius nomen significat materiam determinatam, nulla habita ratione formae, identitas rei sumitur ab identitate materiae. Sin nomen significet formam determinatam nulla habita ratione materiae nisi quatenus ad formam necessario aliqua requiritur, identitas rei aestimatur per identitatem formae.'

meaning than that of the Aristotelian tradition: here the concept of form coincides with that of figure. In *De corpore*, he returns to the question, maintaining that 'that accident for which we give a certain name to any body, or the accident which denominates its subject, is commonly called the ESSENCE thereof' and gives as an example how *extension* itself is the 'essence of a body.'[66]

The concept of the body is, therefore, inseparably bound to the attribute of *extension*, which is simply our way of conceiving the size of that body. This attribute, therefore, becomes essential to Hobbes' philosophy, which he terms a *science of bodies*. It emerges quite clearly from these ideas that Hobbesian speculation on the nature of bodies and accidents—developed in growing detail, starting from *The Elements of Law* through to *De corpore*—are based on Galileo's inquiries and the distinction he established between the essential *qualities* of bodies and the qualities that reside only in the perceiver.

2.3 Bodies, Accidents, and Causes

The existence of certain accidents that Hobbes deems 'essential' to bodies and their coinciding with the *qualities* that Galileo considered inseparable from the body's very essence invites us to consider the Hobbesian development of the concept of *body*.

In Chapter 27 of *De motu, loco et tempore*, Hobbes pointed out the perfect coincidence of the concepts of *entity* and *body* as he considers *entity*: 'everything that occupies space, or which can be measured as to length, breadth, and depth.'[67] The concept of *body* in Chapter 8 of *De corpore* is extremely rich in philosophical meaning and takes on implications that were latent but not clearly expressed in *De motu*. Hobbes defines *body*, as that which does not only 'fill some part of the space,' and 'also that it has no dependence upon our thought:'

> And this is that which, for the extension of it, we commonly call *body*; and because it depends not upon our thought, we say is a *thing subsisting of itself*; and also *existing*, because without us; and, lastly, it is called the *subject*, because it is so placed in and sub-*jected* to imaginary space, that it may be understood by reason, as well as perceived by sense. The definition, therefore, of *body* may be this, *a body is that, which having no dependence upon our thought, is coincident or coextended with some part of space.*[68]

As has been widely noted,[69] by defining the *corpus* as *suppositum* and *subjectum*, Hobbes intends to suggest that the concept has a dual nature. On one side, it is the substance underpinning *accidents* and on the other, it is something that is considered, *'non dependens a nostra cogitatione.'* This definition of *body* as that which

---

[66] *De corpore*, VIII, 23, *OL*, I, 104; Eng. Trans. *EW*, I, 117.

[67] *MLT*, XXVII, 1; 311; Eng. Trans. *EW*, I, 311 modified.

[68] *De corpore*, VIII, 1, *OL*, I, 90–91; Eng. Trans. *EW*, I, 102.

[69] See Pacchi 1965, 89 ff.; Schuhmann 1992, 73; Paganini 2008, 213.

does not depend on our thought and subsists in and of itself provides a fundamental starting point for a proper evaluation of the extent of Hobbes' materialism (contrasting with overly rationalistic interpretations of his philosophy[70]), and it also invites us to examine the reality of empirical phenomena in the light of the concept of body itself.

From his earliest reflections in the realm of natural philosophy, Hobbes always considered every physical phenomenon solely as the product of the interaction between bodies.[71] Hobbes' reworking of the concept of *cause* in *De motu, loco et tempore*[72] is associated with this idea and he returned to it in *De corpore*. In Chapter 28 of *De motu*, Hobbes argued that the Aristotelian concepts of 'final cause' and 'formal cause' were baseless.[73] He suggested in their place the concept of *integral cause* (or *causa simpliciter*), which was referred to here simply as 'one act or more acts through which another act is produced or destroyed.'[74] In *De corpore*, it is more clearly defined as 'the aggregate of all accidents both of the agents how many soever they be, and of the patient, put together; which when they are all supposed to be present, it cannot be understood but that the effect is produced at the same instant; and if any one of them be wanting, it cannot be understood but that the effect is not produced.'[75]

Every act that individually contributes to producing the integral cause is termed 'causa sine qua non, or something hypothetically necessary.'[76] Furthermore, some causes *sine qua non* or hypothetically necessary 'lie in the agent, someone in the patient,' so 'all the properties present in the agent, if taken collectively, are termed "an efficient cause," and those causes in the patient, taken collectively, are called "a material cause."'[77]

His removal of Aristotelian concepts of 'final' and 'formal cause'[78] and reduction of other causes to 'efficient' and 'material' suggest that Hobbes also translated the Aristotelian vocabulary and concepts of the idea of cause with a mechanical

---

[70] Neo-Kantians, such as Cassirer, suggested there is a split in Hobbes' philosophy between the empirical world of bodies and the hypothetical world of scientific discourse. See Cassirer 1922–1957, II, 46 ff. On this topic, see also Malherbe 1989, 24. For a critical overview of neo-Kantian interpretations of Hobbes' materialism, see Milanese 2016.

[71] See *TO I, OL*, V, 217.

[72] See Leijenhorst 2005, esp. 118 ff.

[73] See *MLT*, XXVII, 2, 314–316. However, in all that pertains to the concept of 'final cause,' we should note that Hobbes includes the concept in the scope of his anthropology, where it is translated in mechanical terms (being physiologically included as an efficient cause). Barnouw 1990 argues that the Hobbesian interpretation of sensation is far from strictly mechanical. See also Barnouw 1989, esp. 55 ff. Lejienhorst gives a valid discussion of this orientation. See Leijenhorst 2007, 90.

[74] *MLT*, XXVII, 2, 315; Eng. Trans. Hobbes 1976, 314, modified

[75] *De corpore*, IX, 3; *OL*, I, 197–198; Eng. Trans. *EW*, I, 121–122.

[76] *MLT*, XXVII, 2, 315.

[77] Ibid., Eng. Trans. Hobbes 1976, 314–315.

[78] On the need for the four causes: 'material,' 'efficient,' 'formal,' and 'final,' see Aristotle, *Physics*, II (B), 3, 194b, 16 ff.

approach in perfect keeping with Galilean precepts.[79] In Hobbesian philosophy, in which 'local motion' is the fundamental principle, there is no room for the concept of 'formal cause' nor 'final cause.' Every action is motion, produced by an efficient cause or a bodily agent on a patient (which must necessarily also be bodily).

In the following paragraphs of Chapter 27 of *De motu, loco et tempore*, Hobbes continues the process of translating Aristotelian vocabulary into mechanical terms, specifying what he meant by 'potential' and 'act.' Here his *Megarian* conception of possibility emerges, according to which, 'when an act (or any number of acts taken collectively), if postulated, makes it necessary for another act to come into being, the original act is called *potential* with respect to the production of the future act'[80] and, as a result, the potential of the agent and efficient cause necessarily coincide. Hobbes conceives 'potential' only as the collection of characteristics, i.e. accidents that contribute to the production of the effect, so it makes sense to speak of potenctial only in relation to the act, as we can speak of a 'cause' only in relation to the effect that this produces, or has produced.

Hobbes' treating the concept of cause as inseparable from that of the body has an interesting aspect in that he considers any change in the physical world solely as an interaction between bodies; nevertheless, he believes that the body cannot be created nor destroyed. But we should keep in mind that he considered the pure materiality or corporeity of the body itself impossible to generate and indestructible.[81] It is only accidents that are born and perish with the death of an individual or the destruction of an object. In this case, therefore, the 'essential' accidents of the bodies, which are in the *figure* and the *extension* and which Galileo considered among the primary qualities of bodies, must also disappear.

In Chapter 9 of *De corpore*, in which Hobbes discusses 'cause' and 'effect,' he says: 'a body is said to work upon or *act*, that is to say *do* something to another body, when it either generates or destroys some accident in it: and the body in which an accident is generated or destroyed is said to *suffer*.'[82] In other words, action and passion are necessarily produced by the interaction between bodies, but that which causes the effect is not its corporeity but the existence of certain *accidents* that produce a certain effect in the patient. He writes:

> An agent is understood to *produce* its determined or certain effect in the patient, according to some certain accident or accidents, with which both it and the patient are affected; that is to say, the agent hath its effect precisely such, not because is a body, but because such a body, or so moved. For otherwise all agents, seeing they are all bodies alike, would produce like effects in all patients. And therefore the fire, for example, does not warm because it is

---

[79] On the difference between the Hobbesian concept of cause and that in the Aristotelian tradition, see Leijenhorst 2002, 203 ff.

[80] *MLT*, XXVII, 3, 316; Eng. Trans. Hobbes 1976, 316. On the *Megarian* concept of possibility in Hobbes, see Leijenhorst 2002, 181 ff.

[81] Ibid., XXVII, 1, 312–314.

[82] *De corpore*, IX, 1, *OL*, I, 106; Eng. Trans. *EW*, I, 120.

a body, but because it is hot, nor does one body put forward another body because it is a body, but because it is moved into the place of that other body.[83]

His discussion shows that the concepts of 'cause,' 'effect,' 'potential,' and 'act,' must be linked by the concept of 'body,' as well as to that of 'accident,' which determines every mutation in the physical world. But a clarification should be made: not only is every change or phenomenon determined by motion and certain accidents, the effect of these changes must necessarily be expressed in the presence or absence of certain accidents, which—at a phenomenal level—manifest the new characteristics that the body takes on, which let us notice the change.

From the perspective of accidents, the ones that prove decisive are those accidents that have 'Galilean' roots: extension and figure. Indeed, motion—which, along with matter, is the key element of Hobbesian philosophy and the driver of all mutations and changes—cannot effect materiality in general but only definite and determinate bodies that show a principle of individuation. The concepts of agent and patient, cause and effect only apply to bodies that are delimited by a figure and have an extension. Accordingly, the concept of *cause*—of prime importance in the development of science according to Hobbes[84]—also only finds a realm of application thanks to Galileo's speculations as it is rooted in the Galilean establishment of primary and secondary qualities.

2.4 'Dominions, Occult Properties, and Similar Childish Ideas:' Galileo, Hobbes, and Criticism of Kepler

As rightly underscored by Jesseph, the main Galilean legacy in Hobbes' philosophy is the idea of conceiving the universe as a mechanical system, within which every phenomenon can be expressed in mathematically quantifiable terms, through the laws of motion.[85] As we know, the concept of motion plays a fundamentally important role within the Hobbesian philosophical paradigm and Hobbes makes explicit reference to a particular type of motion: 'local motion.'

Every action is a local motion in the agent, as well as every passion is local motion in the patient. With the term *agent*, I mean a *body*, whose motion produces an effct in another-body; with *patient* I mean a body in which the motion is produced by another body.[86]

---

[83] Ibid., IX, 3, 107; Eng. Trans. 121.

[84] In Chapter 6 of *De corpore* Hobbes defines philosophy as a solely causal knowledge. See *OL*, I, 59: 'Itaque, scientia τοῦ διότι, sive causarum est; alia cognitio omnis quae τοῦ ὅτιdicitur, sensio est vel a sensione remanens imagination sive memoria.'

[85] See Jesseph 2004, 191.

[86] *TO I*, *OL*, V, 217 (my trans.): 'Omnis actio est motus localis in agente, sicut et omnis passio est motus localis in patiente. *Agentis* nomine intelligo *corpus*, cujus motu producitur effectus in alio corpore; *patientis*, in quo motus aliquis ab alio corpore generatur.' See also *MLT*, V, 1, 128–129; *MLT*, XIV, 2, 202; *De corpore*, IX, 9, *OL*, I, 112. On the contrary, Aristotle conceived four types of change: 1) local motion, 2) increase and decrease, 3) alteration or change of quality, 4) the sub-

Local motion is therefore the principle that generates every natural phenomenon, since nothing else exists in the universe except matter and motion.

In *De corpore*, Hobbes credited Galileo with having 'opened to us the gate of natural philosophy universal' and for having founded a 'new science of motion.' Here he echoed the text of the letter addressed from the printer to the readers, at the beginning of the *Discourses and Mathematical Demonstrations Relating to Two New Sciences*, where the author was presented as the discoverer of 'two entirely new sciences and to have demonstrated them in a rigid, that is, geometric, manner.'[87] What is more, the closing lines stated that:

> Of these two sciences, full of propositions to which, as time goes on, able thinkers will add many more; also by means of a large number of clear demonstrations the author *points the way* to many other theorems as will be readily seen and understood by all intelligent readers.[88]

Of these two sciences, the focus was placed on the former, dedicated to 'a subject of never-ending interest, perhaps the most important in nature, one which has engaged the minds of all the great philosophers and one concerning which an extraordinary number of books have been written. I refer to local motion [*moto locale*].'[89]

Galileo developed the problem of local motion extensively, accompanied by mathematical and physical illustrations, in the third day of his *Discourses*.[90] However, even in his *Dialogue Concerning the Two Chief World Systems*, he conveyed a mechanical 'image of the world.'[91] Here he described a complete and detailed cosmic system, in which the only type of movement to be contemplated was local motion.

Galileo's mechanism particularly emerges in the explanation of the phenomenon that the scientist held to be the crucial proof of the accuracy of the Copernican system: *tides*.[92] As early as 1616, in his *Discourse on the Tides*, he had provided a

---

stantial change that takes place in the process of generation and corruption (see Aristotle, *Physics*, III, 1, 201a). See the footnote by Paganini, in Hobbes 2010, 291. On this subject see also Leijenhorst 2002, 179–181.

[87] Galilei, *Discorsi e dimostrazioni matematiche intorno a due nuove scienze*, *OG*, VIII, 46 (my italics); Eng. trans. Galilei 2003, 20.

[88] Ibid. (my italics); Eng. trans. ibid., 21.

[89] Ibid.; Eng. trans. ibid., 20, modified. The *Discorsi e dimostrazioni matematiche intorno a due nuove scienze* was known in the seventeenth century as *Dialogi de motu*, and Galileo used these title to refer to his last work. In fact, this title derives from the Latin treatise, *De motu locali*, which features in the third and fourth day. Moreover, it was Galileo himself who defined the work with those words, as demonstrated by a letter written in June 1637 to Lorenzo Realio, where Galileo mentions the *Discorsi* as his 'libro *de motu*.' See Galileo to Lorenzo Realio, 5 June 1637, *OG*, XVII, 100.

[90] See Galilei, *Discorsi e dimostrazioni matematiche intorno a due nuove scienze*, *OG*, VIII, 190 ff.

[91] See Dijksterhuis 1961, esp. 348 ff.

[92] Regarding the fact that Galileo considered the tides to be crucial proof of the physical correctness of the Coperican system, see Altieri Biagi 1995, who indicated that the planned *Dialogo sul flusso e reflusso del mare*, which then became the *Dialogo sopra i due massimi sistemi*, should have

solution to the problem of tides that mentioned a combination of annual and daily motion to which the earth is subjected.[93] This reasoning was taken up again in the *Dialogue*, where the fourth and final day is dedicated entirely to this phenomenon.[94] In the first few pages of this final day, Galileo stated that he had identified an analogy between the 'monthly' period of the tides and the 'motion of the Moon,' but he claimed that the latter did not introduce 'other movements, but merely alters the magnitude of those already mentioned,' that is to say the daily movements.[95]

However, as he continues with his argument, he rules out any attraction exercised by the lunar orb,[96] focusing his attention on the combination of the two movements of the earth, daily and annual.

The partial exclusion of the moon from Galileo's analysis of the phenomenon of the tides is anything but coincidental and, as we shall see, has its roots in his science of motion.[97] Despite this, it is worth noting that Hobbes was not only influenced by Galileo's reflections on the composition of earthly motions in relation to the phenomenon of the tides, but he also draws from Galileo the analogy between the movements to which the mass of the oceans on the terrestrial surface is subjected and the fluctuations of the water contained in a pot or basin,[98] which can be found in several of Hobbes' scientific works.

Nevertheless, before highlighting Hobbes' interest in Galileo's theory of tides, it is worth examining the passage from the *Dialogue* in which Galileo expressed his clear-cut opposition to the explanation of the phenomenon put forward by Johannes Kepler. In fact, the rejection of the Keplerian theory is significant, because Kepler had developed a model centered around the relationships between the earth and moon, which was strongly criticized by Galileo.

Galileo started by highlighting that the moon's orbit proceeds in the same direction as the earth's. As a result, the turning of the moon could not produce that variation in motion that Kepler believed to be the cause of the tides and that Galileo

---

opened—as Galileo intended—with a discussion of the phenomenon of the tides. On tides see also Shea 1974, 217 ff.

[93] See Galilei, *Discorso del flusso e del reflusso del mare*, OG, V, 381 ff. It is interesting to observe that the analogy between the motions within a *basin* and the composition of movements to which the terrestrial sphere is subjected also features in one of Paolo Sarpi's *pensieri* (thoughts) dating to c. 1595. See Sarpi 1996, pensiero 569, 424. See also Sosio 1995, esp. 305 ff.; Heilbron 2010, 216; Naylor 2014.

[94] An extensive analysis of the phenomenon of the tides features in Naylor 2007, 1–22. On Galileo's theory of tides see also Festa 2007, 195–200; Henry 2011, on which we will return.

[95] Galilei, *Dialogo sopra i due massimi sistemi*, OG, VII, 444; Eng. trans. Galilei 2001, 486.

[96] In Galileo's *Dialogue* (OG, VII, 445–446), Simplicio recalls a tract by '*un certo prelato*' (a certain prelate), who said that the Moon '*vagando per il cielo, attrae e solleva verso di sé un cumulo d'acqua, il quale la va continuamente seguitando, sì che il mare alto è sempre in quella parte che soggiace alla Luna*', but Sagredo brands these lucibrations as 'fancies' that should not credited.

[97] Ibid., 452.

[98] Ibid., 453–454.

instead attributed to the composition of the two motions inherent solely to the earth, daily and annual motion.[99]

What is more, Galileo also observed that were we to imagine that the moon exercised an influence or attraction over the earth, 'it is neither explained nor self-evident how this is supposed to happen,' that is to say there is no adequate theory to explain a possible attraction exercised by the satellite on the oceans of the terrestrial sphere. His criticism of Kepler is directly connected to this difficulty:

> However, of all great men who have philosophized on such a puzzling effect of nature, I am more surprised about Kepler than about anyone else; although he had a free and penetrating intellect and grasped the motions attributed to the earth, he lent his ear and gave his assent to the dominion of the moon over the water, to occult properties, and to similar childish ideas.[100]

According to Galileo, Kepler, who had claimed the moon exercised an attraction over the waters of the sea, was unable to provide adequate and comprehensive explanations in this regard, simply referring to the 'dominion of the moon over the water, to occult properties, and to similar childish ideas.' Instead, he claimed that 'this movement of the oceans is a local and sensible one, made in an immense bulk of water.' He openly denied that the moon had a direct effect on the production of the tides, instead stating that this idea was 'repugnant' to his mind, which cannot 'give credence to such causes as lights, warm temperatures, predominances of occult qualities, and similar idle imaginings.'[101]

Galileo's criticism of his 'colleague' Kepler is interesting, because it is indicative of the different epistemological approach adopted by the two thinkers and conceals the general principle underlying all of Galileo's natural philosophy. Where it is not possible to include a natural phenomenon in terms of a mechanistic explanation (in this case the effect of the moon on the waters of the sea), Galileo deems it necessary to rule the element out from the theory.[102]

---

[99] Galilei, *Dialogo sopra i due massimi sistemi del mondo*, *OG*, VII, 486: 'Il dire anco (come si riferisce d'uno antico matematico) che il moto della Terra, incontrandosi col. moto dell'orbe lunare, cagiona, per tal contrasto, il flusso e il reflusso, resta totalmente vano, non solo perché non vien dichiarato né si vede come ciò debba seguire, ma si scorge la falsità manifesta, atteso che la conversione della Terra non è contraria al moto della Luna, ma è per il medesimo verso: talché il detto e imaginato sin qui da gli altri resta, al parer mio, del tutto invalido.' As underscored by Naylor, Galileo is aware that the phases of the moon and the tides coincide and tries to explain the phenomenon various times, through the composition of the rotations and revolutions of the planets. See Naylor 2007, 18 ff. See also the letters by Galileo to Fulgenzio Micanzio, dated 7 November 1637 and 30 January 1638 (*OG*, XVII, 214–215 and 269–271).

[100] Galilei, *Dialogo sopra i due massimi sistemi del mondo*, *OG*, VII, 486; Eng. trans. Finocchiaro 2008, 268.

[101] Ibid., 470; Eng. trans. Galilei 2001, 516.

[102] Galileo returned to the subject several times and, in a letter written to Micanzio in 1637, he even contemplates the movement of the Moon as being one of the causes of the phenomenon. See Galileo to Fulgenzio Micanzio, 7 November 1637, *OG*, XVII, 214–215). See Naylor 2007, 18 ff. See also Bucciantini 2003, 306 ff.

Astronomical problems also offer us a privileged viewpoint of the development of Hobbes' mechanism. In Chapter 24 of *De motu, loco et tempore*, Hobbes sought to identify the forces that link the motions of the earth to those of the moon and, in this passage, he did not seem to agree with Galileo's criticism of Kepler. Although he supposed that the earth's effect upon the moon mainly entailed the transmission of a movement originating in its daily rotation, Hobbes did not rule out the action of 'a kind of influence-as if a magnetic one,'[103] which presented more than a few analogies with Kepler's ideas in *Astronomia nova*.[104]

Hobbes evidently realized that Galileo was unable to clarify how the lunar phases and terrestrial motions were related,[105] while Kepler—despite referring to somewhat vague magnetic attractions—had instead sketched out an explanation in this regard.

In *De corpore*, on the other hand, Hobbes reconsidered the terms of the question and categorically rejected any occult influence or quality. In this sense, his analysis of the magnetic phenomena proves to be paradigmatic. Although the problem is not addressed explicitly, in Chapter 26 of the work, when discussing the revolution of the moon around the earth, Hobbes seemed to adopt a position that was diametrically opposed to the claims he had made in *De motu, loco et tempore*:

> And as for them that suppose this may be done by magnetical virtue, or by incorporeal and immaterial species, they suppose no natural cause; nay no cause at all. For there is no such thing as an incorporeal movent, and magnetical virtue is a thing altogether unknown; and whensoever it shall be known, it will be found to be a motion of body.[106]

Hobbes refused to resort to a magnetic force and, when exploring the cause of the eccentricity of the earth's revolution, he commented on Kepler's position in almost identical terms with regard to Galileo:

> I am, therefore, of Kepler's opinion in this, that he attributes the eccentricity of the earth to the difference of the parts thereof, and supposes one part to be affected, and another disaffected to the sun. And I dissent from him in this, that he thinks it to be by magnetic virtue, and that this magnetic virtue, or attraction and thrusting back of the earth is wrought by immateriate species[107]; which cannot be; because nothing can give motion, but a body

---

[103] *MLT*, XXIV, 1, 289, Eng. Trans. Hobbes 1976, 278: "...sol quidem motu primum illo quo & terram, & lunam quotannis simul circumfert, deinde etiam motu illo quo omnia recta quadam undequaque eiaculatione illuminat. Terra autem in eandem non modo motu suo diurno aliquid potest, sed etiam influentia quadam tanquam magenticam, & praeterea vaporibus fortasse eousque elevatis'.

[104] See Horstmann 1998, 142 ff.

[105] *MLT*, XVI, 2, 211.

[106] *De corpore*, XXVI, 7, *OL*, I, 351; Eng. Trans. *EW*, I, 430. The interesting aspects of Hobbes' mechanical explanation of magnetism are picked up by Leijenhorst 2002, 188 ff. Henry instead highlights the inconsistencies and the criticism directed at Boyle. See Henry 2016, esp. 30 ff.

[107] Cf. *Astronomia nova*, pars III, chap. XXXIII, in Kepler 1937—, III, 240: 'Effluxus igitur, quemadmodum et lucis, immateriatus est; non qualis odorum cum diminutione substantiae, non qualis caloris ab aestuante fornace, et si quid est simile, quibus media implentur. Relinquitur igitur, ut quemadmodum lux, omnia terrena illustrans, species est immateriata ignis illius, qui est in corpore solis: ita virtus haec, planetarum corpora complexa et vehens, sit species immateriata ejus virtutis,

moved and contiguous. For if those bodies be not moved which are contiguous to a body
unmoved, how this body should begin to be moved is not imaginable.[108]

Hobbes could not be clearer in his mechanistic interpretation of astronomical
phenomena. In perfect agreement with the principle of inertia, a body in a state of
rest cannot start to move without the intervention of another external and contiguous
body. He rules out the possibility of any remote action that cannot be conceived
in mechanistic terms of action and reaction between bodies, either directly or
through the medium.

In the cited passages, Hobbes excludes magnetism from his explanation of lunar
motion due to the considerable difficulties encountered when examining this matter
in accordance with the principles of mechanical philosophy. It is no coincidence
that in the 'third day' of his *Dialogue*, Galileo's criticism of William Gilbert focused
precisely on his failure to explore the phenomenon. Despite praising Gilbert for
having 'framed such a stupendous concept,' Galileo also expressed his regret that he
was not 'more of the mathematician, and especially [with] a thorough grounding in
geometry:'

> ... a discipline which would have rendered him less rash about accepting as rigorous proofs
> those reasons which he puts forward as *verae causae* for the correct conclusions he himself
> had observed. His reasons, candidly speaking, are not rigorous, and lack that force which
> must unquestionably be present in those adduced as necessary and eternal scientific
> conclusions.[109]

Over the following pages, Galileo tackled the problem in part, discussing 'arma-
tured lodestones' and referring to 'contacts,'[110] that is to say to microparticulate
bonds that develop between the surfaces of two ferrous materials, bonds that exist
thanks to the perfect adherence of the 'innumerable tiny particles if not by the infinity of points on both surfaces.'[111]

In addition to the particular explanation of the phenomenon, it is interesting to
underscore Sagredo's reaction to the argument presented by his Galilean interlocutor, Salviati. He declared that the reason given 'satisfies me little less than it would
if it were a pure geometrical proof' and this is despite the fact that despite it being a
'physical problem,' concerning 'natural science,' where—as Simplicio was also

---

quae in ipso sole residet, inaestimabilis vigoris, adeoque actus primus omnis motus mundani.' See
also Gargani 1971, 274–278.

[108] *De corpore*, XXVI, 8, *OL*, I, 354; Eng. Trans. EW, I, 434.

[109] Galilei, *Dialogo sopra i due massimi sistemi del mondo*, *OG*, VII, 432; Eng. trans. Galilei
2001, 471.

[110] Ibid., pp. 432–436; Eng. trans. Galilei 2001, 470–475.

[111] Ibid., 436; Eng. trans. Galilei 2001, 475. See the discussion of this phenomenon in Galluzzi
2011, 82–84.

well aware[112]—'geometrical evidence cannot be demanded.'[113] Nevertheless, Galileo maintained that, in order to rise to the level of scientific theory, every explanation of natural phenomena always had to be expressed in strictly mathematical terms. As a result, the 'sympathy,' to which the Aristotelian Simplicio referred, 'which is a certain agreement and mutual desire,' in no way represented an explanation of magnetism. On the contrary, Sagredo ironically stated: 'this method of philosophizing seems to me to have great sympathy with a certain manner of painting used by a friend of mine,' who limited himself to writing the names of what he wanted to paint on the canvas, then leaving the painter with the task of producing the work, but still believing that 'he himself had painted the story of Acteon – not having contributed anything of his own except the title.'[114]

Likewise, in *De corpore*, Hobbes maintained that Kepler's theory—based on concepts of 'attraction' and 'thrusting back'—consisted of 'connexions of words ἀδιανοήτῳ,'[115] meaning that it was unconceivable, while with regard to 'similitude' (Lat. *cognatio*), which Kepler believed to be the cause of their mutual attraction, Hobbes stated sarcastically: 'if it were so, I see no reason why one egg should not be attracted by another.'[116]

These passages clearly show that Hobbes kept faith with the lesson imparted by Galileo. However, in reference to magnetism we must observe that not even the third day of the *Dialogue* offered a clear and comprehensive explanation of the attraction exercised by the magnet from a distance. Hobbes must evidently have realized the problem. In fact, in Chapter 30 of *De corpore*, he developed an alternative to Galileo's theory, which is particularly interesting in any case as it is imbued with 'Galilean spirit.'

The concept of internal movements within bodies, formulated by Hobbes in order to explain their hardness, is at the basis of this elucidation.[117] He compares magnetism to the phenomena of static electricity and gives the example of 'jet,' which is able to attract straw. However, in order for this effect to be activated, it has to be rubbed first, while 'the loadstone hath sufficient excitation from its own nature, that is to say, from some internal principle of motion peculiar to itself.'[118] Nevertheless, Hobbes continues:

---

[112] See Aristotle, *Metaphysics*, II, 3, 995a, 15–18; Eng. trans. Aristotle 1989, 95: 'Mathematical accuracy is not to be demanded in everything, but only in things which do not contain matter. Hence this method is not that of natural science, because presumably all nature is concerned with matter. Hence we should first inquire what nature is; for in this way it will become clear what the objects of natural science are.'

[113] Galilei, *Dialogo sopra i due massimi sistemi del mondo*, *OG*, VII, 436; Eng. trans. Galilei 2001, 475.

[114] Ibid.; Eng. trans. Galilei 2001, 475–476.

[115] *De corpore*, XXVI, 8, *OL*, I, 354.

[116] Ibid., Eng. Trans. *EW*, I, 434.

[117] See above, Chap. 1, and Baldin 2017.

[118] Ibid., XXX, 15, *OL*, I, 427; Eng. Trans. *EW*, I, 526.

Now whatsoever is moved, is moved by some contiguous and moved body, as hath been formerly demonstrated.[119] And from hence it follows evidently, that the first endeavour which iron hath towards the loadstone, is caused by the motion of that air which is contiguous to the iron: also that this motion is generated by the motion of the next air, and so on successively, till by this succession we find that the motion of all the intermediate air taketh its beginning from some motion which is in the loadstone itself; which motion, because the loadstone seems to be at rest, is invisible. It is therefore certain, that the attractive power of the loadstone is nothing else but some motion of the smallest particles thereof.[120]

Hobbes supposes that 'lodestones' are endowed with an internal circular motion at the level of its smallest particles, which is transmitted first to the surrounding air before then reaching the ferrous materials. A sort of internal counter-reaction develops inside them, which makes it possible for the metal to be drawn to the lodestone, since, through their movement, these bodies are able to progressively move the air that separates them, thereby coming together.[121] In order to clarify his explanation, Hobbes also uses the example of a lyre string that, by vibrating, is able to generate motion in another string with an analogous vibration. Over the following pages, he discusses various phenomena associated with magnetism,[122] while the same theory, albeit with slight variations, also features in his later scientific works.[123]

Nevertheless, at the end of this paragraph of *De corpore*—which also marks the end of the work—Hobbes lingers on a number of noteworthy methodological considerations. After specifying that the first, second and third part of the text are—with the exception of errors of any kind—'rightly demonstrated,' because founded *a priori* and developed solely on the basis of deductive reasoning,[124] he then states that, on the contrary, the fourth part (concerning *Physics*) 'depends upon

---

[119] As we will see later on (see below, chap. III) Hobbes continuously reiterates his adhesion to an inertial concept of movement throughout the work. See, for example., *De corpore*, IX, 7, *OL*, I, 110–111; ibid., XXVI, 8, 354.

[120] Ibid. Eng. Trans. *EW*, I, 526–527.

[121] As we will see in the next chapter, Hobbes developed an analogous argument to explain the phenomenon of gravity. See below, chap. III, § 6.

[122] An interesting discussion also features further on in this paragraph to explain the attraction of the pole of the magnet to the terrestrial magnetic North. *De corpore*, XXX, 15, *OL*, I, 428–429, Eng. Trans. *EW*, I, 528–529.

[123] See *Problemata Physica*, *OL*, IV, 357–359 (Eng. Trans. *Seven Philosophical Problems*, *EW*, VII, 56–59); *Decameron Physiologicum*, ibid., 154 ff., although here Hobbes' claims are much more hypothetical and less assertive compared to the earlier works.

[124] *De corpore*, XXX, 15, *OL*, I, 430; Eng. Trans. *EW*, I, 531: 'In the first, second and third parts, where the principles of ratiocination consist in our own understanding, that is to say, in the legitimate use of such words as we ourselves constitute, all the theorems, if I be not deceived, are rightly demonstrated.'

Hypotheses.'[125] And yet, he also suggests that not all hypotheses were the same,[126] deeming that the right hypotheses had to be '*cogitabiles*' (Eng. *conceivable*)[127]:

> For as for those that say anything may be moved or produced by *itself*, by *species*, by *its own power*, by *substantial forms*, by *incorporeal substances*, by *instinct*, by *antiperistasis*, by *antipathy, sympathy, occult quality*, and other empty words of schoolmen, their saying so is to no purpose.[128]

Hobbes also goes back to the accusation regarding the vacuous nature of the scholastic discourse, as formulated by Galileo in his *Dialogue*, without however making any specifications regarding the required 'conceivability' of these hypotheses.[129] The previously cited passages make it is easy to deduce that Hobbes was alluding to something very similar to the mathematical proof desired by the author of the *Dialogue*, but this is perhaps also confirmed by a passage from the *Decameron physiologicum* (published in 1678). Here, Hobbes once again returned to the cause of the eccentricity of terrestrial motion, ascribed by Kepler to a magnetic virtue. The comment made by Hobbes' character regarding Kepler's explanation is extremely eloquent in this regard: 'I am not satisfied with that. It is *magical* rather than *natural*, and unworthy of Kepler.'[130] Hobbes' criticism of Kepler literally takes up the objection made against him by Galileo and is based on the same theoretical instance. Kepler's explanation presents magical rather than scientific connotations, because it is not expressed on the basis of the laws of motion, that is to say in mechanical terms.[131]

---

[125] Ibid., 430–431; Eng. Trans. 531: 'The fourth part depends upon hypotheses; which unless we know them to be true, it is impossible for us to demonstrate that those causes which I have there explicated, are the true causes of the things whose productions I have derived from them.'

[126] In this regard, see: Horstmann 2001, esp. 492–496; As we will see further under, Giudice underscored the convergence between the scientific research method specific to Galileo and that adopted by Hobbes. See Giudice 2016, 92–95.

[127] *De corpore*, XXX, 15, *OL*, I, 431; Eng. Trans. *EW*, I, 531: 'Nevertheless, seeing I have assumed no *hypothesis*, which is not both possible and easy to be comprehended; and seeing also that I have reasoned aright from those assumptions, I have withal sufficiently demonstrated that they may be the true causes; which is the end of physical contemplation. If any other man from other *hypotheses* shall demonstrate the same, or greater things, there will be greater praise and thanks due to him then I demand for myself, provided his *hypotheses* be such as are conceivable'.

[128] Ibid., 431; Eng. Trans. *EW*, I, 531. Strangely, this sort of 'indirect criticism' targeted at Kepler was ignored by Horstmann in his well documented and interesting article on the relationship between Kepler and Hobbes. See Horstmann 1998, 149 ff.

[129] The theme of the conceivability of the hypotheses is also taken up again in the *Dialogus physicus* (*OL*, IV, 247; Eng. In Shapin and Schaffer 1985, 356), where Hobbes considers 'is his not the rule for all hypotheses, that all things that are supposed must all be of a possible, that is, conceivable, nature?'

[130] *Decameron Physiologicum*, *EW*, VII, 102.

[131] In an interesting article on the importance of Galilean kinematics, as an element that made a decisive contribution to the development of the scientific revolution and that would be inherited by the great philosophers of the seventeenth century (including Hobbes), John Henry emphasized the central role played by Galileo's theory of tides, which features extremely mechanistic connotations, associated with the rejection of Kepler's theory (and of every notion of occult qualities). It is

Throughout the entire work, Hobbes continuously opposes arguments that refer to sympathy, influence, substantial forms, or incorporeal effluvia, instead proposing only those explanations centered around movement or, better still, around the only Aristotelian category of movement contemplated by Galilean philosophy: local motion.

> You know I have no other cause to assign but some *local motion*, and that I never approved of any argument drawn from sympathy, influence, substantial forms, or incorporeal effluvia. For I am not, nor am accounted by my antagonists for a witch.[132]

Only theories expressed through mathematically quantifiable laws of movement can achieve the status of scientific hypotheses. Any other explanation—founded on 'sympathy, influence, substantial forms, or incorporeal effluvia'—is worthy of a 'child' or a 'witch.'

2.5   Tides

The Hobbesian theories of terrestrial motion and magnetism are evident examples of Hobbes' stringent mechanism. Nevertheless, we must linger a while longer on the issue of tides, because Galileo's explanation of the phenomenon leaves evident marks on Hobbesian natural philosophy. We know that the Italian scientist had resolutely rejected Kepler's theory, based on the idea of an influence of the moon over the waters that cover the earth, and had come up with an alternative explanation, centered around the imperfect coincidence of the motions to which the earth is subject. We should also remember that, since 1616 (the year that the *Discourse on the Tides* was written), Galileo considered the phenomenon of tides to be evident proof of the accuracy of the Copernican system. In the *Dialogue Concerning the Two Chief World Systems*, the theory resorts to an analogy between the water present on the surface of the earth and that contained in a vessel or basin. The mass of water, if subjected to different movements that are not perfectly synchronous and concordant, undergoes jerks and motions that tend to drive part of the liquid out of the receptacle.[133] Galileo rules out any attraction exercised by the moon, because his

---

of fundamental importance because it is rooted in Galileo's mechanism. The convergence with Hobbes' scientific writings on the subject seems to confirm this idea in my opinion. See Henry 2011, 3–36 (meanwhile, the observations advanced by the author in the second part of the article and inspired by Koyré's interpretation are less convincing).

[132] *Decameron Physiologicum*, EW, VII, 155 (my italics).

[133] Galilei, *Dialogo sopra i due massimi sistemi del mondo*, OG, VII, 453–454: 'Concludiamo per tanto, che sì come è vero che il moto di tutto il globo e di ciascuna delle sue parti sarebbe equabile ed uniforme quando elle si movessero d'un moto solo, o fusse il semplice annuo o fusse il solo diurno, così è necessario che, mescolandosi tali due moti insieme, ne risultino per le parti di esso globo movimenti difformi, ora accelerati ed ora ritardati, mediante gli additamenti o sottrazioni della conversion diurna alla circolazione annua. Onde se è vero (come è verissimo, e l'esperienza ne dimostra) che l'accelerazione e ritardamento del moto del vaso faccia correre e ricorrere nella

mechanical physics—which only contemplates the action of moving bodies—is unable to explain that particular and curious effect of the moon on the surface of the earth. On the contrary, the earth is subjected both to daily rotation and annual revolution, and the combination of these motions produces the phenomenon of the rising and falling sea.

As we know, a number of letters from Hobbes dating to 1636 suggested that he advocated a mechanism of Galilean inspiration, which is first set out in full in his *Elements of Law* and *Tractatus Opticus I*, where Hobbes openly expresses the idea that every physical phenomenon is only explained through the action and reaction of moving bodies.[134] Despite this, when it comes to discussing the phenomenon of tides in *De motu, loco et tempore*, Hobbes realized that Galileo's refusal to examine the possible relationship between lunar motions and the tides invalidated his entire theory, to the point that the position taken by the Italian scientist was not acceptable.[135] On the contrary, Hobbes assumes the concurrence between the phases of the moon and the tidal variations to be a matter of fact,[136] peremptorily stating that these phenomena are 'accepted by popular opinion and proved by experience.'[137] His comment on Galileo's explanation is to the point: 'since not even Galileo has considered this correspondency between the tide and the moon, there is no reason why one should agree with what Galileo has written on the subject.'[138]

However, Hobbes grasps another problem underlying Galileo's theory, and that is that he failed to consider the fact that the oceans and terrestrial sphere form part of the same physical system, according to the theory developed by Galileo himself.[139] On the other hand, Hobbes is unable to provide an alternative theory to Galileo and simply admits that he had insufficient knowledge to settle the question definitively.

He returns to the issue in *De corpore*, where he comes up with an explanation that takes up the model developed by Galileo in the 'fourth day' of his *Dialogue*.[140] In Chapter 26, Hobbes has the following to say about the subject of tides:

---

sua lunghezza, alzarsi ed abbassarsi nelle sue estremità, l'acqua da esso contenuta, chi vorrà por difficultà nel concedere che tale effetto possa, anzi pur debba di necessità, accadere all'acque marine, contenute dentro a i vasi loro, soggetti a cotali alterazioni, e massime in quelli che per lunghezza si distendono da ponente verso levante, che è il verso per il quale si fa il movimento di essi vasi? Or questa sia la potissima e primaria causa del flusso e reflusso, senza la quale nulla seguirebbe di tale effetto.'

[134] See *EL*, pars I, chap. II, §10, 7; *TO I, OL*, V, 217.

[135] The phenomenon of the tides occupies a good part of Chapt. XVI and all of Chapt. XVII in *De motu, loco et tempore*. Here, Hobbes primarily puts forward his counter-objections to White's criticism of Galileo's *Dialogue*. See *MLT*, 210–214.

[136] Ibid., XVI, 2, 210–211.

[137] Ibid., 211; Eng. Trans. Hobbes 1976, 172, modified.

[138] Ibid.

[139] Ibid., XVII, 6; 220.

[140] Galileo deemed that the moon formed part of a system hinged upon the revolution of the earth (and the other planets). Within this macro-system, the earth and moon constitute a sort of micro-system, so that the moon follows the earth in its revolutions (daily and annual), which are not

To these three simple motions, one of the sun, another of the moon, and the third of the earth … together with the diurnal conversion of the earth, by which conversion all things that adhere to its superficies are necessarily carried about with it, may be referred the three phenomena concerning the tides of the ocean.[141]

The three phenomena that Hobbes attributes to the tides are the rising and falling of the waters along the coast, the coincidence of the new moon with the highest tide and, lastly, an additional increase in the height of the tide during the equinox. Nevertheless, in addition to the particular motions to which the tides are subject, it is important to underscore that the explanation developed by Hobbes is analogous to that of Galileo. What is more, it is also strictly mechanical, just like that of the Italian scientist. Hobbes considers the tides to be the result of the combination of three movements, albeit caused by a single mover: the sun. Like Kepler and Galileo,[142] he believes that the sun rotates around its own axis, to which Hobbes adds an additional movement that he calls 'simple circular motion.'[143]

This model features in almost all Hobbesian works and, despite the different variants, one can perceive the common matrix, which brings together elements drawn primarily from Copernicus' *De Revolutionibus orbium coelestium* and arguments taken from Galileo's *Dialogue*.

In his *Tractatus Opticus II*, Hobbes supposes that the two annual motions of the earth, described by Copernicus, could be perfectly represented by his theory of *motus cribrationis*, based on the analogy between the rotation and revolution of the planets and the movements to which grains of wheat are subjected within a sieve as it is turned.[144] He also cites Galileo's observations regarding the motion of the sun,

---

perfectly synchronous. According to Galileo, the fact that the movement of the moon coincides with the movements of the tides derives from this fact. See Galilei, *Dialogo sopra i due massimi sistemi del mondo*, *OG*, VII, 477 ff.

[141] *De corpore*, XXVI, 10, *OL*, I, 356, Eng. Trans. *EW*, I, 437.

[142] When discussing sunspots in his *Dialogue*, Galileo advanced the theory that the sun rotated around its own axis (See Galilei, *Dialogo sopra i due massimi sistemi del mondo*, *OG*, VII, 79), as he had claimed more explicitly in his *History and Demonstration concerning Sunspots* (Galilei, *Istoria e dimostrazioni intorno alle macchie solari e loro accidenti*, Letter II, *OG*, V, 117), where he wrote that 'il corpo del Sole è assolutamente sferico; secondariamente, ch'egli in sé stesso e circa il proprio centro si raggira portando seco in cerchi paralleli le dette macchie, e finendo una intera conversione in un mese lunare in circa, con rivolgimento simile a quello de gli orbi de i pianeti, cioè da occidente verso oriente.' In his *Astromia Nova* (in Kepler 1937—, III, 34), Kepler had already supposed that the sun rotates around its axis and that this movement, together with the action of the *species*, produced the revolution of the planets, but he explained it by referring to 'speciem immateriatam corporis sui, analogam speciei immateriatae lucis suae'. On the analogies between Kepler and Galileo regarding this subject see: Applebaum and Baldasso 2001. On the influence of these Galilean ideas on the genesis of 'simple circular motion' in Hobbes, see Henry 2016, 25 ff.

[143] See *De corpore*, XXVI, 6, *OL*, I, 349. Horstmann 1998, 139 ff. supposes that the source of this Hobbesian theory is Kepler, but I think it is more likely that here Hobbes is referring to the texts of Galileo, because the Italian scientist's name is cited directly.

[144] See *TO II*, fol. 198v/153.

in order to support his theory that the movement of the star causes the earth's motion.[145]

On the one hand, therefore, Hobbes refers to the demonstrations contained in Chapter 11 of book I of *De Revolutionibus Orbium Coelestium*,[146] where Copernicus, in order to explain the parallelism of the terrestrial axis during revolution, had supposed the existence of two annual motions attributable to the earth: the first its revolution around the sun along the ecliptic, according to the order of the signs, the other *in declinatione* around its own axis, in the opposite direction to the former.

However, Hobbes' argument seems to be modelled on the 'fourth day' of the *Dialogue*, where Galileo had drawn upon Copernican speculations in his turn:

> We said that there are two motions attributed to the terrestrial globe: the first is the annual motion performed by its center along the circumference of the annual orbit in the plane of the ecliptic and in the order of the signs of the zodiac, namely, from west to east; the other is performed by the same globe rotating around its own center in twenty-four hours, likewise from west to east, but around an axis somewhat inclined and not parallel to that of the annual revolution. From the combination of these two motions, each of which is in itself uniform, there results, I say, a variable motion for the parts of the earth.[147]

Although the term *motus cribrationis* is missing from *De motu, loco et tempore*, as well as from his later works, this model still persists in Hobbesian astronomy, just as the references to Copernicus' writings also remain.[148]

Hobbes also draws upon Copernican theories in Chapters 21 and 26 of *De corpore*, basing his argument on the idea that 'the several points ... do in several equal times describe several equal arches,' in the movement that he describes as 'simple circular motion.'[149] Hobbes supposes that these points do not rotate around a common axis, but that each individual point is positioned on the surface of an imaginary circumference and rotates around its own center. This reasoning means that segment AB (*see figure*) can always remain parallel in DE, just as the axis of the earth always stays parallel to itself in the earth's annual revolution.

---

[145] Ibid.: 'Non erit igitur absurdum existimare, siquidem sol causa sit conversionis terrae annuae, eiusmodi quoque motum esse in Sole, praesertim cum motum quendam esse in sole circa Axim Ecclipticae observaverit Galilaeus neque putandum sit temere esse quod in tanta coelorum amplitudine nullum inveniatur Corpus quod circumvolvatur per aliam orbitam praeter Ecclipticam vel ab Ecclipica parum digredientem.'

[146] See Copernicus 2015, II, 74 ff.

[147] Galilei, *Dialogo sopra i due massimi sistemi del mondo*, *OG*, VII, 452; Eng. trans. Finocchiaro 2008, 257.

[148] In section 4 of Chapter 6 of *De motu, loco et tempore*, Hobbes includes the theory of *simple circular motion* deriving from the writings of Copernicus, which returns in a more complete and comprehensive form in *De corpore*. See *MLT*, VI, 4, 137–139.

[149] *De corpore*, XXI, 1, *OL*, I, 258 Eng. Trans. *EW*, I, 317–318.

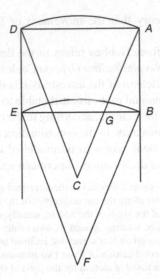

Hobbes applies the same model to the movement of the moon, which, according to him, is to be attributed to the rotation of the earth, which behaves towards its satellite as the sun does with our planet. Hobbes develops various versions of this theory, from the second optical treatise to *De corpore* and later works, always claiming that the rotation of the sun (connected, initially, to an additional movement of systoles and diastoles) is able to move the other planets, generating two movements in them, rotation and revolution, transmitted by them to their satellites, just as the earth does with the moon.

In order to explain the phenomenon of the tides, Hobbes also supposes a combination of terrestrial and lunar movements, although he fails to provide any explanation as to how the tide is caused by these movements.

Incidentally, this problem is taken up again in *Dialogus physicus sive de natura aeris* (1661), returning to the theory already put forward in *De corpore* and defined as 'simple circular motion.' Nevertheless, compared to *De corpore*, Hobbes uses another extremely significant example if compared to Galileo's argument. He describes a 'basin' containing a small quantity of water, undoubtedly referring to the argument in Galileo's *Dialogue*.[150]

The following year, in his *Seven Philosophical Problems*, Hobbes tackles the problem of the tides directly, using Galileo's example of the basin of water to

---

[150] *Dialogus physicus sive de natura aeris*, OL, IV, 251–252; Eng. trans. Shapin and Schaffer 1985, 360: 'he [Hobbes] assumes another hypothesis, this one, that the earth has its own motion, due to its own nature or creation, which is also simple, circular. And from this supposition he demonstrates clearly many things about natural causes; and indeed, what that may be you will understand thus. Take up a basin in your hand, in whose bottom is a little water … Could you not move the water by moving the basin, so that it ran in a circle, raising itself a little around the concave surface of the basin?'.

describe the phenomenon.[151] In the Latin version of the work (*Problemata physica*), he expressly names Galileo and Bacon,[152] claiming that both had attributed the cause of tides to the motion of the earth. However, Hobbes states that this movement could only be contemplated by conceiving the sun, earth, and moon as conjoined by some firm bond, similar to that of a ball of lead hanging from a string.[153]

The same explanation and the same image of the basin return in the *Decameron physiologicum*,[154] but, above and beyond the fascination exercised by the example taken from the *Dialogue*, it is interesting to underscore the Galilean legacy that emerges in Hobbes' astronomical theories. Both authors reject any kind of explanation that is not expressed in terms of action and reaction between bodies, that is to say every interpretation of physical phenomena that transcends the limits of mechanism. The methodological criteria imposed upon the new science by Galileo are absorbed and redeveloped by Hobbes, who projected them well beyond the limits and objectives that Galileo had set himself, to the point of constructing an entire philosophical system with Galilean premises.[155]

---

[151] *Seven Philosophical Problems, EW*, VII, 13. See also *Problemata Physica, OL*, IV, 313. Here the matter is discussed in a different and more comprehensive way than in *De corpore* and *Dialogus physicus*, because Hobbes explored the issue of the coincidence of the high tide with the phenomena of the new moon and the full moon, that is to say direct interaction between the moon and earth. See *Seven Philosophical problems, EW*, VII, 15.

[152] Hobbes refers to a passage from *De fluxu et refluxu mari*. See Bunce 2003, 57–58.

[153] *Problemata Physica, OL*, IV, 317: '…sed (*Bacon*) motus aquae causam adscribit motui diurno primi mobilis, qui motus primi mobilis, cum sit in circulo cujus centrum est centrum terrae, propellere aquam non potest. Etiam Galileus causam aestuum horum terrae motui cuidam adscribit: quem motum terrae habere non potest, nisi sol, terra, et luna solido aliquo vinculo connecterentur, tanquam in fune pendulo totidem pilae plumbeae.' Interestingly, Hobbes seems to echo an explanation put forward by Galileo himself, in the fourth day of his *Dialogue*, where he used the example of the weight hanging from a string to explain the phenomenon of the tides. See Galilei, *Dialogo sopra i due massimi sistemi, OG*, VII, 475 ff.

[154] *Decameron physiologicum, EW*, VII, 109–111. See also ibid., 100–102, where Hobbes goes back to the model he had put forward in *De corpore*, in which the sun, by moving around its axis, causes the motion of all the other planets and, as an indirect result, their satellites too.

[155] In the light of the themes analysed in the previous chapters, I feel that the roots of Hobbes' mechanism should be sought in the 'Galilean' premises of his philosophy. Because of this, I disagree with Garber 2013, 13–5, who considers Galileo as essentially extraneous to mechanism. Garber bases this claim on two arguments: a supposed debt of the Italian scientist to Aristotelian mechanics and the infinitesimal conception of matter sketched out by Galileo in his *Discourses*, which would necessarily make him extraneous to corpuscularism. As regards the former, it should be observed that Galileo tackles a number of Aristotelian themes, but he certainly does not come up with the same responses. Meanwhile, as regards the latter, Galileo has a very complex concept of matter, although, in any case, the 'infinitesimal' structure of matter in the *Discourses* is far from invalidating the arguments on the qualities of bodies in *The Assayer* and is based on a 'corpuscular' (broadly speaking) concept of matter.

2.6 Things Produced 'Mechanically by Nature'

The passage from *The Assayer* that proposed a distinction between qualities inherent to the bodies and qualities, which could instead be linked solely to manifestations induced in the sentient being by the movement of the smallest parts present in the objects, is one of the few ideas dedicated to the nature of sensation in Galileo's texts. The lack of focus on this phenomenon has sometimes been interpreted as being due to a greater propensity for purely physical study or, even, as a defect attributable to the absence of a comprehensive and systematic philosophy in Galileo, which underlies all properly scientific speculations.[156] In reality, in Galileo's texts a well outlined natural philosophy emerges, even if sometimes not clearly and distinctly. In this regard, it is interesting to underscore how, in a letter to Diodati dated 1638, Galileo clearly expressed his wish to study animal sensation, especially those things 'that are produced mechanically by nature.'[157]

For his part, Hobbes instead took his first steps in the world of natural philosophy driven by his interest in discovering the causes and nature of sensation and intellection, and by his wish to interpret these phenomena by referring solely to matter and motion.[158] In fact, the first part of his *Elements* is dedicated to anthropology and, in it, he emphasizes how the analysis of sense, passions, and the formation of concepts, does not require resorting to alternative explanations compared to those we refer to in order to explain any other phenomenon that occurs in the natural world.[159] The idea of contemplating animal sensations and passions solely as an effect of moving matter allows Hobbes to conceive the natural universe by applying a single, uniform explanatory criterion to it.[160] This is also expressed very clearly in *De motu,*

---

[156] This, for example, was how Descartes interpreted it. See Descartes to Mersenne, 11 October 1634, *AT*, II, 380–402. On Descartes's interpretation of Galileo see Shea 1978, and Bucciantini 2007.

[157] Galileo Galilei to Elia Diodati, 23 January 1638, *OG*, XVII, 262 (my italics): '… io ho un buon numero di problemi e questioni spezzate, tutte, al mio consueto, nuove e con nuove dimostrazioni confermate. Sono ancora sul tirare avanti un mio concetto assai capriccioso; e questo è il portar, pur sempre in dialogo, una moltitudine di postille fatte intorno a' luoghi più importanti di tutti i libri di coloro che mi ànno scritto contro et anco di qualch'altro autore e in particolare di Aristotele, il quale nelle sue Questioni Meccaniche mi dà occasione di dichiarare diverse proposizioni belle, ma molto più ancora me ne dà nel trattato *De incessu animalium*, materia piena di cose ammirabili, come quelle che *son fatte meccanicamente dalla natura*.' See Camerota 2004, 561. This passage leads me to disagree with Alan Gabbey, who claims that in Galileo there is no sense of *mechanism* in the sense of the term used by Boyle. See Gabbey 2004, 15.

[158] For a global overview of mechanism in the seventeenth century it is always useful to refer to: Dijksterhuis 1961, 287–491, who devotes plenty of space to Galileo, while overlooking Hobbes. See also Garber and Roux 2013.

[159] See *EL*, part 1, chapt. 2–9, 3–48.

[160] Despite Hobbes' intention to present his thinking as a unitary system, we need to underscore the problems associated with this approach. Hobbes' thinking cannot be considered as a perfectly uniform and joined structure, which unfolds seamlessly from philosophy all the way through to politics. In this regard, see Malcolm 1990. Now in Malcolm 2002, 146–155. See also Gert 1996, who shows very clearly how Hobbes' concept of matter is not really at the basis of his political philosophy, but itself a 'psychological' analysis of human passions.

loco et tempore, where the author disputes the need to resort to the concept of soul to explain animal sensation.[161]

It is noteworthy that Hobbes' handling of the phenomenology of sensation did not undergo any radical changes as regards its general principles and, in Chapter 25 of *De corpore*, he once again took up the speculations featured in Chapters 27 and 30 of *De motu, loco et tempore*. Chapter 30 is particularly interesting, especially when compared to Hobbes' later speculations. Here, he develops the series of problems concerning sensation and imagination, before coming to reasoning and the birth of science or philosophy. These pages recall his argument in the *Elements* and anticipate the detailed analysis in the first part of the *Leviathan*, where the link between 'natural philosophy' and 'political' or 'civil philosophy,' which lies in the notion of passion, becomes clear to see.[162] In fact, the notion of passion dominates Hobbesian anthropology and its importance as a link between natural exploration and the artificial construction specific to political philosophy has been extensively underscored. Hobbes always dedicates plenty of space to the phenomenology of passions, as is particularly apparent in Chapter 6 of his *Leviathan* (but also Chapter 12 of *De homine*).[163] Here he claimed that the 'Imagination is the first internall beginning of all Voluntary Motion,'[164] and that, as we know, it 'is nothing but decaying sense.'[165] The expressions and concepts set out in the *Leviathan* recall those already featured in the *Elements*,[166] and in *De motu, loco et tempore*: 'imagination and perception (Lat. *sensione*) are the same thing. As long as it concerned with the object, it is called sense; if [the object] moves away, it is called imagination, by a name taken from the images of the vision.'[167] The phenomenology of the passions is therefore placed in natural continuity with respect to sensation and imagination, which, as we know, are determined by movements of external objects on our sensory organs. In Chapter 6 of his *Leviathan*, Hobbes used the same explanation once again, extending both to *vital* motion (that is to say involuntary motion) and to animal motion (or voluntary acts), referring to a key notion of his philosophical system,

---

[161] *MLT*, XXVII, 326–327. In his *Elements* and *Tractatus Opticus I* Hobbes only ascribed sensation to the brain, while in his *Tractatus Opticus II* (see *TO II*, fol. 240v/199) and in *De motu, loco et tempore* the heart constitutes the primary origin of the reaction to sensible motion.

[162] For an analysis of the concept of *passion* in Hobbesian philosophy see Gert 1989, and, above all Tricaud 1992. The importance of the *passions* as a link between anthrolopology and politics in Hobbes has primarily been underscored by Strauss (see Strauss 1952, 11–27) and Pacchi 1987 (now in Pacchi 1998, 79–95). See also Sorell 1986, 87 ff. The importance attributed by Hobbes to the notion of *passion* is demonstrated not only by Hobbes' extensive discussion of it in Chapter 6 of *Leviathan* (see Hobbes 2012, 79–97), but also by a manuscript dating to around 1644, which constitutes a sort of draft of the aforementioned chapter in *Leviathan*. See Hobbes 1988b.

[163] *De homine*, *OL*, II, 103–110.

[164] Hobbes 2012, 78.

[165] Ibid., 26.

[166] See *EL*, part I, chapt. VIII, § 1, 31.

[167] *MLT*, XXX, 4, 350–351; Eng. Trans. Hobbes 1976, 366, largely modified.

both in terms of natural and political philosophy: the concept of 'endeavour' (or *conatus*)[168]:

> And because *going*, *speaking*, and the like Voluntary motions, depend alwayes upon a precedent thought of *whither*, *which way*, and *what*; it is evident, that the Imagination is the first internall beginning of all Voluntary Motion. And although unstudied men, doe not conceive any motion at all to be there, where the thing moved is invisible; or the space it is moved in, is (for the shortnesse of it) insensible; yet that doth not hinder, but that such Motions are. For let a space be never so little, that which is moved over a greater space, whereof that little one is part, must first be moved over that. These small beginnings of Motion, within the body of Man, before they appear in walking, speaking, striking, and other visible actions, are commonly called ENDEAVOUR.[169]

Conatus (or 'endeavour' in English) 'when it is toward something which causes it, is called APPETITE, or DESIRE.'[170] It is, therefore, at the origin of the passions and every voluntary act, and Hobbes explains it exclusively in mechanical, dynamic, and causal terms.

The term *conatus*, which Hobbes uses to indicate the internal reaction of the living body (and that has already been used in the *Elements*, the *Tractatus Opticus I* and *De motu, loco et tempore*), is closely connected to the specifically physical meanings in which it recurs (as we shall see in the next chapter).[171] The definition of conatus in *De corpore*, as 'motion made in less space and time than can be given,'[172] in fact, encompasses all the multiple variations and aspects of the term.

Nevertheless, what needs to be underscored here is the continuity that links the process of sensation and the phenomenon of the passions in Hobbesian natural philosophy: a single explanatory model has to be used for both. In fact, the natural development of the perceptive phenomenon, from sensation to passions, which represents the crucial link between the various parts of Hobbes' philosophical system, is still centered around concepts of matter and motion.

2.7 Cause and Necessity

The particular conception of sensation in Hobbes leads us to consider his reflections on voluntary acts and will in general. If we can identify certain mechanical reactions within the sentient being at the basis of every voluntary act, generated by the action of external bodies upon the sensory organs, it is difficult to conceive how human actions are not necessarily determined by a series of causal factors. In other words, Hobbesian theory regarding the origin of voluntary acts leads us to consider Hobbes' determinism (or necessitarianism).

---

[168] See Jesseph 2016, esp. 78 ff.
[169] Hobbes 2012, 78.
[170] Ibid.
[171] See below, chap. 3.
[172] *De corpore*, XV, 2, *OL*, I, 177; Eng. Trans. *EW*, I, 206.

Chapters 30 and 33 of *De motu, loco et tempore* contain invaluable indications in this regard, reflected in later texts. The first of these two chapters contains a definition of 'will' as 'the final act of him who deliberates,'[173] and this expression becomes much clearer in the light of the next paragraph, dedicated to that which can be defined as voluntary.[174] In fact, the term voluntary can only be referred to actions, not to the process of deliberation, and the concept of freedom illustrated by Hobbes in the next paragraph is in keeping with his definition of will:

> As regards all the motions other than animal motion, something whose movement is unhindered is termed 'free', i.e. it is free on that path [to action] where it encounters no obstacle. A river, for instance, runs freely where unchecked by its banks, though its freedom is not due to any [particular] spot on them. But as regards animal motion, a thing is 'free' whose motion suffers no check save that arising from its own will.[175]

The definition of freedom as negative liberty, characteristic of Hobbesian philosophy,[176] conceives freedom as an absence of external impediments. Nevertheless, here Hobbes seems to exclude will from the chain of causes and instead considers it to be that which determines the realization or non-realization of an event.

On the contrary, if we focus on Chapter 33, which discusses the liberty of God to create the world, even voluntary human acts seem to be inevitably traced back to the concept of cause.[177] Here, Hobbes seems to admit—albeit very surreptitiously—that the definition of a free agent (provided that we accept his negative conception of liberty) does not rule out events being determined by a cause or series of causes. In

---

[173] *MLT*, XXX, 27, 361; Eng. Trans. Hobbes 1976, 380: 'Voluntas est ultimus actus deliberantis; qui quidem ultimus actus, si appetitus sit, est *voluntas faciendi*. Si fuga sit, est voluntas non faciendi. Et sicut non intelligitur voluntas in eo qui adhuc deliberat, sic neque deliberatio in eo qui voluit.'

[174] Ibid., XXX, 29, 361: 'Volontarium est nomen quod competit solis actionibus. Actio voluntaria ea. est ad cuius causam requiritur agentis voluntas; voluntas ipsa ergo voluntaria non est, quia non est actio; neque nisi absurde & ridicule potest aliquis dicere "Volo cras sic velle," vel "Volo cras habere voluntatem faciendi hoc aut illud," si enim dicere possum "Volo velle," quidem, & "Volo velle velle" & sic in infinitum.'

[175] Ibid., XXX, 30, 362; Eng. Trans. Hobbes 1976, 381–382, modified.

[176] As regards the development and limitations of Hobbesian necessitarianism see Zarka 2004, 249–262.

[177] *MLT*, XXXIII, 2, 377: '*Liberum* proprie dicitur in animalibus, quod cum habeat coeteram potestatem agendi omnem, voluntatem tamen nondum habet. In rebus autem inanimatis liberum est & dicitur quod non impeditur quominus faciat quicquid ex natura eius facere potest, sic aqua libere decurrit quae ripis, neque obstaculo alio externo fluere prohibetur. Neque tamen nesciunt qui sic loquuntur, decurrere aquam vi gravitatis suae, id est per necessitatem naturalem. Confitentur ergo libertatem non opponi necessitati internae, sed impedimento externo: sic etiam ii qui hominibus libertatem agendi tribuunt, non ignorant quaedam esse quae homines non possunt velle, qualia sunt quae videntur pessima sibi, & quaedam, quae non posssunt nolle, qualia sunt quae videntur optima sibi facta esse. Non tamen ideo negant eos libere & per electionem agere; nam & electionis (sicut aliarum rerum omnium) causa aliquis existit, eaque necessaria, neque qui necessario eligit, ob id minus eligit, nisi etiam dicamus lapidem non cadere quia necessario cadit.' These reflections are taken up again in Chapter 37, see *MLT*, 401 ff.

Chapter 34, this idea is translated into a presentation of a *Megarian* conception of possibility that conceives Aristotelian 'potential' solely as a cause and considers every potential or possibility that is not translated into action to be in vain.[178] The same problems are also extended to Chapter 35, in relation to the fundamental aspects of historical fate, where Hobbes also defines his concepts of necessary and contingent. He starts by reiterating that '*cause*, then, and *potential* are the same.'[179] What is more, he explains what he means by *necessary*: 'which is unable not to be'[180] and emphasizes that the Stoics used the word 'fate' to describe what is really necessity.[181] Consequently, a particular event is determined by a series of preceding events that cause it and from which it necessarily results. So, the definition of *contingent* is equivalent to that of 'unknown' and, in fact, people refer to everything for which they do not know the cause as contingent.[182] Further on, Hobbes seems to clarify this concept even more, underscoring once more that the definition of contingent is linked to ignorance of the necessary cause.[183]

These reflections suggest that, as early as 1643, Hobbes had already come up with many of the ideas that characterize his later speculations: the necessitarianism, or determinism, that was at the center of the dispute with the Arminian bishop John Bramhall, already appeared in clear-cut and specific terms.

In his short work *Of Liberty and Necessity*, written in 1645 in response to a number of objections made against him by Bramhall and published in 1654, Hobbes went back to the considerations featured in *De motu, loco et tempore*. Here he explained many of his theories concerning will and determinism. An analysis of some short passages enables us to consider the essential coherence of the contents in the evolution of Hobbesian thought. Here he defines liberty as 'the absence of all the impediments to action that are not contained in the nature and intrinsical quality of the agent,'[184] and, in describing voluntary phenomena, he claims that 'when first a man has an appetite or will to something, to which immediately before he had no appetite nor will, the cause of his will is not the will itself, but something else not in his own disposing.'[185] As a result:

---

[178] This concept can also be noted in *De corpore*, *OL*, I, 113–115.

[179] *MLT,* XXXV, 4, 388; Eng. Trans. 423.

[180] Ibid., XXXV, 6, 389; Eng. Trans. 424.

[181] Ibid.

[182] Ibid., XXXV, 10, 391.

[183] Ibid., XXXVIII, 1, 412: 'Vera & unica causa quare res humanas casu geri homines arbitrantur, videtur esse haec, quod ignorant earum causas integras & necessarias; si enim qui causarum omnium συντυχίαν, sive concursum ad effectum aliquem futurum producendum praesciret, nunquam illum effectum fortuito, sed ex necessitate, eventurum affirmaret; quod enim aliquis futurum certo scit ex certitudine scientiae, certe, id est non fortuito, eventurum pronuntiaret … Cum itaque eventus omnes propter causas suas necessarii sint, sequitur ut non alia de causa fortuiti videantur, quam quod eorum causas omnes non percipimus.'

[184] *Of Liberty and Necessity*, *EW*, IV, 273.

[185] Ibid., 274.

...whereas it is out of controversy that of voluntary actions the will is the necessary cause, and by this which is said the will is also caused by other things whereof it disposes not, it follows that voluntary actions have all of them necessary causes and therefore are necessitated.[186]

In the same way, Hobbes reiterated the same *Megarian* concept of possibility he had set out in *De motu, loco et tempore* and that implies the absurdity of the concept of the 'free agent,' according to the accepted definition supported by Bramhall:

Lastly, I hold that ordinary definition of a free agent, namely that a free agent is that which, when all things are present which are needful to produce the effect, can nevertheless not produce it, implies a contradiction and is nonsense; being as much as to say the cause may be sufficient, that is necessary, and yet the effect shall not follow.[187]

The conclusions reached by Hobbes in *Of Liberty and Necessity* are nothing other than the natural conclusion of the premises already contained in *De motu, loco et tempore* and are grounded in his natural philosophy, in which the concept of *cause* plays a leading role. It represents the 'dogmatic bulwark' of Hobbes' philosophy[188]: not only does he never question it, but it actually forms the link between the subjective dimension of sensible perceptions and the external reality of moving bodies.[189] Just as he had attributed all Aristotelian motions to that single movement contemplated by Galilean physics, that is to say local motion, so in Chapter 9 of *De corpore*, Hobbes reduces the four Aristotelian causes to two[190]: the efficient cause and the material cause, which together compose what he defines as the entire cause.[191] It is at the same time necessary and sufficient, because it is 'the aggregate of all the accidents both of the agents how many soever they be, and of the patient, put together; which when they are all supposed to be present, it cannot be understood but that the effect is produced at the same instant; and if any one of them be wanting, it cannot be understood but that the effect is not produced.'[192] Every action and, therefore, every element that forms part of the causal chain must necessarily be expressed in kinetic terms and, as a result, 'there is no cause of motion in any body, except it be contiguous and moved.'[193]

The same concept emerges when Hobbes discusses sensation. In Chapter 25, he claims that motion cannot be generated except by a moved and contiguous body and, as a result, there is nothing other than the action of an external object on a sensory organ at the origin of sensation and our every concept.[194] What is more, even

---

[186] Ibid.

[187] Ibid., 275.

[188] See Leijenhorst 2005, 79–119.

[189] See Paganini 2004, 323.

[190] See Aristotle, *Physics,* II (B), 194b ff.

[191] See *De corpore*, IV, 4, *OL*, I, 104.

[192] Ibid., IX, 3, 107–8; Eng. Trans. *EW*, I, 121–122.

[193] Ibid., IX, 7, 110; Eng. Trans. *EW*, I, 125.

[194] Barnouw 1990 maintains that, despite his causal and kinetic conception of sensation, Hobbes does not rule out the possibility of including the concept of intentionality. See, however: Leijenhorst 2007, 90.

the succession of sensations and images, recalled by the imagination and memory, responds exactly to the same principles:

> Now it is not without cause, nor so casual a thing as many perhaps think it, that phantasms in this their great variety proceed from one another; and that the same phantasms sometimes bring into the mind other phantasms like themselves, and at other times extremely unlike. For in the motion of any continued body, one part follows another by cohesion; and therefore, whilst we turn our eyes and other organs successively to many objects, the motion which was made by every one of them remaining, the phantasms are renewed as often as any of those motions comes to be predominant above the rest.[195]

Above and beyond the subjective, varied and diverse reality of sensory perception, there therefore exists an objective reality of material entities, of moving bodies, to which not only does the concept of cause apply, but which are themselves the cause of our perception, and even of our imagination. Far from occurring randomly, all these phenomena are the product of a causal chain that unfolds with unavoidable necessity.

2.8   Mathematics and the Book of Nature

One of the most problematic elements of both Hobbes and Galileo's natural philosophy is the issue of the epistemological status of scientific inquiry. Additional issues are connected to this topic, specifically the common denominator of the two thinkers' scientific and philosophical ideas: the mathematization of the natural world,[196] a particular feature of Galileo's thinking inherited by Hobbes.[197]

Regarding Galileo, this matter is one of the most widely debated in the critical literature because, though Galileo did not address the question clearly, specifically, and comprehensively, there are references to it throughout his writings that give a very specific, well-structured view of what he meant by science and scientific research.

Nevertheless, scholars who have addressed this issue have often emphasized only one element of Galileo's epistemology, neglecting others, and in some cases offering a partial (or even distorted) view of them, giving us an even more obscure, vague impression of Galileo's science. After a period of 'Platonic' or 'Platonizing' interpretations of Galileo,[198] historiographies investigated the debt Galileo owed to

---

[195] *De corpore,* XXV, 8, *OL,* I, 324; Eng. Trans. *EW,* I, 397–398.

[196] On the importance of the mathematization of nature with regards to the genesis of mechanism, see Koyré, 'Du monde de l'"à-peu-près" à l'univers de la precision'. Now in Koyré 1981, 341–362. See also Crombie 1953, 274 ff. Alfred Rupert Hall and Marie Boas Hall's historical text was one of the first critical studies that stressed the importance of the *Assayer* for the mathematization of nature; see Hall and Boas Hall 1964, 166 ff. See also Palmerino 2010.

[197] See Gargani 1971, 53 ff. and 170–173; Jesseph 2004, Paganini 2010, 24 ff.

[198] The intellectual debate that has grown around this question started from Alexandre Koyré's historical essay published in French in 1939 (Koyré 1966), which focused on Platonic elements in Galileo's work. According to Koyré, in Galileo's writings, deductive, speculative reason dominates

earlier scientific and philosophical traditions, comparing Galileo's methods of scientific inquiry to those fashionable among the sixteenth-century School of Padua[199]; or they identified a particular influence on the young Galileo in authors and texts used at the Jesuit's *Collegio Romano*.[200] But a proper analysis of the actual relationships that Galileo had with earlier philosophical movements and traditions revealed the original elements of his position.[201]

Beyond these historiographical interpretations, it may nonetheless be useful to ask what view of science emerged from Galileo's texts that Hobbes could have read, and the influence these may have had on Hobbes' scientific method.[202] At first glance, Galileo and Hobbes' epistemological positions seem to run on divergent tracks: Galileo stressed the difference separating the 'mathematical astronomer,' from the 'philosopher astronomer'[203] (and there is no doubt that he considered him-

---

a great deal over the experimental element. Koyré's thesis, though widely debated and critically analyzed, still has an enormous impact.

[199] On the 'compositive/resolutive method,' used by Agostino Nifo and especially Jacopo Zabarella, see Randall 1961, 15–68, who suggests a comparison with its application in Galileo (Ibid., 55 ff.). Galileo refers specifically to compositive and resolutive methods in *Considerazioni sopra 'l discorso del Colombo*, *OG*, IV, 520. It should, however, be noted that Neal Gilbert had ruled out Galileo's close adherence to Paduan Aristotelianism and suggested a substantial difference between Zabarella's and Galileo's methods (see Gilbert 1963, 223–231. See also Schmitt, 'Esperienza ed. esperimento: un confronto tra Zabarella e il giovane Galilei' (1969). Now in Schmitt 2001, 25–64; Pastore Stocchi, 1976 1986, esp. 58–60). Antonio Banfi had long before distinguished Galileo's speculative nature from the work of 'erudite exegesis of Aristotelian texts and their commentaries' found in authors like Zabarella and Cremonini. See Banfi 1964, 280.

[200] Edwards and Wallace have published one of Galileo's early texts (see Galilei 1988), which addresses epistemological problems and includes comments to Aristotle's *Posterior Analytics* in use at the Jesuit's Collegio Romano. The text is, however, for the most part, a transcription of the teachings given in the prestigious College and does not express Galileo's mature epistemology. On this topic, see also Crombie 1975, and Carugo and Crombie 1983. More recently, the topic has been revisited in Laird 1997. Wisan 1978 considered different stages of Galileo's epistemology. He argues that Galileo was originally influenced by late scholasticism, and that in a later period, he expressed a position similar to that of the 'mitigated sceptics.' He would later turn to an epistemological dimension shaped by the dialogue between empirical inquiry and rational speculation to then return to a 'rationalist' position in the *Discourses*.

[201] On this topic, a key reference is Galluzzi 1973, which specifically analyzes texts by Mazzoni, Biancani, Clavius, and other late sixteenth-century authors who were available to Galileo and who influenced him. This study clearly reveals Galileo's *true* relationship with the earlier Platonism and Aristotelism, and his originality is evident when he decided to distance himself markedly from these philosophical traditions.

[202] This topic has been addressed by Minerbi Belgrado 1993, 58–60 and most recently by Giudice 2016, 92 ff.

[203] For instance, in *Istoria e dimostrazioni intorno alle macchie solari* (1612), Galileo had already established a clear dividing line between the figure of the 'pure astronomer' and that of the 'philosopher astronomer.' Galilei, *Istoria e dimostrazioni intorno alle macchie solari e loro accidenti*, Letter I, *OG*, V, 102: '… quei deferenti, equanti, epicicli etc., posti da i puri astronomi per facilitar i lor calcoli, ma non già da ritenersi per tali da gli astronomi filosofi, li quali, oltre alla cura del salvar in qualunque modo l'apparenze, cercano d'investigare, come problema massimo ed. ammirando, la vera costituzione dell'universo, poi che tal costituzione è, ed. è in modo solo, vero, reale

self more truly a philosopher[204]). He specified that the task of the philosopher astronomer was to discover 'the true constitution of the universe,' a constitution that 'is, and is in one way only, true, and impossible to be differently.'[205] Such ideas recur throughout almost all of Galileo's writings, not only in his *Copernican Letters* (where he states with certainty that our knowledge of reality is ensured by nature being 'inexorable' and 'immutable'[206]), but also, and importantly, in the *Dialogue*, where he expresses the same idea in a particularly significant passage in which he distinguishes knowledge acquired by human study from that of scientific knowledge:

> If what we are discussing were a point of law or of the humanities, in which neither true nor false exists, one might trust in subtlety of mind and readiness of tongue and in the greater experience of the writers, and expect him who excelled in those things to make his reasoning most plausible, and one might judge it to be the best. But in the *natural sciences*, whose conclusions are *true* and *necessary* and have nothing to do with human will. One must take care not to place oneself in the defense of error; for here a thousand Demostheneses and a thousand Aristotles would be left in the lurch by every mediocre wit who happened to hit upon the truth for himself.[207]

In contrast, Hobbes not only strove to provide an indisputable foundation for human law and political science,[208] but he particularly emphasized that the world of natural philosophy is the realm of hypotheses and that researchers of it must be satisfied with discovering possible causes of a particular phenomenon.

However, we must investigate what epistemological status Hobbes gave to these hypotheses, and, importantly, understand *what* kind of hypotheses can reach the status of scientific theories. As we have noted several times, Hobbes insisted (like Mersenne and Clavius before him[209]) on the importance of the geometric-demonstrative aspect of philosophy, and in *De cive*, he proposed a deductive method shaped on the model of the geometric sciences as the unifying criteria of the vast 'ocean' of philosophy.[210] As we know, this idea is found in all of Hobbes' philo-

---

ed. impossibile ad esser altramente, e per la sua grandezza e nobiltà degno d'esser anteposto ad ogn'altra scibil questione de gl'ingegni specolativi.'

[204] Galileo is known to have especially coveted the title of natural philosopher, which was attributed to him after he moved from Padua to Florence. See Galileo Galilei to Belisario Vinta, May 7, 1610, *OG*, X, 353: '... quanto al titolo et pretesto del mio servizio, io desidererei, oltre al nome di Matematico, che S. A. ci aggiugnesse quello di Filosofo, professando io di havere studiato più anni in filosofia, che mesi in matematica pura.' See also Camerota 2004, 186.

[205] Galilei, *Istoria e dimostrazioni intorno alle macchie solari e loro accidenti*, Letter I, *OG*, V, 102 (my italics). On Galileo's interpretation of Copernicanism as the *true* and *real* description of the universe, see Torrini 1993. On Galileo's awareness of himself as a *natural philosopher*, see Bucciantini 2003, 53 ff. Clavelin also notes that 'nombreux sont les textes où Galilée affirme un réalisme sans equivoque.' (see Clavelin 2001, 29 note).

[206] See Galilei, *Lettera a Madama Cristina di Lorena, Granduchessa di Toscana* (1615), *OG*, V, 316.

[207] Galilei, *Dialogo sopra i due massimi sistemi*, *OG*, VII, 78; Eng. trans. Galilei 2001, 53–54, my italic.

[208] See, in particular, the epistle dedicatory and the preface to the readers in *De cive*, *OL*, II, 135 ff.

[209] See *supra*, chap. I.

[210] Hobbes (epistle dedicatory of 1647 ed.) *De cive*, *OL*, II, 136–138.

sophical and scientific works: starting with *Elements*,[211] and including *De motu, loco et tempore*,[212] his later *De corpore*,[213] and the *Six Lessons*, where he maintains that the geometric method is what lets us go beyond the 'cavils of the sceptics.'[214]

On this subject, we can consider what Galileo wrote in a text Hobbes is sure to have known both in the original,[215] and in the French version by Marin Mersenne[216]: *Le Mecaniche*. The *incipit* of this short treatise focuses on the importance of the deductive method, specific to the demonstrative sciences:

> In this essay we should describe what is necessary to observe in *demonstrative sciences*; namely propose the definitions of terms which are appropriate to this faculty, and the first suppositions, from which, as from *fertile seeds*, there *arise and germinate causes and demonstrations* of the properties of all mechanical instruments.[217]

The method that Galileo recommends in studying mechanical tools is the typical method of *demonstrative sciences*: a strictly deductive process in which, starting from definitions or principles—the 'fertile seeds'—we arrive by deduction at the conclusions and the true demonstrations. The demonstrative sciences that Galileo is referring to are, of course, mathematics, and, specifically for engineering instruments, geometry. Galileo reiterates this faith in mathematical knowledge in almost all of his writings, especially in his *Dialogue*, where he distinguishes two ways of knowing: intensive and extensive; the former refers to the 'multitude of intelligibles'[218] i.e. the breadth of our knowledge, and the latter refers to the certainty of that knowledge. Based on this distinction, he maintains that in the realm of the 'pure mathematical sciences, namely, geometry, and arithmetic,' the human intellect comes to the point of matching intensive, divine intellect 'for it is capable of grasping their necessity, which seems to be the greatest possible assurance there is.'[219]

---

[211] See *EL*, part I, chapt. XIII, § 3, 65–66.

[212] *MLT*, XXIII, 1, 269–270.

[213] See the epistle dedicatory in *De corpore*, *OL*, I, not numbered; 61 ff.

[214] *Six Lessons*, *EW*, VII, (the epistle dedicatory), 184–185.

[215] There is an English translation of the text by Robert Payne, titled *'Of the Profitt wch is drawn from the Art Mechanics & it's Instruments. A Tract of Sig.^r Galileo Galilei, Florentine'* (see *infra*, Appendix). The text is dated November 1636 and is in the British Library, MS Harley 6796, fol. 317–337 *v* et *r*. On the last page, it reads: 'Raptim ex Italico in Anglicum sermonem transfusam. Novemb. 11, 1636, By Mr Robert Payen', which means that Payne had received the Italian original, very likely through William Cavendish and Thomas Hobbes, who obtained it from Marin Mersenne (it is highly unlikely that they would have received it from Galileo himself, who was working diligently at the time on *Discourses*).

[216] Mersenne 1634. Now in Mersenne 1985, 443.

[217] Galilei, *Le Mecaniche*, *OG*, II, 159; Eng. Trans. Galilei 1960, 149 (my italics).

[218] Galilei, *Dialogo sopra i due massimi sistemi del mondo*, *OG*, VII, 128.

[219] Ibid., 128–129. Eng. trans. Finocchiaro 2008, 113–114 (my italics). Answering Simplicio's perplexity, who stressed the recklessness of Salviati's statement, this last one says: 'Però, per meglio dichiararmi, dico che quanto alla verità di che ci danno cognizione le dimostrazioni matematiche, ella è l'istessa che conosce la sapienza divina; ma vi concederò bene che il modo col. quale Iddio conosce le infinite proposizioni, delle quali noi conosciamo alcune poche, è sommamente più eccellente del nostro, il quale procede con discorsi e con passaggi di conclusione in conclusione,

Hobbes also considered mathematical knowledge to be absolutely certain: in mathematics, the degree of truth is absolute and determined by a logical deductive system within these disciplines, therefore defined by a purely aprioristic and conventional dimension. Based on principles established by convention, a syllogism is developed that leads to logically necessary conclusions: Hobbes states this clearly in the letter to Charles Cavendish, where he explains that in the realm of 'mathematicall sciences wee come at last to a definition which is a beginning or Principle, made true by pact and consent amongst ourselves'[220] and this same idea is also found in the almost coeval *Tractatus Opticus II*.[221] However, he states with equal clarity—both in his correspondence and in his treatise on optics—that in the realm of natural philosophy we must limit ourselves to the possibility of developing probable hypotheses.

Arrigo Pacchi has thoroughly studied the hypothetical elements of Hobbesian natural philosophy, and by focusing on this aspect, he tended to attribute a conventionalist position to Hobbes.[222] Shapin and Schaffer more recently emphasized Hobbes' distance from the emerging culture of experiments,[223] which contributed to his developing an absolute distaste for *empiria*, in its two senses: observational and experimental.[224] The roots of Hobbesian scientific method are, however, to be found in the range of problems both methodological and specifically content-based. Since his earliest thinking, Hobbes proposed a dichotomy in the realm of philosophical knowledge: certain and aprioristic in the mathematical sciences and merely

---

dove il Suo è un semplice intuito: e dove noi, per esempio, per guadagnar la scienza d'alcune passioni del cerchio, che ne ha infinite, cominciando da una delle più semplici e quella pigliando per sua definizione, passiamo con discorso ad un'altra, e da questa alla terza, e poi alla quarta, etc., l'intelletto divino con la semplice apprensione della sua essenza comprende, senza temporaneo discorso, tutta la infinità delle passioni; … Il che né anco all'intelletto umano è del tutto incognito, ma ben da profonda e densa caligine adombrato, la qual viene in parte assottigliata e chiarificata quando ci siamo fatti padroni di alcune conclusioni fermamente dimostrate e tanto speditamente possedute da noi, che tra esse possiamo velocemente trascorrere.' Ibid., 129.

[220] Hobbes to Sir Charles Cavendish, from Paris, Jan. 29, [Feb 8.] 1641, *CH*, I, 83. Interestingly, in this same letter, Hobbes analyzes and comments on several passages of *Discourses* (ibid., 82).

[221] After making a distinction between the realms of scientific and mathematical knowledge, in *TO II*, Chap. I, § 1, f. 193r/147, Hobbes writes: 'In caeteris [*i.e. the mathematical sciences*] enim fundamenta sive principia prima demonstrandi alia neque requiruntur, neque admittuntur, quam definitiones vocabulorum, quibus excludatur Amphibologia. Eae primae veritates sunt, est enim definitio omnis, vera propositio; & prima; propterea quod definiendo, id est consentiendo circa vocabulorum usum, ipsi inter nos veram esse facimus. Si quidem enim nobis inter nos libuerit figuram hanc Δ Triangulum appellare, verum erit, figura illa Δ est Triangulum.'

[222] See Pacchi 1965. In my opinion, Gargani offers a more balanced interpretation of this issue, demonstrating the difference between the scientific perspective of authors like Hobbes and Descartes compared to that of the thinkers known as the 'mitigated sceptics.' See Gargani 1971, 173 ff. See also Malherbe 1984, 79 ff.; Sorell 1986, 46–47; Leijenhorst 2005; Paganini 2003.

[223] See Shapin and Schaffer 1985, esp. 80–153.

[224] For a more balanced interpretation of Hobbes' relationships with Royal Society and the English scientific milieu of the time, see Malcolm, 'Hobbes and the Royal Society'. Now in Malcolm 2002, 317–335; Terrel 2009, and Terrel 2013 (on the dual meaning of experience, as consequential experience based on the use of names and de facto experience, 68–69); Médina 2013, 160–162.

probable in the realm of natural philosophy. Furthermore, in *The Elements of Law*, Hobbesian science already had a nominalist, logical and axiomatic[225] quality. Hobbes saw 'truth' and 'falseness' as pertaining only to names and language and the ability to reason as coinciding with the syllogistic faculty.[226] In this aspect, a substantial difference emerges from Galileo's thinking: Galileo maintained that truths were not 'names' but 'things.'[227]

De motu, loco et tempore gives a definition of philosophy that is completely coherent with the ideas of *Elements* and takes the formal and logical-linguistic characteristics even further: 'philosophy is the science of general theorems, or of all the universals of any matter whatsoever, whose truth can be demonstrated by natural reason.'[228] Philosophy is therefore to be addressed in a 'logical form,'[229] as 'its aim is to know the necessity of consequences and the truth of universal propositions.'[230] In this work, Hobbes clearly expresses the deductive conception of philosophical science and stresses its nominalistic character, equating it with 'a faithful, correct, and accurate nomenclature of things.'[231] In Chapter 30 of the book he specifically returns to his 1640 thoughts and maintains that true and false are concepts that should only be given to names and affirmations and that the tool we need to reason correctly, and thereby reach the necessary conclusions, is syllogism.[232]

In addition, in *The Elements of Law*[233] Hobbes had already made a distinction between 'experience of fact' and 'evidence of truth,' which was returned to in *De motu, loco et tempore*, and particularly in *Leviathan*,[234] implying a split between

---

[225] *EL*, I, Chap. 5, § 4, 19: 'By the advantage of names it is that we are capable of science'. Zarka 1999, 85–93 focuses a great deal on the nominalistic aspects of Hobbesian thinking (emphasizing the similarities and divergences of Hobbes' philosophy of language with that of William of Ockham).

[226] *EL*, I, Chap. 5, §4, 22.

[227] See Galilei, *Dialogo sopra i due massimi sistemi del mondo*, *OG*, VII, 218.

[228] *MLT*, I, 1, 105; Eng. Trans. Hobbes 1976, 23, modified. The topic appeared frequently in Descartes' *Objections* and *Meditations*. See *Objectiones Tertiae*, *AT*, VII, 178.

[229] A discussion on the need for logic had already been addressed by Galileo at the beginning of the 'first day' of the *Dialogue*. Here Salviati emphasized the superiority of mathematical-geometric demonstration over Aristotelian metalogic. In *De motu, loco et tempore*, Hobbes appears to agree with Simplicio, saying that philosophy should be treated in a logical form. However, we should remember that Hobbes associated Euclidean geometry and logical demonstration (as Mersenne). In the epistle dedicatory of *De corpore* Hobbes presents geometry as the 'true logic.' *De corpore, Epistola dedicatoria*, *OL*, I, not numbered (first page of the dedicatory letter): 'Scio philosophiae partem illam, quae versatur circa lineas et figuras, traditam nobis esse bene cultam a veteribus simulque verae logicae.'

[230] *MLT*, I, 3, 107; Eng. Trans. Hobbes 1976, 26, modified

[231] Ibid., XIV, 1, 201; Eng. Trans. 157.

[232] Ibid., 355 ff.

[233] See *EL*, Part I, Chap. VI, §§ 1–4, 24–6.

[234] Hobbes 2012, 124: 'There are of KNOWLEDGE two kinds; whereof one is *knowledge of Fact*: the other *knowledge of the Consequence of one Affirmation to another.* The former is nothing else, but Sense and Memory, and is *Absolute Knowledge*; as when we see a Fact doing, or remember it done; And this is the Knowledge required in a Witnesse. The later is called *Science*; and is *Conditionall;*

historical and scientific knowledge. In Chapter 4 of *Leviathan*, in which Hobbes presents an interesting tree diagram representing the entire realm of science, (meaning: 'knowledge of consequences,' which is also named 'science or philosophy'),[235] he excludes from this domain any kind of history, whether natural or civil.[236]

On this subject, it is worth looking at the preface to the readers of *De corpore*, which includes the famous tribute that Hobbes made to Galileo and Harvey. He, of course, argues that before these scientists, there was nothing certain in natural philosophy but individual experiments, and the natural histories.[237] The reference to natural histories and individual isolated experiments was a quite explicit critical reference to the experimental approaches lacking in theory that Hobbes had found in some members of the nascent Royal Society, perhaps even in Francis Bacon himself, the new scientific academy's 'patron saint.'[238]

On the contrary, Hobbes considered Galileo's main achievement as having been the first to have applied—particularly in *Discourses and Demonstrations*—mathematics to the study of the science of motion.

This issue is discussed more in depth in *Dialogus physicus* in which Hobbes addresses the question of the correct scientific method. It should, however, be noted that the handling of this topic in this work is partially 'marred' by the diatribe against the natural philosophers of the Royal Society. After repeating that it is right not to believe in histories blindly,[239] Hobbes expresses particular scepticism about the nature of experiments. He particularly criticizes experiments conducted individually and asks that they be done collectively instead.[240]

---

as when we know, that, *If the figure showne be a Circle, then any straight line through the Center shall divide it into two equall parts.* And this is the Knowledge required in a Philosopher; that is to say, of him that pretends to Reasoning.'

[235] Ibid., 130–131. Hobbes' definition of science, as we know, was influenced by the earlier philosophical tradition (Aristotelian-Scholastic), where *scientia* was understood as a strictly deductive knowledge. Yet, in Hobbes, as in other authors of the time, a new concept of science took shape, which escaped solely deductive reasoning, encompassing several aspects of the modern concept of science. He does not favor the experimental dimension but insists on the mathematization of nature. On the concept of *scientia* in Hobbes, see Jesseph 2010. On *scientia* in the early modern age, see   Sorell et al. 2010, vii–viii. See also Garber 2010; Gaukroger 2010.

[236] According to Hobbes 2012, 130–131, history is in the realm of the *knowledge of fact* and differs from *natural history*, 'which is the History of such Facts, or Effects of Nature, as have no Dependance on Mans Will; such as are the Histories of *Metalls*, *Plants*, *Animals*, *Regions*, and the like. The other, is *Civill History*; which is the History of the Voluntary Actions of men in Common-wealths'.

[237] *De corpore*, *Epistola dedicatoria*, *OL*, I, not numbered.

[238] In the *Advancement of Learning* Bacon actually argued that there was a fundamental need to study mathematics in the field of 'mixed mathematics'. See Bacon, 1858–1874, III, 360. However, Bacon had a well-known general distaste for the constant application of mathematics to the study of natural philosophy. See Gaukroger 2001, 20–27.

[239] *Dialogus physicus*, *OL*, IV, 241.

[240] Ibid. Eng. Trans. Shapin and Schaffer 1985, 351: 'Indeed, it is right not to believe in histories blindly. But are not those phenomena, which can be seen daily by each of you, suspect, unless all of you see them simultaneously?'. As Shapin and Schaffer have correctly pointed out, the distinc-

Hobbes repeats there are two criteria to define a scientific hypothesis: 'of which the first is, that it be conceivable, that is, not absurd; the other, that, by conceding it, the necessity of the phenomenon may be inferred.'[241] Here, Hobbes particularly stresses that no inquiry in the world of physics can be without theory and that, indeed, it must draw on true scientific research, only after having developed an adequate, systematic theory of motion, which Hobbes felt that the followers of the Gresham College had not succeeded in doing:

> Since Aristotle had rightly said that *to be ignorant of motion is to be ignorant of nature*,[242] how did you dare take such a burden upon yourselves, and to arouse in very learned men, not only of our country but also abroad, the expectation of advancing physics, when you have not yet established the doctrine of universal and abstract motion (which was easy and *mathematical*)?[243]

As is the case in many passages of Hobbes, here abstract reasoning and the deductive element seem to have absolute dominance in the realm of science and scientific knowledge, but Hobbes, nevertheless, repeats once again that the universal doctrine of motion must be mathematical: a mathematical description of reality following Galileo's model.

Indeed, what Galileo wanted and suggested was the mathematicization of the natural world. This emerges particularly from a famous passage of *The Assayer*, referencing the 'book of nature,' which would have considerable influence on Hobbes' thinking:

> Philosophy is written in this all-encompassing book that is constantly open before our eyes, that is the universe; but it cannot be understood unless one first learns to understand the language and knows the characters in which it is written. It is written in mathematical language, and its characters are triangles, circles, and other geometrical figures; without these it is humanly impossible to understand a word of it, and one wanders around pointlessly in a dark labyrinth.[244]

To understand the structure of the universe, we must make use of the language equipped to interpret it correctly and this language is mathematical. Nonetheless, though the structure of the universe relies on a necessary mathematical order,[245] according to Galileo, it is anything but fully deducible in advance without needing an experimental verification. This concept is far from Galileo's thinking, as is seen clearly in several passages of his writings and some of his interesting letters. For instance, in *Dialogue*, he expresses disappointment in the lack of a mathematical approach in Gilbert's work, affirming the importance of two basic elements in sci-

---

tion between experiment and observation seems outside of Hobbes' mindset. Ibid., 139–143, and 146 ff.

[241] *Dialogus physicus*, *OL*, IV, 247; Eng. Trans. 362

[242] Hobbes indirectly references Aristotle, *Physics*, III, I, 200 b, 12 ff.

[243] *Dialogus physicus*, *OL*, IV, 273; Eng. Trans. Shapin and Schaffer 1985, 379.

[244] Galilei, *Il saggiatore*, *OG*, VI, 232. Eng. trans. Finocchiaro 2008, 183.

[245] On the mathematical order of the universe in Galileo, see Galluzzi 1979, which analyzes the topic with critical precision, including relating the topic of 'essences.'

entific research: 'sensible experience' and 'the necessary mathematical demonstration.'[246]

2.9 Galileo and Hobbes Between 'Sensible Experiences' and 'Necessary Demonstrations'

At the center of Galileo's epistemology is the question of the dialectic between necessary demonstrations and the empirical dimension.[247] Galileo repeats many times throughout his writings his wish for inquiry that seeks to discover the true constitution of the universe, and he is absolutely sure that 'Among the safe ways to pursue truth is the putting of experience before any reasoning.'[248] Empirical observation is the fundamental method for acquiring certain and true knowledge of the world: by applying this method we are certain to avoid committing fallacies, 'it not being possible that a sensible experience is contrary to truth.'[249] Therefore, in the final analysis, the criterion for reaching the truth in scientific inquiries is exclusively through experience.[250]

However, Galileo certainly does not neglect the question of rational discourse, so decisive in his epistemology. Often accused of being rigidly anti-Aristotelian, on the contrary, Galileo maintains that in his own research he meticulously observed the method suggested by Aristotle: '...everything that Aristotle teaches us in his logic, pertaining to care in avoiding fallacies in discourse, using reason well so as to

---

[246] Galilei, *Dialogo sopra i due massimi sistemi del mondo*, *OG*, VII, 432.

[247] For the dialectic between *ratio* and *experience* in Galileo, a fundamental text is Galluzzi 1994. See also Stabile 2002, and Bucciantini 2003, 210. In some letters contemporary to the publication of the *Discorsi e dimostrazioni necessarie intorno a due nuove scienze*, Galileo suggested that the necessary demonstrations would be no less conclusive (demonstratively), even if empirical verification did not corroborate rational speculation. For example, in a letter to Pietro Carcavy (5 June 1637, *OG*, XVII, 90–91), about the fall of bodies, Galileo expresses himself as follows: 'se l'esperienza mostrasse che tali accidenti si ritrovassero verificarsi nel moto dei gravi naturalmente descendenti, potremmo senza errore affermare questo essere il moto medesimo che da me fu definito e supposto, quando che no, le mie dimostrazioni, fabricate sopra la mia supposizione, niente perdevano della sua forza e concludenza; sì che come niente progiudica alle conclusioni dimostrate da Archimede circa la spirale il non ritrovarsi in natura mobile che in quella maniera spiralmente si muova'. This idea returns in another letter sent to Giovan Battista Baliani (7 January 1639, *OG*, XVIII, 12–13): 'Ma tornando al mio trattato del moto, argomento *ex suppositione* sopra il moto, in quella maniera diffinito; siché quando bene le conseguenze non rispondessero alli accidenti del moto naturale de' gravi descendenti, poco a me importerebbe, siccome nulla deroga alle dimostratione di Archimede il non trovarsi in natura alcun mobile che si muova per linee spirali.'

[248] Galileo Galilei to Fortunio Liceti, September 15, 1640, *OG*, XVIII, 249. Eng. trans. Drake 1978, 409.

[249] Ibid.

[250] Geymonat has discussed this topic at length. See Geymonat 1957, 224–229. Clavelin also emphasized the fundamental importance of experience in Galileo's epistemology, see Clavelin 1968, 415 ff.

syllogize properly and deduce from the conceded premisses the necessary conclusion.'[251] Galileo claims to have learned this method well, which coincides with the demonstrative and deductive methods of pure mathematics, and that applying this process enabled him to avoid errors, save in very few instances.[252]

Galilean scientific research is oriented by these two heterogeneous, yet inseparable elements: on the one hand, the deductive reasoning that originates from the mathematical sciences, and on the other hand, the necessary contribution provided by experience, the repeated observations of physical phenomena, as well as by experiments. Galileo himself explained how these factors merge and interact with scientific inquiry in a passage of *Dialogue* in which Salviati expresses his opposition to Simplicio's apriorism. The Aristotelian character maintains that Aristotle 'first laid the basis of his argument *a priori*, showing the necessity of the inalterability of heaven by means of natural, evident, and clear principles. He afterward supported the same *a posteriori*, by the senses and by the traditions of the ancients.'[253] Nonetheless, Salviati felt the need to correct his interlocutor and reiterate the importance of empirical analysis and of the 'resolutive method:'

> What you refer to is the method he uses in writing his doctrine, but I do not believe it to be that with which he (*i.e.* Aristotle) investigated it. Rather, I think it certain that he first obtained it by means of the senses, experiments, and observations, to assure himself as much as possible of his conclusions. Afterward he sought means to make them demonstrable. That is what is done for the most part in the demonstrative sciences; this comes about because when the conclusion is true, one may by making use of analytical methods hit upon some proposition which is already demonstrated, or arrive at some axiomatic principle; but if the conclusion is false, one can go on forever without ever finding any known truth—if indeed one does not encounter some impossibility or manifest absurdity.[254]

Galileo maintains that the method Aristotle used—as for any natural philosopher—necessarily came from empirical observation, and only after having ascertained the truth of the conclusion would Aristotle formulate it through a rational discourse, i.e., through the reasoning of demonstrative sciences—those necessary demonstrations that are so important to science.

---

[251] Galileo Galilei to Fortunio Liceti, 15 September 1640, *OG*, XVIII, 248; Eng. trans. Drake 1978, 409

[252] *OG*, XVIII, 248: 'Io stimo (e credo che essa ancora stimi) che l'esser veramente Peripatetico, cioè filosofo Aristotelico, consista principalissimamente nel filosofare conforme alli Aristotelici insegnamenti, procedendo con quei metodi e con quelle vere supposizioni e principii sopra i quali si fonda lo scientifico discorso, supponendo quelle generali notizie il deviar delle quali sarebbe grandissimo difetto. Tra queste supposizioni è tutto quello che Aristotele c'insegna nella sua Dialettica, attenente al farci cauti nello sfuggire le fallacie del discorso, indirizzandolo et addestrandolo a bene sillogizzare e dedurre dalle premesse concessioni la necessaria conclusione; e tal dottrina riguarda alla forma del dirittamente argumentare. In quanto a questa parte, credo di havere appreso dalli innumerabili progressi matematici puri, non mai fallaci, [tal] sicurezza nel dimostrare, che, se non mai, almeno rarissime volte io sia nel mio argumenta[re] cascato in equivoci. Sin qui dunque io sono Peripatetico.'

[253] Galilei, *Dialogo sopra i due massimi sistemi del mondo*, *OG*, VII, 75. Eng. trans. Galilei 2001, 50.

[254] Ibid.

Significantly, Hobbes echoed the above passage almost word for word in a letter to William Cavendish, Earl of Newcastle in the summer of 1636 (the same letter in which—on the basis of *The Assayer*—he reduced the visual sensation first to motion and then to the sense organs). Hobbes followed Galileo's assertions exactly though giving them a more probabilistic slant. He states that in natural philosophy 'the most that can be atteyned unto is to have such opinions, as no certayne experience can confute, and from which can be deduced by lawfull argumentation, no absurdity.'[255] This position was repeated, in the same way, in *Tractatus Opticus II*.[256]

In the quoted passage, Galileo included an important reference to the resolutive method. Gargani has already noted that the Hobbesian conception of *resolutio* was influenced by Galileo.[257] In *De motu, loco et tempore*, Hobbes also maintains that the resolutive method is the correct process to apply to the 'science of causes,' i.e. natural philosophy.[258]

However, we should pay attention to the emphasis that Hobbes put on the hypothetical character of these opinions about the natural world, which he repeated in *Tractatus Opticus II* as well: he expressly suggests that the hypotheses should be developed around events manifest to the senses and empirically verifiable.[259] A small, but certainly not negligible, point should be added on this matter: Hobbes did not consider natural philosophy to be unprovable in its entirety, but rather 'for the most part,' because it depends on the motion of bodies so small they are invisible, like the air and the spirits,[260] a position he also maintained much later in *De corpore*.[261]

However, Galileo was also well aware that it can be very difficult, or even impossible, to observe phenomena because they are completely beyond experimental verification. On this subject, there are two interesting questions that Galileo

---

[255] Hobbes to William Candish, Earl of Newcastle, from Paris, 29 July/8 August 1636, *CH*, I, 33.

[256] *TO II*, Chapt. I, § 24, ff. 203r–203v/159: '… quaerenda erat hypothesis Conveniens, id est supponendus motus possibilis imaginabilis, et cui nullum absurdum consequeretur …'

[257] See Gargani 1971, 50 ff.

[258] *MLT*, XXX, 10, 352–353: '… siquidem processus fiat ab imaginatione causae ad imaginationem effectus, & ita deinceps versus finem (qui est semper effectus ultimus), vocatur animi discursus *compositio*, σύνθεσις; quod si contra procedatur ab effectu ad causam, & deinceps versus priora, vocatur *resolutio*, ἀνάλυσις; utraque autemvocatur reminiscentia … Eadem haec reminiscentia mediorum ad finem, si quoties finem imaginamur, toties eundem mediorum ordinem percurreret imaginatio procedendo a causa ad effectum, *ars* diceretur & contra procedendo ab effectu ad causam, *scientia* causarum'. Hobbes attributes the compositive method to art, and the resolutive method to the 'science of causes.' Ibid., II, 8, 115.

[259] *TO II*, Chap. I, § 1, f. 193r/147: 'Cum enim quaestio instituta sit, de alicuius eventus sensibus manifesti (quod Phaenomenon appellari solet) causa efficiente.'

[260] Hobbes to William Cavendish, Earl of Newcastle, from Paris, 29 July/8 Aug. 1636, *CH*, I, 33.

[261] *De corpore*, XXX, 14, *OL*, I, 424–425: 'Quis enim est ex sensibus nostris per quem judicamus esse aerem, quem neque videmus, neque audimus, neque gustamus, neque olfacimus, neque tangentes quidem cognoscimus esse aliquid? … Ratione autem corpus esse aliquod quod aerem dicimus, cognosci potest, sed unica, nimirum, quia sine medio corpore, corpora procul posita in sensoria nostra agere non possint, neque omnino sentiremus nisi contigua. Naturae ergo corporeae, absque ratiocinatione ab effectu, soli sensus idonei testes non sunt.'

addressed regarding the dispute with Orazio Grassi. Grassi had reproached Mario Guiducci, and through his disciple, Galileo himself, of having proposed dubious theories about the phenomenon of comets, which, according to Grassi, lacked the necessary observational confirmation. In the opening of the *Discourse on Comets*, (*Discorso delle comete*, 1619), through his student's pen, Galileo had maintained that we can argue, 'not affirmatively, but only *probably* and *dubiously*,'[262] and we can produce only '*conjectures*' about particular astronomical phenomena such as comets.

In *The Assayer*, Galileo reiterates the doubting, undogmatic nature of his speculations and those of Guiducci, because we 'let it be seen that we are not departing from our custom of declaring nothing as certain except what we know beyond doubt, as our philosophy and mathematics teach us.'[263] The asserting nature of rational demonstrations of the mathematical sciences can not always be the legacy of natural philosophy. Due to the lack of observational data, in these instances, we must formulate hypotheses or conjectures that do not have the traits of truth and certainty of mathematics.[264] Galileo uses a fable of considerable philosophical interest to address the issue: he references the speculative interest of an 'isolated man' who is obsessed with finding the origins of sound. But he is ultimately forced to accept the impossibility of identifying the cause of each sound with absolute accuracy.[265] Galileo's text concludes that the assayer of natural phenomena is sometimes precluded from a certain, objective knowledge of reality, and in such cases, must state a mere hypothesis, determining only one of the possible causes of the phenomenon.

> I might by many other examples make clear the bounty of nature in producing her effects by means which we would never think of if our senses and experience did not teach us of them, though even these are sometimes insufficient to remedy our incapacity. Therefore, I should not be denied pardon if I cannot determine precisely the manner in which comets are produced, especially as I never boasted that I could, knowing that it may occur in some way far beyond our power to imagine. The difficulty of understanding how the cicada's song is formed even when we have it singing to us right in our hands, is more than enough to excuse us for not knowing how a comet is formed at such immense distance.[266]

Even though we are aware that the confirmation of observational data ultimately comes only from experience, in the physical disciplines we must nonetheless make do with formulating hypotheses, i.e., not express ourselves 'affirmatively,' but only 'probably' and 'doubtfully.'[267]

---

[262] Galilei, *Discorso delle comete*, *OG*, VI, 47 (my italics). My translation.

[263] Galilei, *Il saggiatore*, *OG*, VI, 279. Eng. trans. Drake 1960, 233 ff.

[264] On the difficulty of translating highly mathematical arguments into the field of physics, see Giusti 1995. Regarding the 'mathematization' of nature, applied to the structure of matter, see the discussion in Chap. 4.

[265] Galilei, *Il saggiatore*, *OG*, VI, 279–281.

[266] Ibid., 281. Eng. trans. Drake 1960, 236–237.

[267] On this topic, see also Piccolino 2005, 27–37.

This is particularly true of some celestial phenomena, about which on the first day of the *Dialogue*, Galileo says 'there are more ways known to us that could produce the same effect, and perhaps others that we do not know of.'[268] On this topic, an argument made by Sagredo is interesting, which Hobbes later quoted almost exactly:

> I do not know nor do I suppose that herbs or plants or animals similar to ours are propagated on the moon, or that rains and winds and thunderstorms occur there as on the earth; much less that it is inhabited by men. Yet I still do not see that it necessarily follows that since things similar to ours are not generated there, no alterations at all take place, or that there cannot be things there that do change or are generated and dissolve; things not only different from ours, but so far from our conceptions as to be entirely unimaginable by us.[269]

According to Sagredo, this could be so particularly for the moon:

> ... Thus, and more so, might it happen that in the moon, separated from us by so much greater an interval and made of materials perhaps much different from those on earth, substances exist and actions occur which are not merely remote from but completely beyond all our imaginings, lacking any resemblance to ours and therefore being entirely unthinkable. For that which we imagine must be either something already seen or a composite of things and parts of things seen at different times; such are sphinxes, sirens, chimeras, centaurs, etc.[270]

In Chapter 7 of *De motu, loco et tempore*, Hobbes makes almost exactly the same argument, also referring to *chimera* and monstrous beings, specifically arguing the difficulty of providing exact, accurate observations in some realms of natural philosophy.

> Therefore, body or materia prima can be changed, and its parts moved in innumerable ways; and by means of motion of this sort it can arouse innumerable phantasms (*phantasmata*) in the minds of the percipients. i.e. numerous kinds of images. Granted that is impossible to know what motions the separate particles of the whole world have, it follows that we cannot know how many varieties of things there are and hence *whether or not there are in the heavens bodies like ours*. It may be there are; it may be that all the *chimeras* and *monsters of human imagination have their counterparts in the heavens; it also may be that there is not in the heavens any heavy or light object, any* man *or* animal *or* tree. As being in

---

[268] Galilei, *Dialogo sopra i due massimi sistemi del mondo*, *OG*, VII, 124. Eng. Trans. Galilei 2001, 99. Giudice pointed out an analogy and a quotation of this passage in § 2 of Chap. 1 of *Tractatus Opticus II*, where Hobbes compares the main cosmological theories (see Giudice 2016). I think Giudice was generally correct about Galileo's influence on Hobbes' method; however, in my opinion, that analogy is not very evident (for instance, a quite similar argument was made by Descartes and Gassendi. See Roux 1998, 20). On the contrary, I find striking similarities between an argument in another part of the same day of the *Dialogue*, and the argument in a passage of Chapter 7 of *De motu, loco et tempore* (see next page). Giudice also cites a passage of the optical treatise about the principle of the relativity of motion on a ship, (*TO II*, chap. IV, § 56, f. 264*v*/225) which refers to a comparable place on the *second day* of the *Dialogue* (Galilei, *Dialogo sopra i due massimi sistemi del mondo*, *OG*, VII, 141–143).

[269] Ibid., *OG*, VII, 86. Eng. trans. Galilei 2001, 61.

[270] Ibid.

fact the things we cannot know at all, because they do not work on our senses from so great a distance.[271]

However, we should note that in Chapter 26 of the same book, Hobbes reiterates that 'to prove that something exists, there is need of the senses, or experience,'[272] while maintaining that 'even so, the demonstration is not established,' because 'for someone rigidly demanding the truth from people who say that Socrates lived or existed will tell them to add: "Unless I have seen a spirit, or a ghost, or I have dreamed, I have seen Socrates, so Socrates existed, etc."'[273] Empirical observation remains a fundamental element of scientific inquiry, but nonetheless, we should emphasize that in some of his works, particularly the *Leviathan*, Hobbes insists on the aprioristic and conventional element of philosophy, seeming to propose a rather inflexible vision of a single large deductive system, in which all the necessary conclusions develop as a consequence of the premises. This, however, is not the conception that emerges from his scientific works, and we should consider briefly the Hobbesian classification of the sciences, which proves more complex than it may seem at first glance.

2.10 Hobbes' Classification of Sciences

The diagram of the sciences in *Leviathan* is much influenced by the book's primarily political intent, which seeks to present philosophy as a unified body of knowledge, dominated by demonstrative reasoning, free from paralogisms and interpretative difficulties. As a result, here Hobbes' philosophy has more clearly conventionalistic features than in his works about natural philosophy. Resuming some ideas already presented in *The Elements of Law*, Hobbes restates some positions outlined in his earlier works,[274] thoroughly and precisely expressing his conventional and linguistic conception of science (and philosophy).[275] He stresses the

---

[271] *MLT*, VII, 4, 147–148 Eng. Trans. Hobbes 1976, 81, modified. See also ibid., XXIV, 1, 289.

[272] *MLT*, XXVI, 2, 309; Eng. Trans. Hobbes 1976, 305 (my italics).

[273] Ibid.

[274] In *Leviathan*, Hobbes reiterates that *truth* and *falsehood* only exist in relationship to discourse. See Hobbes 2012, chap. IV, 54–56: 'For *True* and *False* are attributes of Speech, not of Things. And where Speech is not, there is neither *Truth* nor *Falsehood* … Seeing then that *truth* consisteth in the right ordering of names in our affirmations, a man that seeketh precise *truth*, had need to remember what every name he uses stands for, and to place it accordingly.'

[275] Ibid., chap. V, 72: 'By this it appears that Reason is not, as Sense and Memory, borne with us; nor gotten by Experience onely, as Prudence is; but attayned by Industry; first in apt imposing of Names; and secondly by getting a good and orderly Method in proceeding from the Elements, which are Names, to Assertions made by Connexion of one of them to another; and so to Syllogismes, which are the Connexions of one Assertion to another, till we come to a knowledge of all the Consequences of names appertaining to the subject in hand; and that is it, men call SCIENCE'.

distinction (already made in *The Elements*[276]) between knowledge limited to the realms of sensation and memory (which are only '*de facto* knowledge') and knowledge that consists of 'knowledge of the consequences of names.' Only the latter achieves the level of philosophy because it lets us discover a possible cause of phenomena and lets us act on natural processes to benefit humanity: 'Because when we see how any thing comes about, upon what causes, and by what manner; when the like causes come into our power, wee see how to make it produce the like effects.'[277] In Chapter 9, Hobbes reiterates the conventional nature of scientific knowledge, returning to the split that includes two different kinds of knowledge, maintaining that *science* is a '*Conditionall knowledge* … as when we know, that, *If the figure showne be a Circle, that any straight line through the Center shall divide it into two equall parts*.'[278] On the contrary, *history* is 'The Register of Knowledge of Fact' and it divides into *natural history* and *civil history*.[279] However, we should note that the distinction that Hobbes makes does not imply a devaluing of empirical knowledge; this is, indeed, the type of knowledge required of an eye witness, and is therefore termed: absolute knowledge.

The diagram that Hobbes drew in the same chapter is the most complete, clearest image of Hobbes' aspiration to conceive the entire philosophical panorama in the framework of a single system. *Science* or *philosophy* are presented as a syllogistic system within which all branches of knowledge are meant to have the criteria of truth and a uniform methodology.[280] In this classification, mathematical and physical sciences are both included together in the larger whole that makes up natural philosophy. However, if we examine the diagram more carefully, we see that some disciplines that we would normally consider in the realm of physics Hobbes considers siblings of geometry. Hobbes included among these: astronomy, geography, and science of engineers (i.e. architecture and navigation).[281] Of course, these were the so-called 'mixed mathematics,' which Hobbes included among the 'Sciences or Knowledge of consequences.'[282] They refer to 'Consequences from the accidents Common to all Bodies Naturall which are *Quantity* and *Motion*, whereas physics studies the 'Consequences from the *Qualities of Bodies*.'[283]

In this book, Hobbes suggests translating certain sciences that we consider in the realm of physics (such as mechanics) into specifically geometric terms, based on interpreting geometry as a science of motion, founded only on two elements: quantity and motion. This result also came, ultimately, from applying the method that

---

[276] *EL*, Part I, chap. VI, §§ 1–4, 24–6.

[277] Hobbes 2012, 72.

[278] Ibid., 124.

[279] Ibid.

[280] See Watkins 1965, 61 and Malherbe 1984, 90.

[281] See Hobbes 2012, chap. IX, 130–132.

[282] As we know, Hobbes' definition of philosophy in *Leviathan* is that of the 'knowledge of consequences' The study of the consequences of natural bodies, is 'natural philosophy,' while the study of the consequences of the political bodies is 'civil, or political philosophy.' (ibid).

[283] Ibid.

Galileo had proposed, which considered as the true properties of bodies only 'shapes, numbers, and slow or rapid movements.'[284]

Yet, in *Leviathan*, the demonstrative element of knowledge gains a total, absolute dominance, and, as we noted, Hobbes' goal was to present an ideal single framework of science. Nonetheless, to fully understand the diagram presented here, we must refer to *De corpore*, which discusses in depth the method of knowledge acquisition, addressed in less detail in his 1651 book.

In the introduction, Hobbes already presented the image of philosophy as a *reminiscence*, which was already part of *De motu, loco et tempore* and in Galileo's writings.[285] Yet, rather than a Platonic reminiscence, the concept that Hobbes illustrates evokes the idea of the philosopher's revelation and discovery of the natural world through a suitable method: speculative inquiry. Philosophy is 'such knowledge of effects or appearances, as we acquire by true ratiocination from the knowledge we have first of their causes or generation: And again, of such causes or generations as may be from knowing first their effects.'[286] In *De corpore*, Hobbes also reiterates his computational concept of reasoning,[287] and his notion of true and false.[288] Furthermore, in Chapter 4 (about method), his ideas complete and at times alter the methodological framework that emerged from *De motu, loco et tempore*. Hobbes repeats that philosophical or scientific knowledge is merely causal knowledge. Any other kind of knowledge that is not $\tau o\hat{v}$ $\delta\iota\acute{o}\tau\iota$, but only $\tau o\hat{v}$ $\acute{o}\tau\iota$, is perception by sense, or the trace remaining after such a perception, i.e. imagination, or memory.[289]

Yet, (as he had in previous works) Hobbes restates the empirical genesis of all of our knowledge: the first principles of science are the 'phantasms of sense and imagination': of which we naturally know *quod sunt*; but to know 'why they be' (*quare sunt*), or 'from what causes they proceed, is the work of ratiocination.' He also reaffirms that philosophical knowledge is necessarily demonstrative. However, unlike in *De motu, loco et tempore*—where knowledge of causes is exclusively resolutive—in *De corpore*, he maintains that both methods: *resolutive* and *compositive*, i.e. *analytical* and *synthetic* are necessary to science.

> ... seeing universal things are contained in the nature of singular things, the knowledge of them is to be acquired by reason, that is, with the resolutive method. For example, if there be propounded a conception or idea of some singular thing, as of a square, this square is to be resolved into a *plain, terminated with a certain number of equal and straight lines and right angles*. For by this resolution we have these things universal or agreeable to all matter, namely, line, plain, (which contains superficies) *terminated, angle, straightness, rectitude,*

---

[284] Galilei, *Il saggiatore*, *OG*, VI, 350. Eng. trans. Drake 1957, 275.

[285] *De corpore, Ad lectorem, OL*, I, not numbered (first page of the dedicatory letter): 'Mentis ergo tuae et totius mundi filia Philosophia in te ipso est; nondum fortasse figurata, sed genitori mundo qualis erat in principio informi similis. Faciendum ergo tibi est, quod faciunt statuarii qui materiam exculpentes supervacaneam, imaginem non faciunt, sed inveniunt.'

[286] Ibid., I, 2, *OL*, I, 2.

[287] Ibid., *OL*, I, 3.

[288] Ibid., III, 7, *OL*, I, 31–32.

[289] Ibid., VI, 1, *OL*, I, 58–59.

and equality; and if we can find out the causes of these, we may compound them altogether into the causes of a square.[290]

In Chapter 2 (about *words*), he states that the *universal* 'is never the name of any thing existent in nature, nor any idea or phantasm formed in the mind, but always the name of some word or name.'[291] The operation of extrapolating universals from every *res* should, therefore, be interpreted as an exclusively mental process, which has some creative elements. Indeed, though Hobbes maintains that universal causes are manifest to themselves 'known to nature' and that the only universal cause of them is, ultimately, motion[292]; he specifies that the things known to nature (*naturae notiora*) ought to be understood only as knowledge acquired through reason as distinguished from purely empircal experience.[293] There is no reality known to nature, things commonly defined as *naturae notiora* are to be considered as acquired through reason, which is to say demonstratively. On the other hand, *nobis notiora* are realities perceived exclusively through the senses, and this type of knowledge, of course, does not reach the level of science or philosophy.[294]

Chapter 25 of *De corpore* opens the fourth part of the book, focused on 'Physics,' i.e. 'natural phenomena.'[295] Hobbes distinguishes the following chapters from the previous ones based both on the subject of study and the method to be applied. He says specifically that whereas he was able to use a strictly deductive method in earlier chapters, a particular methodology had to be used here:

> The principles, therefore, upon which the following discourse depends, are not such as we ourselves make and pronounce in general terms, as definitions; but such, as being placed in the things themselves by the Author of Nature, are by us observed in them; and we make use of them in single and particular, not universal propositions. Nor do they impose upon us any necessity of constituting theorems; their use being only, throughout not without such general propositions as have been already demonstrated, to show us the possibility of some production or generation. Seeing, therefore, the science, which is here taught, hath its principles in the appearances of nature, and endeth in the attaining of some knowledge of natural causes, I have given to this part the title of PHYSICS, or the *Phenomena of Nature*.[296]

It is worth emphasizing that, although the analysis of physics starts from observing phenomena, i.e. the effects, to arrive at possible causes, Hobbes nonetheless considers it necessary for them to be based on 'general propositions' and physical science is also expressed as a causal knowledge.

---

[290] Ibid., VI, 4, 61; Eng. Trans. *EW*, I, 68–69.

[291] Ibid., II, 9, *OL*, I, 17–18; Eng. Trans. *EW*, I, 20. On this topic, see Minerbi Belgrado 1993, 122.

[292] *De corpore*, VI, 5, *OL*, I, 62; Eng. Trans. *EW*, I, 69: "…causa enim eorum omnium universalis una, est motus'.

[293] Ibid., 61.

[294] On the *nobis notiora/naturae notiora* distinction and the difference between Hobbes' interpretation and that which we find in Zabarella's *De Regressu*, see Prins 1990, 34–6. On the *nobis notiora/naturae notiora* distinction, see also Lupoli 2006, 74 ff.

[295] *De corpore*, XXV, 1, *OL*, I, 334–335.

[296] Ibid., 335; Eng. Trans. *EW*, I, 388.

In order to understand Hobbesian epistemology, we need to turn to Chapter 6 to explore the relationship that he establishes between geometry and physics. He started by defining the concepts of *place* and *motion*, closely bound with Hobbesian philosophy's cardinal principle: the idea of *body*. *Place* is 'that space which is possessed or filled adequately by some body' and motion is nothing more than 'the privation of one place, and the acquisition of another.'[297] On the basis of these definitions, Hobbes says that, through the *compositive method*, a researcher may investigate what generates the body's motion: line, length, and area are produced by the moved body and the part of philosophy that addresses operations applied to motions (additions, subtractions, multiplications, and divisions of motions) is geometry.[298]

Based on these considerations, geometry would, strictly speaking, be nothing more than a 'science of motion,' and, as such, Hobbes would have had to consider Galileo the founder of not only physics but of geometry as well. Nonetheless, of course, Galileo is celebrated for having revealed the 'nature of motion,'[299] and in Chapter 6 of *De corpore*, Hobbes suggests a less rigid, dogmatic classification of sciences than that in *Leviathan*, which also implies a split between geometry and physics that is less neatly delineated. After defining geometry, Hobbes investigates the science that deals with the motion of bodies and the interaction between bodies, defining this discipline, or area of research, broadly as 'that part of Philosophy which treats of Motion.'[300] He then addresses the 'search that comes from the motion of parts,' which are the motions that produce changes in sense organs. His inquiry focuses on 'sensible qualities, such as *light, colour, transparency, opacity, sound, odour, savour, heat, cold*, and the like.' According to Hobbes, this is the true realm of physics, which, though it addresses the qualities, does not do so as if they were true accidents of the bodies, but as being produced by their actual underpinnings: *bodies* and *motion*.[301]

Hobbes' undertaking leads to forming a kind of bridge, or point of passage, from geometry to physics, creating a dual result of geometrizing physics and, inversely, physicizing geometry (as geometric entities are the product of the motion of bodies). This process is based on fundamental notions of *quantity* and *motion*: these are the accidents that Galileo considered the 'primary qualities' of bodies. and these are the fundamental principles of Hobbesian philosophy. Not coincidentally, Hobbes significantly identifies *extension* and *figure* as the essential accidents of the body, inseparable from the very concept of *body*.[302] These accidents can be considered the

---

[297] Ibid. VI, 6, *OL*, I, 62–63; Eng. Trans. 70.

[298] Ibid., 63.

[299] Ibid., not numbered, 62. Jesseph has investigated Hobbes' mechanics, the foundation and premise of Hobbes' philosophical system whose key elements are motion and mathematically quantifiable bodies. See Jesseph 2006.

[300] *De corpore*, VI, 6, 63.

[301] Ibid.

[302] Ibid., VIII, 3, 92–93. These thoughts draw from ideas found in the manuscript of *De principiis*, where he identifies an accident that was the essence of a specific body. See *De Principiis* (National Library of Wales, Ms. 5297), *MLT*, Appendix II, 457: 'That accident for which we impose a certain

bricks of the building of human knowledge, and, as such, are the basis of not only the science of motion, but all of Hobbes' philosophy, starting from his 'first philosophy,' which is focused, as noted, on the concept of body. We can infer from this that Hobbes had absorbed and re-worked the attempt to interpret the objects that fill the natural world into mathematically quantifiable entities, a particular aim of Galileo's natural philosophy (on which modern science was founded). However, in the 'search for causes'[303]—the domain of natural philosophy—there must be a joint application of *compositive* and *resolutive* methods. The resolutive method lets us deconstruct the phenomenon in simple data, tying it to the principles of quantity and motion. The compositive method then performs the unified recomposition of the individual data analyzed.[304]

A year after *De corpore* was published, in *Six Lessons*, Hobbes started with the same vision of geometry based on the concept of quantity[305] (and therefore—consistent with the ideas illustrated in *De corpore*, that of body[306]). *De homine* includes some interesting developments on the relationship between physics and geometry. In Chapter 10, on *Discourse on the Sciences*, Hobbes again proposes a constructivist conception of mathematics as we have seen in his earlier works: mathematics completely rules the cause and genesis of the geometric figure, whereas the principles of nature are subject to natural philosophy and are exclusively in the hands of the creator. Mathemetization does not, however, create the geometric entities out of nothing but, as in something of a demiurgic operation, builds geometric arguments on the basis of tools that already exist, and these tools are the cardinal principles of geometric science.[307] Hobbes also argues (as he had 20 years earlier in his letter to William Cavendish, Earl of Newcastle) that the 'greater part of natural realities is invisible, which means that it is not possible to conduct a direct empirical observation of these phenomena. This kind of demonstration is called *a posteriori* and its science, physics.' But at times we can trace the causes of events based on observed effects, from the properties we see, 'we can, by deducing as far as possible the consequences of those qualities we do see, demonstrate that such and such *could* have been their causes.'[308] The discussion's continuation is equally interesting as it leads Hobbes to include *physics* among the mixed mathematics:

---

name upon any body: or that accident which does denominate his subject is called the *essence* thereof.'

[303] *De corpore*, VI, 10, *OL*, I, 70; *EW*, I, 77.

[304] Ibid.: 'Interea manifestum est quod in causarum investigatione partim methodo analytica partim synthetica opus est. Analytica, ad effectus circumstantias sigillatim concipiendas, synthetica ad ea. quae singulae per se efficiunt in unum componenda.'

[305] *Six Lessons*, *EW*, VII, 191: 'Geometry is the science of determining the quantity of anything, not measured, by comparing it with some other quantity or quantities measured.'

[306] Ibid.: 'I suppose, most egregious professors, you know already that by geometry, though the word import no more but the measuring of land, is understood no less the measuring of all other quantity than that of bodies.'

[307] See *De homine*, X, 5, *OL*, II, 93.

[308] Ibid., 93.

And since one cannot proceed in reasoning about natural things that are brought about by motion from the effects to the causes without a knowledge of those things that follow from that kind of motion; and since one cannot proceed to the consequences of motions without a knowledge of the quantity, which is geometry; nothing can be demonstrated by physics without something also being demonstrated *a priori*. Therefore, physics (I mean true physics), that depends on geometry, is usually numbered among the mixed mathematics.[309]

Hobbes maintains that physics, unlike mathematics, is a science structured *a posteriori*. Only after having seen the effects of a specific phenomenon through the *resolutive* method can we discover its possible causes. However, physics is based on principles outside of our power which are not fully under the assayer's control. Hobbes argues at the same time that without cognition of motions and quantities, the domain of geometry, no demonstration can be produced in the realm of physics. He is quite emphatic on this point: without that process of mathematizing the natural world, of which Galileo is the 'patron saint,' no rigorous physical inquiry can be made. Correctly applying an aprioristic method in physics becomes possible by being able to express the objects of physics in mathematically-quantifiable geometric terms, through the method introduced by Galileo.[310]

The Italian scientist had never given up on the idea that the book of nature, i.e. universe, be written in mathematical characters, i.e. 'triangles, squares, circles, spheres, cones, pyramids, and other mathematical figures' (as he repeated in a letter to Fortunio Liceti in January 1641[311]). Hobbes takes up and re-elaborates the Galilean legacy, which emerges primarily in the geometricization of reality. Exactly like Galileo, Hobbes maintains that the only criteria for correctly interpreting physical phenomena is codifying in a mathematical-geometric language the phenomenal entities that are made evident to our sense organs. Researchers who attempt to know the causes of natural phenomena must translate the world of perceptions acquired through the senses into mathematically and geometrically quantifiable entities.

In Hobbes' work, this geometrization of the natural world effectively takes on much more hypothetical connotations, but this does not at all imply that he discredited scientific knowledge over mathematics and geometry. On the contrary, Hobbes affirms in *Seven Philosophical Problems* (1662), that although he was aware that the

---

[309] Ibid. My translation.

[310] Zvi Biener has investigated Galileo's attempt to apply his method of mathematization to reality as well as to matter, found in the first day of *Discourses* where Galileo focuses on the possibility (and need) to extend mathematical speculations to physics (and therefore the science of matter). Biener has argued that Galileo had a clear polemical target: the ideas that Niccolò Tartaglia had developed in *Quesiti et inventioni diverse* (1546). See Biener 2004, 267–269. Regarding Galileo's relationships with the so-called mixed mathematics, I disagree with the conclusion reached by Machamer 1978, which would reduce Galileo the philosopher to the role of a mixed mathematician. The image that emerges from Galileo' texts is that of a natural philosopher, as is clear from his conception of matter (see below, Chap. 4), which is the basis for his distinction between primary and secondary qualities of bodies. In my opinion, an analysis of Hobbes' texts also suggests that this was Hobbes' interpretation, and this is why he considered Galileo the greatest philosopher of all time.

[311] Galileo Galilei to Fortunio Liceti, January 1641, *OG*, XVIII, 295. Eng. Trans. Palmerino 2002, 30

doctrine of natural causes had no infallible and evident principles,[312] the study of nature was 'the most noble employment of the mind that can be.'[313]

References

Works of Thomas Hobbes:

Hobbes, Thomas. 1976. *Thomas White's De Mundo Examined*, ed. Harold Whitmore Jones. Bradford: Bradford University Press.

———. 1988a. (probably by Robert Payne), *Court traité des premiers principes*, ed. Jean Bernhardt. Paris: Presses Universitaires de France.

———. 1988b. *Of Passions* (British Library: Ms. Harl. 6083), ed. Anna Minerbi Belgrado. *Rivista di Storia della Filosofia* (4): 729–738.

———. 2010. *Moto, luogo e tempo*, ed. Gianni Paganini, Utet, Torino.

———. 2012. *Leviathan*, ed. Noel Malcolm, 3 Vols. Oxford, Clarendon Press.

Other Works:

Aristotle. 1986. *De Caelo*, Eng. Trans. W.K.C. Guthrie. Cambridge, MA: Harvard University Press.

———. 1989. *The Metaphysics*, Eng. Trans. Hugh Tredennick. London: William Heinemann.

Bacon, Francis. 1858–1874. *The Works of Francis Bacon*, 14 Vols, ed. James Spedding, Robert Leslie Ellis and Douglas Denon Heath. London: Longman, Green and Roberts.

Boyle, Robert. 1999–2000. *The Works of Robert Boyle*, 14 Vols, ed. Michael Hunter and Edward B. Davis. London: Pickering and Chatto.

Copernicus, Nicolaus. 2015. *De Revolutionibus Orbium Coelestium*, ed. Michel-Pierre Lerner, Alain-Philippe Segonds and Jean-Pierre Verdet, 3 Vols. Paris: Les Belles Lettres.

Galilei, Galileo. 1960. *Le Mecaniche*, Eng. Trans. Stillman Drake. Madison: University of Wisconsin-Madison Press.

———. 1988. *Tractatio de praecognitionibus et praecognitis* and *Tractatio de demonstratione*, ed. William F. Edwards and William A. Wallace. Padua: Antenore.

———. 2001. *Dialogue Concerning the Two Chief World Systems*, Eng. Trans. Stillman Drake. New York: The Modern Library (1st ed. Los-Angeles-Berkeley: University of California Press, 1967).

———. 2003. *Dialogues Concerning Two New Sciences*, Eng. Trans. Henry Crew and Alfonso de Salvio. Mineola: Dover Publications (1st ed.: New York: The Macmillan Company, 1914).

Kepler, Johannes. 1937—. *Gesammelte Werke*, ed. Walther von Dick, Max Caspar and Franz Hammer. München: Beck.

Mersenne, Marin. 1634. *Les Mechaniques de Galilee, Mathematicien et Ingenieur du Duc de Florence*. Paris: Henry Guenon (Reprint in Mersenne. 1985: 427–513).

———. 1985. *Questions Inouyes*, (Corpus des œuvres de philosophie en langue française) (Paris: Fayard, 1985).

---

[312] *Seven Philosophical Problems*, EW, VII, 3: 'The doctrine of natural causes hath not infallible and evident principles.'

[313] Ibid., 4.

Sarpi, Paolo. 1996. *Pensieri, naturali, metafisici e matematici*, ed. Luisa Cozzi and Libero Sosio. Milan-Naples: Ricciardi.

Sarsi, Lotario. 1619. (*alias* Orazio Grassi). *Libra Astronomica ac Philosophica*. In *OG*, VI: 109–179.

Secondary Sources:

Altieri Biagi, Maria Luisa. 1995. L'incipit del *Dialogo sopra i due massimi sistemi*. In *Galileo e la cultura veneziana*, 351–361. Venice: Istituto Veneto di Lettere ed Arti.

Anstey, Peter. 2000. *The Philosophy of Robert Boyle*. London/New York: Routledge.

———. 2013. The Theory of Material Qualities. In *The Oxford Handbook of British Philosophy in the Seventeenth Century*, ed. Peter Anstey, 240–260. Oxford: Oxford University Press.

Applebaum, Wilbur, and Renzo Baldasso. 2001. Galileo and Kepler on the Sun as Planetray Mover. In *Largo campo di filosofare*, ed. José Montesinos and Carlos Solís, 381–390. La Orotava: Fundaciòn Canaria Orotava de Historia de la Ciencia.

Aquinas, Thomas. 1989. *Summa Theologiae*. Bologna: Edizioni Studio Domenicano.

Baldin, Gregorio. 2017. Archi, spiriti e *conatus*. Hobbes e Descartes sui principi della fisica. *Historia Philosophica* 15: 147–165.

Banfi, Antonio. 1964. *Galileo Galilei*. Milan: Il Saggiatore (1st ed. 1949).

Barnouw, Jeffrey. 1989. *Respice Finem!* The Importance of Purpose in Hobbes's Psychology. In *Thomas Hobbes: de la métaphysique à la politique*, ed. Martin Bertman and Michel Malherbe, 47–59. Paris: Vrin.

———. 1990. Hobbes's Causal Account of Sensation. *Journal of the History of Philosophy* 18 (n 2): 115–130.

Bernhardt, Jean. 1990. Grandeur, substance et accident: une difficulté du *De corpore*. In *Thomas Hobbes. Philosophie première, théorie de la science et politique*, ed. Yves-Charles Zarka and Jean Bernhardt, 39–46. Paris: Presses Universitaires de France.

Biener, Zvi. 2004. Galileo's First New Science: The Science of Matter. *Perspectives on Science* 12 (3): 262–287.

Brandt, Frithiof. 1928. *Thomas Hobbes' Mechanical Conception of Nature*. Copenhagen/London: Levin & Mungsgaard-Librairie Hachette (or. Ed. 1921).

Bucciantini, Massimo. 2003. *Galileo e Keplero. Filosofia, cosmologia e teologia nell'Età della Controriforma*. Turin: Einaudi.

———. 2007. Descartes, Mersenne e la filosofia invisibile di Galileo. *Giornale critico della filosofia italiana* 86 (1): 38–52.

Bunce, Robin. 2003. Thomas Hobbes' Relationship with Francis Bacon – An Introduction. *Hobbes Studies* 16: 41–83.

Camerota, Michele. 2004. *Galileo Galilei e la cultura scientifica nell'età della Controriforma*. Rome: Salerno Ed.

Carugo, Adriano, and Alistair C. Crombie. 1983. The Jesuits and Galileo's Ideas of Science and of Nature. *Annali dell'Istituto e Museo di Storia della Scienza di Firenze* 8 (2): 3): 3–3):68.

Cassirer, Ernst. 1922–1957. *Das Erkenntnisproblem in der Philosophie und Wissenschaft der neuren Zeit*, 4 Vols. Berlin: Bruno Cassirer Verlag.

Clavelin, Maurice. 1968. *La philosophie naturelle de Galilée*. Paris: Librairie Armand Colin.

———. 2001. Galilée astronome philosophe. In *Largo campo di filosofare. Eurosymposium Galileo 2001*, ed. José Montesinos and Carlos Solís, 19–39. La Orotava: Fundaciòn Canaria Orotava de Historia de la Ciencia.

Crombie, Alistair C. 1953. *Augustine to Galileo. The History of Science A.D. 400–1650*. Cambridge, MA: Harvard University Press.

————. 1975. Sources of Galileo's Early Natural Philosophy. In *Reason, Experiment, and Mysticism in the Scientific Revolution*, ed. Maria Luisa Righini Bonelli and William Shea, 157–175. New York: Science History Publ.

————. 1990. *Science, Optics and Music in Medieval and Early Modern Thought*. London/Ronceverte: The Hambledon press.

Dijksterhuis, Eduard J. 1961. *The Mechanization of the World Picture: From Pythagoras to Newton*. Oxford: Clarendon Press.

Drake, Stillman. 1957. *Discoveries and Opinions of Galileo*. New York: Doubleday & Co..

————. 1960. *The Controversy on the Comets*. Philadelphia: University of Pennsylvania Press.

————. 1978. *Galileo at Work: His Scientific Biography*. Mineola: Dover Publications, Inc..

Festa, Egidio. 2007. *Galileo. La lotta per la scienza*. Rome/Bari: Laterza.

Finocchiaro, Maurice A. 2008. *The Essential Galileo*. Indianapolis/Cambridge: Hackett Publishing Company.

Gabbey, Alan. 2004. What Was 'Mechanical' About 'The Mechanical Philosophy'. In *The Reception of Galilean Science of Motion in Seventeenth-Century Europe*, ed. Carla Rita Palmerino and J.M.M.H. Thijssen, 11–23. Dordrecht/Boston/London: Kluwer.

Galluzzi, Paolo. 1973. Il 'Platonismo' del tardo Cinquecento e la filosofia di Galileo. In *Ricerche sulla cultura dell'Italia moderna*, ed. Paola Zambelli, 39–79. Rome/Bari: Laterza.

————. 1979. Il tema dell''ordine' in Galileo. In *Ordo. Atti del II colloquio internazionale del Lessico Intellettuale Europeo*, ed. Marta Fattori and Massimo L. Bianchi, 235–277. Rome: Ateneo & Bizzarri.

————. 1994. Ratio/Ragione in Galileo. Del dialogo tra la ragione e l'esperienza. In *Ratio (VII Colloquio Internazionale del Lessico Intellettuale Europeo)*, ed. Marta Fattori and Massimo L. Bianchi, 379–401. Florence: Olschki.

————. 2011. *Tra atomi e indivisibili. La materia ambigua di Galileo*. Florence: Olschki.

Garber, Daniel. 2010. Philosophia, Historia, Mathematica: Shifting Sands in the Disciplinary Geography of the Seventeenth Century. In *Scientia in Early Modern Philosophy. Seventeenth-Century Thinkers on Demonstrative Knowledge from First Principles*, ed. Sorell Tom, G.A.J. Rogers, and Jill Kraye, 1–17. Dordrecht: Springer.

————. 2013. Remarks on the Pre-History of Mechanical Philosophy. In *The Mechanization of Natural Philosophy*, ed. Daniel Garber and Sophie Roux, 3–26. Dordrecht: Springer.

Gargani, Aldo G. 1971. *Hobbes e la scienza*. Einaudi: Turin.

Gaukroger, Stephen. 2001. *Francis Bacon and the Transformation of Early-Modern Philosophy*. Cambridge: Cambridge University Press.

————. 2006. *The Emergence of a Scientific Culture*. Oxford: Oxford University Press.

————. 2010. The Unity of Natural Philosophy and the End of *Scientia*. In *Scientia in Early Modern Philosophy. Seventeenth-Century Thinkers on Demonstrative Knowledge from First Principles*, ed. Sorell Tom, G.A.J. Rogers, and Jill Kraye, 18–33. Dordrecht: Springer.

Gert, Bernard. 1989. Hobbes's Account of Reason and the Passions. In *Thomas Hobbes: de la métaphysique à la politique*, ed. Martin Bertman and Michel Malherbe, 83–92. Paris: Vrin.

————. 1996. Hobbes's Psychology. In *The Cambridge Companion to Hobbes*, ed. Tom Sorell, 157–174. Cambridge: Cambridge University Press.

Geymonat, Lodovico. 1957. *Galileo Galilei*. Turin: Einaudi.

Gilbert, Neal W. 1963. Galileo and the School of Padua. *Journal of History of Philosophy* 1: 223–231.

Giudice, Franco. 1999. *Luce e visione. Thomas Hobbes e la scienza dell'ottica*. Florence: Olschki.

————. 2011. Echi del caso Galileo nell'Inghilterra del XVII secolo. In *Il caso Galileo. Una rilettura storica, filosofica, teologica. Convegno internazionale di studi. Firenze, 26–30 maggio 2009*, ed. Massimo Bucciantini, Michele Camerota, and Franco Giudice, 277–287. Florence: Olschki.

————. 2016. Optics in Hobbes's Natural Philosophy. *Hobbes Studies* 29 (1): 86–102.

Giusti, Enrico. 1995. Il ruolo della matematica nella meccanica di Galileo. In *Galileo Galilei e la cultura veneziana*, 321–337. Venice: Istituto Veneto di Lettere e Arti.

Hall, Alfred R., and Marie Boas Hall. 1964. *A Brief History of Science*. New York: The New American Library of World Literature Inc.

Heilbron, John L. 2010. *Galileo*. Oxford: Clarendon Press.

Henry, John. 2011. Galileo and the Scientific Revolution: The Importance of His Kinematics. *Galilaeana* 8: 3–36.

———. 2016. Hobbes, Galileo and the Physics of Simple Circular Motion. *Hobbes Studies* 29: 9–38.

Horstmann, Frank. 1998. Ein Baustein zur Kepler-Rezeption: Thomas Hobbes'. *Physica Coelestis Studia Leibnitiana* 30 (2): 135–160.

———. 2001. Hobbes on Hypotheses in Natural Philosophy. *The Monist* 84 (4): 487–501.

Jesseph, Douglas M. 2004. Galileo, Hobbes and the Book of Nature. *Perspectives on Science* 12 (2): 191–211.

———. 2006. Hobbesian Mechanics. *Oxford Studies in Early Modern Philosophy* 3: 119–152.

———. 2010. *Scientia* in Hobbes. In *Scientia in Early Modern Philosophy. Seventeenth-Century Thinkers on Demonstrative Knowledge from First Principles*, ed. Sorell Tom, G.A.J. Rogers, and Jill Kraye, 117–127. Dordrecht: Springer.

———. 2016. Hobbes on 'Conatus': A Study in the Foundations of Hobbesian Philosophy. *Hobbes Studies* 29 (1): 66–85.

Kargon, Robert H. 1966. *Atomism in England from Hariot to Newton*. Oxford: Clarendon Press.

Koyré, Alexandre. 1966. *Études galiléennes*. Paris: Hermann (or. Ed. 1939).

———. 1981. *Études d'histoire de la pensée philosophique*. Paris: Gallimard.

Laird, Walter Roy. 1997. Galileo and the Mixed Sciences. In *Method and Order in Renaissance Philosophy of Nature (The Aristotle Commentary Tradition)*, ed. Daniel A. Di Liscia, Eckhard Kessler, and Charlotte Methuen, 252–270. Ashgate: Aldershot.

Leijenhorst, Cees. 2002. *The Mechanisation of Aristotelianism. The Late Aristotelian Setting of Thomas Hobbes' Natural Philosophy*. Leiden/Boston/Köln: Brill.

———. 2004. Hobbes and the Galilean Law of Free Fall. In *The Reception of the Galilean Science of Motion in Seventeenth-Century Europe*, ed. Carla Rita Palmerino and J.M.M.H. Thijssen, 165–184. Dordrecht/Boston/London: Kluwer.

———. 2005. La causalité chez Hobbes et Descartes. In *Hobbes, Descartes et la métaphysique*, ed. Dominique Weber, 79–119. Paris: Vrin.

———. 2007. Sense and Nonsense About Sense: Hobbes and the Aristotelians About Sense Perception and Imagination. In *The Cambridge Companion to Hobbes's Leviathan*, ed. Patricia Springborg, 82–108. Cambridge: Cambridge University Press.

Lupoli, Agostino. 2006. *Nei limiti della materia. Hobbes e Boyle: materialismo epistemologico, filosofia corpuscolare e Dio corporeo*. Milan: Baldini, Castoldi, Dalai.

Machamer, Peter. 1978. Galileo and the Causes. In *New Perspectives on Galileo*, ed. Robert E. Butts and Joseph C. Pitt, 161–180. Dordrecht/Boston: Reidel Publishing Company.

Malcolm, Noel. 1990. Hobbes's Science of Politics and His Theory of Science. In *Hobbes Oggi*, ed. Bernard Willms et al., 145–159. Milan: Franco Angeli.

———. 2002. *Aspects of Hobbes*. Oxford: Clarendon Press.

Malherbe, Michel. 1984. *Thomas Hobbes ou l'oeuvre de la raison*. Paris: Vrin.

———. 1989. Hobbes et la fondation de la philosophie première. In *Thomas Hobbes: de la métaphysique à la politique*, ed. Martin Bertman and Michel Malherbe, 17–32. Paris: Vrin.

Médina, José. 1997. Nature de la lumière et science de l'optique chez Hobbes. In *Le siècle de la lumière 1600–1715*, ed. Christian Biet and Vincent Jullien, 33–48. Paris: ENS Éditions Fonteany-St. Cloud.

———. 2013b. Physiologie mécaniste et mouvement cardiaque: Hobbes, Harvey et Descartes. In *Lectures de Hobbes*, ed. Jauffrey Berthier, Nicolas Dubos, Arnaud Milanese, and Jean Terrel, 133–162. Paris: Ellipses.

———. 2015. Le traité d'optique. In *Thomas Hobbes, De l'Homme*, ed. Jean Terrel et al., 87–146. Paris: Vrin.

Milanese, Arnaud. 2011. *Principes de la philosophie chez Hobbes*. Paris: Classiques Garnier.

————. 2013. Philosophie première et philosophie de la nature. In *Lectures de Hobbes*, ed. Jauffrey Berthier, Nicolas Dubos, Arnaud Milanese, and Jean Terrel, 35–62. Paris: Ellipses.

————. 2016. Le matérialisme de Hobbes en question: la critique du matérialisme et la réception de Hobbes depuis l'École de Marbourg. In *Hobbes et le matérialisme*, ed. Jauffrey Berthier and Arnaud Milanese, 91–109. Paris: Éditions matériologiques.

Minerbi Belgrado, Anna. 1993. *Linguaggio e mondo in Hobbes*. Rome: Editori Riuniti.

Naylor, Ron. 2007. Galileo's Tidal Theory. *Isis* 98: 1–22.

————. 2014. Paolo Sarpi and the First Copernican Tidal's Theory. *British Journal for the History of Science* 47 (4): 661–675.

Pacchi, Arrigo. 1965. *Convenzione e ipotesi nella filosofia naturale di Thomas Hobbes*. Florence: La Nuova Italia.

————. 1987. Hobbes and the Passions. *Topoi* 6: 111–119.

————. 1998. *Scritti hobbesiani (1978–1990)*, ed. Agostino Lupoli. Milan: Franco Angeli.

Paganini, Gianni. 2003. Hobbes Among Ancient and Modern Sceptics: Phenomena and Bodies. In *The Return of Scepticism from Hobbes and Descartes to Bayle*, ed. Gianni Paganini, 3–35. Dordrecht/Boston/Leiden: Kluwer.

————. 2004. Hobbes e lo scetticismo continentale. In *Nuove prospettive critiche sul Leviatano di Hobbes*, ed. Luc Foisneau and George Wright, 303–328. Milan: Franco Angeli.

————. 2008. *Skepsis. Le débat des modernes sur le scepticisme*. Paris: Vrin.

————. 2010. Introduzione. In *Moto, luogo e tempo*, ed. Thomas Hobbes, 9–104. Turin: Utet.

————. 2015. Hobbes's Galilean Project. Its Philosophical and Theological Implications. *Oxford Studies in Early Modern Philosophy* 7: 1–46.

Palmerino, Carla Rita. 2002. The Mathematical Characters of Galileo's Book of Nature. In *The Book of Nature in Early and Modern History*, ed. Klaas Van Berkel and Arjo Vanderjagt, 27–44. Peeters: Leuvens.

————. 2010. The Geometrization of Motion: Galileo's Triangle of Speed and Its Various Transformations. *Early Science and Medicine* 15: 410–447.

Pastore, Stocchi. 1976–1986. Manlio. *Il periodo veneto di Galileo Galilei*. In *Storia della cultura veneta*, ed. Girolamo Arnaldi and Manlio Pastore Stocchi, vol. 4, T. 2, 37–66. Neri Pozza: Vicenza.

Pécharman, Martine. 1992. Le vocabulaire de l'être dans la philosophie première: *ens, esse, essentia*. In *Hobbes et son vocabulaire*, ed. Yves-Charles Zarka, 31–59. Paris: Vrin.

Piccolino, Marco. 2005. *Lo zufolo e la cicala. Divagazioni galileiane tra la scienza e la sua storia*. Bollati Boringhieri: Turin.

Prins, Jan. 1990. Hobbes and the School of Padua: Two Incompatible Approaches of Science. *Archiv für Geschichte der Philosophie* 72: 26–46.

Randall, John H. 1961. *The School of Padua and the Emergence of Modern Science*. Padua: Antenore.

Redondi, Pietro. 2009. *Galileo eretico*. Rome/Bari: Laterza (or. Ed. 1983).

Roux, Sophie. 1998. Le scepticisme et les hypothèses de la physique. *Revue de Synthèse* 4e s (2–3): 211–255.

Schmitt, Charles B. 2001. *Filosofia e scienza nel Rinascimento*. Florence: La Nuova Italia.

Schuhmann, Karl. 1992. Le vocabulaire de l'espace. In *Hobbes et son vocabulaire*, ed. Yves-Charles Zarka, 61–82. Paris: Vrin.

————. 1998. *Hobbes. Une chronique*. Paris: Vrin.

Shapin, Steven, and Simon Schaffer. 1985. *Leviathan and the Air-Pump*. Princeton: Princeton University Press.

Shea, William R. 1970. Galileo's Atomic Hypothesis. *Ambix* 17: 13–27.

————. 1974. *La rivoluzione intellettuale di Galileo*. Florence: Sansoni (or. Ed. 1972).

————. 1978. *Descartes as Critic of Galileo*. In *New Perspectives on Galileo*, ed. Robert E. Butts and Joseph C. Pitt, 139–159. Dordrecht/Boston: Reidel Publishing Company.

Sorell, Tom. 1986. *Hobbes*. London/New York: Routledge.

Sorell, Tom, G.A.J. Rogers, and Jill Kraye, eds. 2010. *Scientia in Early Modern Philosophy. Seventeenth-Century Thinkers on Demonstrative Knowledge from First Principles.* Dordrecht: Springer.

Sosio, Libero. 1995. Galileo Galilei e Paolo Sarpi. In *Galileo Galilei e la cultura veneziana. Atti del convegno di studio*, 269–311. Venice: Istituto Veneto di Lettere ed Arti.

Spragens, Thomas A. 1973. *The Politics of Motion. The World of Thomas Hobbes.* London: Croom Helm.

Stabile, Giorgio. 2002. Il concetto di esperienza in Galilei e nella scuola galileiana. In *Experientia. X Colloquio del Lessico Intellettuale Europeo*, ed. Marco Veneziani, 217–241. Florence: Olschki.

Strauss, Leo. 1952. *The Political Philosophy of Hobbes. Its Basis and Its Genesis.* Chicago/London: University of Chicago Press (or. Ed. 1948).

Terrel, Jean. 1994. *Hobbes. Matérialisme et politique.* Paris: Vrin.

———. 2009. Hobbes et Boyle: enjeux d'une polémique. In *La philosophie naturelle de Robert Boyle*, ed. Myriam Dennehy and Charles Ramond, 277–293. Paris: Vrin.

———. 2013. Hobbes: définition et rôle de l'expérience. In *Lectures de Hobbes*, ed. Jauffrey Berthier, Nicolas Dubos, Arnaud Milanese, and Jean Terrel, 66–83. Paris: Ellipses.

Torrini, Maurizio. 1993. Galileo copernicano. *Giornale critico della filosofia italiana* 13 (A. 72, Iss. 1, January–April): 26–42.

Tricaud, François. 1992. Le vocabulaire de la passion. In *Hobbes et son vocabulaire*, ed. Yves-Charles Zarka, 139–154. Paris: Vrin.

Watkins, John W.N. 1965. *Hobbes's System of Ideas.* London: Hutchinson.

Wisan, Winifred L. 1978. Galileo's Scientific Method: A Reexamination. In *New Perspectives on Galileo*, ed. Robert E. Butts and Joseph C. Pitt, 1–57. Dordrecht/Boston: Reidel Publishing Company.

Zarka, Yves-Charles. 1999. *La décision métaphysique de Hobbes.* Paris: Vrin (1st Ed.. 1987).

———. 2004. Liberté, nécessité, hasard: la théorie générale de l'événement chez Hobbes. In *Nuove prospettive critiche sul Leviatano di Hobbes*, ed. Luc Foisneau and George Wright, 249–262. Milan: Franco Angeli.

Chapter 3
Galileo's *Momentum* and Hobbes' *Conatus*

3.1 Inertial motion

Despite his inclination for the Geometrical Method and the mathematization of nature, the most markedly mathematical and geometrical works of Hobbes reveal his gaps of knowledge in these areas, earning him harsh criticism from British mathematicians and scientists.[1]

Nevertheless, physics (in the modern sense of the term) accounts for a significant part of Hobbesian thought and Galileo Galilei plays a decidedly important role here.

In the previous chapters, we broadly explored the roots of Hobbes' philosophy: the notions of body and movement, which reveal an evident Galilean legacy. This legacy emerges clearly when analyzing Hobbes' physics in more detail, since many of his physical concepts present evident conceptual analogies, and sometimes even lexical similarities, with Galileo's physics-related terms. *De motu, loco et tempore* is paradigmatic from this standpoint, since it enables us to grasp the distance between the physics of Thomas White, of an Aristotelian stamp, and that of Hobbes, who has learned from Galileo and absorbed his key concepts.

A number of interesting elements emerge when analyzing Hobbes' critique of the demonstrations contained in *nodus* VIII of White's dialogue I, entitled: *Aerem motu telluris circumferri*. The English priest puts forward the idea that the air that surrounds the terrestrial sphere, right up to the level of the lunar orbit, is a sort of effusion produced by the earth and, because of this, contains particles of earth that make it heavy. What is more, the fact that this part of the air is involved in the daily rotation of the terrestrial sphere can also be attributed to this heaviness.[2]

---

[1] On this aspect see primarily Jesseph 1999; Sergio 2001.

[2] See White 1642, 204 ff. For a biographical and intellectual profile of White see Southgate 1993 (on White's tendency to reconcile Aristotelianism and certain aspects of modern science, ibid., 5 ff.). On the circumstances surrounding the writing and publication of White's *De Mundo* see Jacquot and Jones 1973, 9–70. On the Erastian spirit shared by Hobbes and the 'Blackloists,' that

© Springer Nature Switzerland AG 2020
G. Baldin, *Hobbes and Galileo: Method, Matter and the Science of Motion*,
International Archives of the History of Ideas Archives internationales
d'histoire des idées 230, https://doi.org/10.1007/978-3-030-41414-6_3

Hobbes responds to this concept in Chapter 20 of *De motu, loco et tempore* and continues into the next chapter. He starts by directly criticizing the Aristotelian theory of 'natural places,' which he counters with a concept based on Archimedean principles.[3] Hobbes persists with his criticism of Aristotelian physical principles in the following paragraphs and, in paragraph eight, he expounds upon a number of fundamental concepts of modern science, which Galileo had introduced in his *Dialogue* and *Discourses*.

Thomas White had explored an intriguing issue: that is to say, whether or not a hypothetical sudden stop in the earth's rotation would cause the bodies on the surface of the earth to fall off. The response that he formulated was based on the principles of Aristotelian physics: once the cause of the phenomenon had stopped, its effect must also stop immediately, so that the air would stop at the same time as the terrestrial sphere.[4] To counter this, Hobbes develops a response based on the concepts of the Galilean science of motion and perfectly in keeping with the principles of modern physics:

> First, it is agreed that nothing begins to move of itself, but that every body has the motion of some adjacent body as the cause and start of its movement ... It follows from that [the body], if it starts to move of itself, it possessed everything necessary for motion, before it starts to move. Therefore, it must have moved previously. Consequently, it does not start to move for the first time, as it was supposed. For the same reason, it cannot be proved that anything that moves can acquire rest of its own accord. But, as in the case of motion in a resting body, so it is also true that in order for a moving body to rest, [the existence] of some external and contiguous agent is necessary.[5]

Two profoundly different conceptions regarding the nature of motion clearly emerge here. White's ideas are linked to the Aristotelian concept of motion, according to which movement needs a mover to maintain it.[6] On the other hand, Hobbes has fully understood the nature of the principle of inertia, as it was sketched out by Galileo in his *Dialogue*. The final part of the cited passage is enlightening: Hobbes

---

is, the English Catholics who revolved around White and wanted to abandon the realist faction to open separate talks with Cromwell, see Collins 2002, and Collins 2005, 139–141, and 178–180.

[3] According to Hobbes, the water of the sea or of a lake does not exercise any weight on an object on the bottom, but this is not because the object is in its natural place, but instead—according to Archimedean principles—due to pressure exercized on the mass of water by the object that has occupied its place. See *MLT*, XX, 3, 248–249.

[4] Ibid., XX, 8, 251–252.

[5] Ibid., 251–252, Eng. Trans. Hobbes 1976, 230, largely modified.

[6] On the Aristotelian idea that the persistence of movement requires a mover, see: Aristotle, *Physics*, VII, 241 b ff. Jean Bernhardt, who was the first to devote a short contribution to the presence of Aristotle physics in Hobbes, maintained that Hobbes' principle of conservation of movement is Aristotelian (see Bernhardt 1989b). Nevertheless, it seems that in this and many other aspects, Hobbes owes a debt to Galileo's physics and is also perfectly aware of the process of translating certain Aristotelian concepts in mechanistic terms, which is something that was undertaken by the Galileo and continued by Hobbes (in this regard, see the suggestion made by Spragens 1973, 53 ff.). On the presence of Aristotle and Aristotelianism in Hobbesian philosophy, see Leijenhorst 2002.

is convinced that a body maintains its state of rest or motion until an external and adjacent agent intervenes to change its state.

The concept of inertial motion is sketched by Galileo in various passages of the *Dialogue*. In the first day we find a brief reference to this issue:

> … In the horizontal plane no velocity whatever would ever be naturally acquired, since the body in this position will never move. But motion in a horizontal line which is tilted neither up nor down is circular motion about the center; therefore circular motion is never acquired naturally without straight motion to precede it; but, being once acquired, it will continue perpetually with uniform velocity.[7]

However, as has already been widely observed,[8] *Le Mecaniche* already contained a reference to inertial motion,[9] which was explained in greater detail in the subsequent *History and Demonstration concerning Sunspots (Istoria e dimostrazioni intorno alle macchie solari)*,[10] using the classic example of a ship: 'If a ship on a calm sea were to receive some impetus just once, it would move continuously around our globe without ever stopping; and if it were placed at rest, it would perpetually remain at rest.'[11] The analogy with the ship on the surface of the sea returns on the second day of his *Dialogue*:

> Therefore, a ship moving in a calm sea is a body going over a surface that slopes neither downward nor upward, and so it has the tendency to move endlessly and uniformly with the impulse once acquired if all accidental and external obstacles are removed.[12]

Nevertheless, it is worth noting that in the cited passages Galileo refers to motion that, although uniform, is applied to a spherical surface, which is to say he is referring to circular motion. Alexandre Koyré came up with a famous interpretation for this passage,[13] deeming the concept of uniform motion expressed by Galileo to imply the impossibility of applying the principle of inertia to rectilinear motion. In fact, according to Koyré, Galileo's argument refers exclusively to the real horizontal plane, considered as a spherical surface: the 'neither downward nor upward' sur-

---

[7] Galilei, *Dialogo sopra i due massimi sistemi del mondo*, *OG*, VII, 52–53; Eng. trans. Galilei 2001, 32.

[8] See Camerota 2004, 446 ff.

[9] See Galilei, *Le Mecaniche*, *OG*, II, 180

[10] As underscored by Bucciantini, the *Istoria* is the 'first' Galilean text on 'natural philosophy.' See Bucciantini 2003, 217.

[11] Galilei, *Istoria e dimostrazioni intorno alle macchie solari*, *OG*, V, 134; Eng. trans. Finocchiaro 2008, 98.

[12] Galilei, *Dialogo sopra i due massimi sistemi del mondo*, *OG*, VII, 174; Eng. trans. Finocchiaro, 232.

[13] See Koyré 1966, 229–230. Westfall also adopts a similar position to Koyré. See Westfall 1971, 18–19. Geymonat's position is different, maintaining that Galileo's failure to explicitly set out the principle of inertia in no way implies that this princple is completely lacking from Galilean physics. On the contrary, it is implicit in his discussions of astronomy in the *Dialogues* and in his writings on physics and mechanics in the *Discourses*. See Geymonat 1957, 322–326. In contrast to Koyré's interpretation, see also Drake, 'Galileo and the Concept of Inertia', in Drake 1970, 240–256; and Drake, 'The Case Against 'Circular Inertia''. Ibid., 257–278.

face, as Galileo describes that surface whose points are all equidistant from the center of the earth.[14]

Despite certain passages in the *Dialogue* seeming to confirm Koyré's theory, there are other points in the same work and, above all, in the *Discourses*, which also seem to suggest the application of an inertial concept to uniform rectilinear motion.[15] In greater detail, the short Latin treatise *De motu proiectorum*, in the opening part of the 'fourth day,' refers to the continuation of rectilinear motion *ad infinitum*,[16] while in the third day, in the treatise *De motu naturaliter accelerato*, the scientist seems to suggest that motion on an infinite horizontal plane would be *eternal* by its very nature.[17]

Independently of the debate surrounding Galileo's full and complete formulation of the principle of inertia, it is evident that Thomas Hobbes grasped the revolutionary potential of the concept of inertial motion,[18] offering numerous examples of its application in his *De motu, loco et tempore*:

> Moreover, very frequent experimental tests show that when the cause of the motion ceased, the effect however does not (the effect, I say, not only the work, for no-one can doubt that, though the painter dies the picture remains); even when the bow ceases [vibrating], the arrow continues to fly; when a ship hits the land, those standing on the deck fall forward immediately; when the gunpowder is consumed, the discharged ball flies onward. Why,

---

[14] Koyré's views are supported by certain passages from the *Dialogue*, such as the following exchange of quips between Salviati and Simplicio. Galilei, *Dialogo sopra i due massimi sistemi del mondo*, *OG*, VII, 173–174; Eng. trans. Galilei 2001, 170–171: '*Salv*. ... Now tell me, what do you consider to be the cause of the ball moving spontaneously on the downward inclined plane, but only by force on the one tilted upward? *Simp*. That the tendency of heavy bodies is to move toward the center of the earth, and to move upward from its circumference only with force; now the downward surface is that which gets closer to the center, while the upward one gets farther away. *Salv*. Then in order for a surface to be neither downward nor upward, all its parts must be equally distant from the center. Are there any such surfaces in the world? *Simp*. Plenty of them; such would be the surface of our terrestrial globe if it were smooth, and not rough and mountainous as it is. But there is that of the water, when it is placid and tranquil.'

[15] See, for example: Galilei, *Dialogo sopra i due massimi sistemi del mondo*, *OG*, VII, 201 and 220. On the inertial concept of motion in Galileo see also: Hooper 1996, esp. 157 ff., where the author discusses Koyré's theory. See also Camerota 2004, 448.

[16] Galilei, *Discorsi e dimostrazioni intorno a due nuove scienze*, *OG*, VIII, 268: '*Mobile quoddam super planum horizontale proiectum mente concipio, omni secluso impedimento: iam constat, ex his quae fusius alibi dicta sunt, illius motum aequabilem et perpetuum super ipso plano futurum esse, si planum in infinitum extendatur.*'

[17] Ibid., 243: '*Attendere insuper licet, quod velocitatis gradus, quicunque in mobili reperiatur, est in illo suapte natura indelebiliter impressus, dum externae causae accelerationis aut retardationis tollantur, quod in solo horizontali plano contingit, nam in planis declivibus adest iam causa accelerationis maioris, in acclivibus vero retardationis: ex quo pariter sequitur, motum in horizontali esse quoque aeternum, si enim est aequabilis, non debilitatur aut remittitur, et multo minus tollitur.*'

[18] The importance of the Galilean concept of inertia in Hobbes was first identified by Spragens, although he did not go into the matter in any great depth. See Spragens 1973, 60 ff.

then, when the Earth stops moving, should the air not contnue its movement if nothing else hinders it?[19]

It is also interesting to observe that the examples adopted by Hobbes are the same as those presented by Galileo in his *Dialogue Concerning the Two Chief World Systems*: the *arrow*, the *ship* and the *cannon ball*.[20]

In the following chapter, Hobbes also analyzes the phenomenon of the tilted plane, which is already included in the *Dialogue*, but acquires fundamental importance in the *Discourses and Mathematical Demonstrations Relating to Two New Sciences*. White criticized Galileo 'for having considered that the motion of the sphere on a plane will continue perpetually, once it has started,'[21] and Hobbes observes that White, just like the Aristotelian Simplicio, failed to understand the nature of the principle of inertia:

> Indeed, he [Galileo] really thinks that any movement, once started, is destined to continue perpetually, unless impeded by another motor moving in the opposite direction. This, I say, is that which our author criticizes, although he does not offer any reasoning to the contrary: he simply says that he cannot understand an impressed quality of this kind any more than Simplicio. Nevertheless, I believe that understanding that a thing once moved moves forever, until some body pushing on it from the opposite direction stops it, is no more difficult than understanding that an object at rest always remains in that state, until some body pushes it out of its place.[22]

White failed to grasp the Galilean revolution regarding the nature of motion, because he maintains, in agreement with Aristotle, that the continuation of movement requires the continuous application of a mover (or agent).[23] The English Catholic priest stuck to his Aristotelian principles and never assimilated the founding principles of the new and revolutionary concept of movement theorized by Galileo, which was instead absorbed and developed by Hobbes. Indeed, while

---

[19] See *MLT*, XX, 8, 252, Eng. Trans. Hobbes 1976, 231.

[20] As Pierre Duhem had already emphasized, some of these examples were present in texts by fourteenth-century Parisian natural philosophers, especially the *Livre du Ciel et du Monde* by Nicole Oresme. See Oresme 1968, 519 ff. See Crombie 1953, 255 ff.

[21] *MLT*, XXI, 13, 262.

[22] Ibid.

[23] Here, as elsewhere, White keeps faith with his Aristotelian principles. His arguments against the existence of the vacuum are also based on Aristotelian and scholastic philosophy. Basing his theory on the Aristotelian definition of place, defined as an immobile surface of the body (Aristotle, *Physics*, IV, 212a, 20), he denied both the possibility of the vacuum and the plurality of worlds. In fact, if there were multiple worlds, we would be faced with the problem of justifying the vacuum *inter mundia*. See White 1642, 29 ff. White's theory particularly echoes an argument linked to the theme of the *annihilatio mundi*, found in Jean Buridan, as in other fourteenth-century Parisian natural philosophers and some Mertonians. When reasoning upon issues concerning the *potentia dei absoluta*, medieval thinkers theorized about a miraculous intervention of God that would annihilate the cosmos, asking themselves whether empty space could be conceived as a *place*. On this subject see Grant 1981, esp. 86 ff.; Grant 1996a, b, 82–83; Funkenstein 1986, 117 ff.; Parodi 1981, 201 ff.; Bianchi 1997, 269–303. On the debate surrounding the vacuum during the Renaissance (esp. in Patrizi and Telesio) see: Schmitt, 'Prove sperimentali sull'esistenza del vuoto'. In Schmitt 2001, 65–83.

Galileo may not have produced a precise and complete definition of inertia, the passage cited from *De motu, loco et tempore* shows us that Hobbes had grasped the revolutionary potential of the concept of inertial motion with respect to the Aristotelian tradition.

In reality, even before the publication of Koyré's study (1939), Frithiof Brandt, in 1921, noted Hobbes' failure to define the principle of inertia in *De corpore*, and also added that Hobbes seemed to apply a concept of inertial movement solely to circular motion.[24] Although we can agree with Brandt regarding the absence of an explicit formulation of the principle of inertia in Hobbes' works (as in those by Galileo), nevertheless Hobbes places inertial motion at the foundation of his mechanistic system of nature, as clearly emerges in the manuscript *De Principiis*.[25]

As Paganini has observed, in Hobbes the principle of inertia is closely tied to another fundamental principle of his mechanics, according to which no body can move independently, but receives the impulse to move from outside (from another moving and adjacent body).[26] As a result, 'inertia takes the form of a principle of sufficient reason, in which the Hobbesian notion of *causality* is reflected.'[27]

This primarily emerges in *De corpore*, which also goes back over issues that had already been discussed in *De motu, loco et tempore*, adding further considerations. Firstly, in Chapter 8, Hobbes claims that a body maintains its state of rest, or of motion, until an external agent intervenes to change its state,[28] and in the next chapter, when exploring the cause of movement, which 'will be some external body,'[29] he finally reaches a fairly clear and articulate explanation of inertial motion:

> Seeing we may conceive that whatsoever is at rest will still be at rest, though it be touched by some other body, except that other body be moved; therefore in a contiguous body, which is at rest, there can be no cause of motion. Wherefore there is no cause of motion in any body, except it be contiguous and moved. The same reason may serve to prove that whatsoever is moved, will always be moved on in the same way and with the same velocity, except it be hindered by some other contiguous and moved body; and consequently that no bodies, either when they are at rest, or when there is an interposition of vacuum, can generate or extinguish or lessen motion in other bodies.[30]

---

[24] See Brandt 1928, 328. As regards the principle of inertia associated with circular motion, Mintz 1952 maintains that Hobbes absorbed a circular concept of inertial motion by reading Galilean texts. Henry has returned to this subject recently, underscoring the analogies between certain passages in Galileo's *Dialogue* and Hobbes' simple circular motion. See Henry 2016, esp. 13 ff.

[25] *De Principiis* (National Library of Wales, Ms 5297), *MLT*, Appendix II, 457: '*Quod quiescit nisi ab externo moveatur semper quiescet. Quod movetur nisi ab externo impediatur semper movebitur.*'

[26] See Paganini 2010, 34–35.

[27] Ibid. On the importance of the concept of *cause* in Hobbes see *supra*, chap. 2, § 7.

[28] See *De corpore*, VIII, 19, *OL*, I, 102.

[29] Ibid., IX, 7, 110; Eng. Trans. *EW*, I, 124.

[30] Ibid., 110–111; Eng. Trans. 125. As he continues, while not referring to it directly, Hobbes goes back to and criticizes White's opinion, according to which rest is 'more opposed' to movement, than movement is opposed to the the rest.

In Chapter 15, Hobbes returns to the problem of the mover, which he had already covered in his objection to White, insisting upon the idea: *'quod cessatio moventis non cogit cessare id quod ab ipso motum est.'*[31] What is more, the same chapter also includes additional interesting observations about the principle of inertia, in which a fundamental concept in Hobbesian physics emerges: *conatus*.

3.2 Inertia and *Conatus*

In Chapter 15 of *De corpore*, Hobbes maintains that in an empty space and in the absence of external impediments, a *conatus* is propagated infinitely:

> All endeavour (Lat. *Conatus*), whether strong or weak, is propagated to infinite distance; for its motion. If therefore the first endeavour of a body be made in space which is empty, it will always proceed with the same velocity; for it cannot be supposed that it can receive any resistance at all from empty space; and therefore, ... it will always proceed in the same way and with the same swiftness.[32]

However, he maintains that *conatus* is also propagated infinitely in the plenum and states that this movement is transmitted instantaneously.[33]

> ...therefore all endeavour, whether it be in empty or in full space, proceeds not only to any distance, how great soever, but also in any time, how little soever, that is in an instant, Nor makes it any matter, that endeavour, by proceeding, grows weaker and weaker, till at last it can no longer be perceived by sense; for motion may be insensible; and I do not here examine things by sense and experience, but by reason.[34]

Hobbes' argument regarding the transmission of the *conatus* is actually problematic: he claims that motion within an empty space is propagated due to the absence of external impediments that oppose resistance; yet, on the other hand, he maintains that the same phenomenon occurs in a plenum, because 'seeing endeavour is motion, that which stands next in its way shall be removed, and endeavour further, and again remove that which stands next, and so infinitely.'[35]

Hobbes' reflection is in no way perspicuous. He proposes a theory of infinite propagation in a plenum, openly contradicting the explanation he had given for the same phenomenon considered in an empty space. In the first case, the air is deemed to be an external impediment that offers resistance and whose absence permits propagation *ad infinitum*. Yet in the second case, the same air seems to have a propulsive

---

[31] Ibid., XV, 3, 180.

[32] Ibid., XV, 7, pp. 182–183; Eng. trans. 216.

[33] Ibid., 183: 'Et siquidem in pleno fiat, tamen cum conatus sit motus, id quod in via ejus proxime obstat removebitur, et conabitur ulterius, et hujus conatus removebit rursus id, quod sibi proxime obstat, et sic in infinitum. Pertingit etiam ad distantiam quantamcumque in instante; nam eodem instante, quo prima pars medii pleni removet partem sibi proximam, pars secunda partem rursus sibi proximam obstantem removet.'

[34] Ibid. Eng. Trans. *EW*, I, 217.

[35] Ibid. *EW*, I, 216

function and becomes the means that allows the *conatus* to be propagated. Another difficulty lies in the fact that Hobbes also supposes that this propagation is instantaneous.

Hobbes must have realized the problems associated with his arguments and, in the final part of the paragraph, he includes a cautionary specification, stating that he is aware that motion in a plenum abates due to the resistance put up by air and peremptorily claiming that he does not want to discuss empirical phenomena, but simply physical properties that can be demonstrated 'by reason.'[36] Nevertheless, even when keeping exclusively to an abstract physical model, the argument still seems to contain a paralogism that is difficult to avoid.

The difficulty encountered by Hobbes in this passage conceals the problematic nature of justifying a series of correlated physical phenomena, which he intended to tackle and which he believed to be linked to the concept of *conatus*.

Meanwhile, as regards the case of movement within a void, it should be emphasized that Pierre Gassendi, in his *Epistolae de motu*, had considered the movement of a stone in empty space and this had led Koyré to suppose that Gassendi was the first to correctly formulate the principle of inertia, of which the void constitutes a fundamental aspect:[37]

> You might ask what happen to that stone which I have imagined to be located in the void space, if it was removed from the rest by being pushed by a certain force? I answer that it would probably *move uniformly and indefinitely*, slow or swifty according to whether the impressed impetus was small or big. I infer this conclusion from the already mentioned uniformity of horizontal motion, which only comes to an end because it is mixed with perpendicular motion; and given that in those [empty] spaces there is no mixture with perpendicular motion, in whatever direction the motion is oriented it will remain horizontal and it would not accelerate, nor decelerate, nor would it come to an end.[38]

---

[36] Ibid.: 'I do not here examine things by sense and experience, but by reason' The position expressed by Hobbes in this passage recalls some of the declarations made by Galileo, who stated that the demonstrations 'regarding my supposition lose nothing of their strength and conclusiveness; just as the demonstration of Archimedes is not diminished by the fact that in nature no moving body is found that moves in spiral lines.' (Galileo to Giovan Battista Baliani, January 7, 1639, OG, XVIII, 12–13, my translation).

[37] See Koyré 1966, 314–317. Nevertheless, Carla Rita Palmerino has claimed, on the contrary, that in Gassendi the abovementioned principle is far from being formulated correctly and explicitly. See Palmerino 2004, esp. 150–151, and Palmerino 2008, esp. 160 ff. Paolo Galluzzi highlighted the central role played by Gassendi in the circulation of the Galilean laws of motion, so much so that his *Epistolae de motu* led to a second '*affaire Galilée.*' See Galluzzi 1993, 86–119. Generally, on the links between the physics of Galileo and Gassendi, see Bloch 1971, 189–194; Festa 1999.

[38] Gassendi, *Epistolae tres. De motu impresso a motore translato*. In Gassendi 1658, III, 495b; Eng. Trans. in Palmerino 2004, 150: 'Quaeres obiter, quidnam eveniret illi lapidi, quem assumpsi concipi posse in spatiis illis inanibus, si a quiete exturbatus aliqua vi impelletur? Respondeo probabile esse, fore, ut aequabiliter, indefinenterque moveretur; et lente quidem, celeriterve, prout semel parvus, aut magnus impressus foret impetus. Argumentum vero desumo ex aequabilitate illa motus horizontalis iam exposita; cum ille videatur aliunde non desinere, nisi ex admistione motus perpendicularis; adeo, ut quia in illis spatiis nulla esset perpendicularis admistio, in quacumque partem foret motus inceptus, horizontalis instar esset, et neque acceleraretur, retardereturve, neque proinde unquam desineret.'

As highlighted by Leijenhorst, Hobbes first came into contact with Gassendi's work in 1640,[39] and it seems likely that he formulated his concept of inertial motion in the empty space on the basis of the speculations put forward by his friend.[40]

Meanwhile, as regards the propagation of the *conatus* in a plenum, Hobbes' handling of the argument is based on the speculations and observations associated with light. Indeed, he maintained that light is propagated instantly, but that it also necessarily implies the existence of a medium. In his *Tractatus Opticus I*, when discussing the instantaneous propagation of light, Hobbes put forward considerations that represent a sort of prelude to his discussion of the conatus in *De corpore*. Here he makes explicit reference to *conatus* and the instantaneous propagation of light through the medium.[41]

However, the Hobbesian concept of the propagation of *conatus* in a plenum is also influenced by Gassendi, as we will have the opportunity to see later on, when exploring the subject of gravity.

3.3 *Momentum* and *Conatus*

The exploration of Hobbes' treatment of inertial motion brings out a fundamental term in Hobbesian physics, which is central to Hobbes' entire philosophy too: the notion of *conatus*. The significant presence of this word in Hobbes' work leads to an in-depth analysis of this concept, which has a number of profound analogies with a term specific to Galilean physics: *momento* or *momentum*.[42]

Brandt was the first to emphasize that the concept of *conatus* is already present in the *Elements of Law*, in the English term 'endeavour.'[43] In this first occurrence,

---

[39] Gassendi, accused of plagiarism by Jean-Baptiste Morin (who claimed that some of the speculations in the *Epistola de motu*, published in 1642, had been taken from one of his own works printed the previous year), called up Hobbes as his witness in his response, specifying that the English philosopher had been aware of his ideas since 1640. See Gassendi, 'Epistola III. In librum qui a viro Cl. Ioanne Morino'. In Gassendi 1658, III, 521b. See Leijenhorst 2004, 169.

[40] Lisa Sarasohn has claimed that the development of a 'mechanical world-view' in Hobbes was determined by the ideas of Pierre Gassendi, with whom the English philosopher came into contact during his Grand Tour. See Sarasohn 1985, esp. 368. What is more, the author supposes that a reciprocal partnership developed between the two philosophers in the 1640s, especially in the field of psychology and physics (ibid., 370–371). However, Paganini produced a detailed analysis of the relationship between Hobbes and Gassendi as regards psychology (see Paganini 1990) and physics (Paganini 2004b, 2008).

[41] *TO I, OL*, V, 219–220. See also Prins 1996, 133 ff.

[42] Leijenhorst can be credited with having identified the influence of Galilean momentum on Hobbes' *conatus*. He produced an interesting and in-depth study dedicated to the reception of the Galilean law of falling heavy objects in Hobbes' natural philosophy. See Leijenhorst 2004, 182–184.

[43] See Brandt 1928, 299–300, which draws upon some of Lasswitz's analyses (see Lasswitz 1963, 214 ff.), who was the first to analyze the importance of the concept of *conatus* in Hobbes, comparing it to the earlier cogitations of Galileo and the later ones of Leibniz and Newton. A reference to

the term is used to express the meaning of 'appetite,' understood as impulse, or first internal movement, targeted at the object that attracts or repels.[44] Even in this case, the terminology adopted by Hobbes refers to movement and this *conatus* (whether appetite or aversion) is explicitly defined as motion (more precisely, as the internal principle of animal motion). What is more, Brandt emphasized that the recurrence of the term *conatus* in physical explorations (in *De corpore*) is not completely disconnected from this initial meaning of the term, as they all have the dynamic nature of the concept in common.[45]

De motu, loco et tempore devotes an entire paragraph of Chapter 13 to the examination of *conatus*,[46] and this discussion features two particularly significant aspects. Firstly, Hobbes questions the position of those who do not believe conatus to be a movement, but instead a tendency to motion present in bodies.

These observations should be considered in the light of the epistolary controversy that unfolded between Hobbes and Descartes in 1641. Hobbes disputed the Cartesian concept of 'inclination towards movement,' present in the *Dioptrique*,[47] instead proposing a radically different position, according to which: 'every principle of motion is motion.'[48] This eminently physical meaning of movement already featured in the *Tractatus Opticus I*, where it was used to describe the phenomenon of vision.[49] The term *conatus* recurred here with the meaning of reception and propagation of a motion or impulse, which reaches the retina and is transmitted by it to the optical nerve. Here too, the principle of movement is always a motion and, as we have seen in relation to inertial motion, Hobbes believed it to be completely impossible for movement to begin without the action of a motor, that is to say without an external and adjacent body.

However, it is in *De motu, loco et tempore* that Hobbes specifically defines *conatus* as a motion, either of the entire body, or solely of the internal parts that comprise it.

---

this can also be noted in Malherbe 1984, 103–109. The concept of *conatus* was studied by Robinet 1990, and especially by Barnouw 1992. More recently, the topic has been returned to in Bertman 2001, Lupoli 2001, Pietarinen 2001, and Jesseph 2016.

[44] *EL*, Part I, chap. VII, § 2, 28: 'This motion, in which consisteth pleasure or pain, is also a solicitation or provocation either to draw near to the thing that pleaseth, or to retire from the thing that displeaseth. And this solicitation is the endeavour or internal beginning of animal motion, which when the object delighteeth, is called APPETITE; when it displeaseth it is called AVERSION'. On the meaning of *conatus* in reference to the phenomenology of sensation, see Barnouw 1990, 116.

[45] See Brandt 1928, 299–300.

[46] See *MLT*, XIII, 2, 194–195.

[47] See Descartes, *La Dioptrique*, in *AT*, VI, 88. On the analysis of the notion of the determination of motion and other fundamental concepts in Descartes' physics, see McLaughlin 2000, esp. 88 ff. For a detailed analysis of Descartes' physics see Shea 1991 (on inclination, 232 ff.); Gaukroger 2002, which, however, focuses almost entirely on the *Principia*. On the connection between physics and mechanics in Descartes see Roux 2004.

[48] See Descartes, *AT*, III, 316. See also Brandt 1928, 301; Baldin 2017.

[49] *TO I*, *OL*, V, 220: '… unde propagabitur motus ad retinam, et inde per *conatum* retinae ad nervum opticus usque ad cerebrum'.

So, *conatus* is nothing but an actual motion, either of the whole body that tends, or of its inner and invisible parts. It is therefore motion in the internal parts of all hard bodies, whose visible parts adhere to one another and resist the agent, enabling us to deduce that all resistance is movement: in fact, resistance is a reaction, reaction is an action and every action is movement.[50]

As he continues, Hobbes recalls a peculiar aspect of his physics and his concept of matter, which he had sketched out in his correspondence with Descartes. He goes back to the issue of the 'return of the bow,'[51] supposing that the phenomenon of falling objects and that of the bow are to be attributed to the same physical principles.[52]

The juxtaposition of the two phenomena, which Hobbes explains by resorting to the same principles, will be discussed later on. However, the most significant element here concerns the reference to conatus as movement of the internal and invisible parts of the body. This indication is important, because in *De corpore* Hobbes will provide a number of definitions of *conatus* that are not immediately clear. A comparison of passages in both works is fundamental when attempting to formulate a comprehensive interpretation.

In Chapter 15 of *De corpore*, conatus is defined as '*motion made in less space and time than can be given; that is, less than can be determined or assigned by exposition or number; that is, motion made through the length of a point*'[53] and this definition is taken up again in Chapter 22, where Hobbes puts forward the same idea again.[54] Aldo Gargani, referring back to the observations made by Lasswitz and Brandt,[55] emphasized that Hobbes' works never clarify the concept of the infinitesimal (one of the founding elements of infinitesimal calculus), but nonetheless he deems this concept to be at the origin of the meaning of the expression used in *De corpore* to indicate *conatus*: '*motum per spatium et tempus minus quam quod datur.*'[56] Gargani also considered the central nature of the term *conatus* in reference to physics and motion, which, while not mathematically formulated by Hobbes, still serves to designate the phenomenon of movement in relation to the infinitesimal parts of space and time. On the other hand, lacking a mathematical formulation, the Hobbesian concept of conatus could not constitute the foundation of a scientific dynamic.[57] Gargani referred to the Galilean origin of the concept and drew directly upon the speculations present in *Discourses and Mathematical Demonstrations Relating to Two New Sciences*, where Salviati discusses the infinite divisibility of

---

[50] *MLT*, XIII, 2, 195 (my trans.).

[51] See Baldin 2017.

[52] *MLT*, XIII, 2, 194–195.

[53] *De corpore*, XV, 2, *OL*, I, 177; Eng. trans. Hobbes 1976, 206. The English translation also adds 'and in an instant or point of time.'

[54] *De corpore*, XXII, 1, *OL*, I, 271: 'CONATUM definivimus supra cap. xv, art. 2, esse motum per longitudinem quidem aliquam, consideratam autem non ut longitudinem, sed ut punctum. Itaque sive quid conanti resistit, sive nil resistit, conatus tamen idem est'.

[55] See: Lasswitz 1963, II, 214; Brandt 1928, 301 ff.

[56] See Gargani 1971, 228.

[57] Ibid. See also Grant 1996a, b, 117.

parts, which, although not expressed by an infinite number (which is non-existent), can still be conceived as a sort of process, where every part is always infinitely divisible.[58]

Nevertheless, despite the accuracy of Gargani's observations, the Hobbesian concept of *conatus* encompasses a range of physical questions that are not exhausted solely in this problem. This semantic plurality primarily emerges in *De corpore*, where Hobbes reflects on a number of 'Galilean' problems. In Chapter 23 of the work, where he takes up some of the themes developed by Galileo in *Le Mecaniche* and the *Discourses*, Hobbes (when discussing scales) defines equilibrium and weight in these terms:

> II. *Equiponderation* (lat. *Æquilibrium*) is when the endeavour of one body, which presses one of the beams, resists the endeavour of another body pressing the other beam, so that neither of them is moved; and the bodies, when neither of them is moved, are said to be *equally posed*.
>
> III. *Weight* (lat. *Pondus*) is the aggregate of all the endeavours, by which all the points of that body, which presses the beam, tend downwards in lines parallel to one another; and the body which presses is called the *ponderant*.[59]

Hobbes uses the term *conatus* with different, albeit linked meanings. This conceptual polysemy for the word and its use in reference to scales leads us to focus our attention on the Galilean concept of *momentum*. In fact, Cees Leijenhorst emphasized that the term *conatus*, present in Chapter 15 of *De corpore*, is clearly inspired by the Galilean concept of *momentum*.[60]

The Galilean term *momentum* was the object of an extensive and well-documented essay by Paolo Galluzzi, who also showed that the term already had a consolidated tradition at the time of Galileo. He primarily used it to mean *instant of time*, but in the sixteenth century it was widely present in translations of Aristotle's *Physics*, where it had already acquired the connotations that would be drawn upon and reworked by Galileo: that of inclination towards movement.[61]

Just like Galileo, Hobbes also attributed a twofold meaning to the concept of *conatus*. On the one hand, he assigned *conatus* the Galilean meaning of *momentum velocitatis*, that is to say an infinitesimal degree of speed. However, at the same time, he also invoked the other meaning of Galilean *momentum*: that of the inclination or tendency for motion.[62]

---

[58] See Galilei, *Discorsi e dimostrazioni intorno a due nuove scienze*, *OG*, VIII, 81 ff. The subject of the infinite division of matter, featured in his *Discourses,* will be discussed in depth in the next chapter.

[59] *De corpore*, XXIII, 1, *OL*, I, 287. Engl. Transl. *EW*, I, 351.

[60] See Leijenhorst 2004, 182.

[61] See Galluzzi 1979a, b, 89 ff.

[62] See Leijenhorst 2004, 183. Leijenhorst had already made some interesting observations on this subject in his monograph dedicated to the links between Hobbes' natural philosophy and the Aristotelian tradition. See Leijenhorst 2002, 188 ff.

The first definition of *conatus* is directly linked to considerations on infinitesimal size, also explored by Gargani, while the second expresses other meanings of Galilean *momentum* that we shall be analyzing.

A number of interesting considerations can be made on the basis of a brief analysis of the Galilean texts known to Hobbes. The first work to attract our attention, because this is the first to contain the word *momentum*, is *Le Mecaniche*. Before developing the principal argument relating to the physical principles underlying the operation of machines, here Galileo presents the definitions of 'gravity' and '*momentum*,' using the first term to mean 'that propensity for naturally downward motion that, in solid bodies, is brought about by the greater or lesser copy of matter, out of which they are constituted.'[63] The meaning of *momentum* is closely connected to the first word, and Galileo explains it in detail, referring to the 'propensity to move downward, caused not so much by the gravity of the moving body, as by the disposition that several falling bodies have with each other; through which movement we will often see a lighter body counterbalance another heavier body.'[64] What is more, Galileo makes use of the example of the 'lever scales' (*stadera*), in which 'one sees a small counterweight lift another very large weight not due to excess gravity, but because of the distance from the point where the lever scales are supported; this, combined with the gravity of the lesser weight, increases downward *momentum* and *impetus*, which can exceed the *momentum* of the other heavier weight.'[65] Following this example, we find a clearer and more succinct definition of *momentum*, described as follows: '*that downward impetus, composed of gravity, position and more, which causes this propensity.*'[66]

The considerations that Galileo goes on to develop are equally important, because they focus on the same problems explored by Hobbes. The Italian scientist defines the *center of gravity*, where the concepts of *momentum* and *impetus* play an essential part, since 'the center of gravity in every heavy body is defined as that point around which parts of equal *momenta* are to be found.'[67]

It is fundamental to emphasize here the central role of the concept of *momentum of gravity* and the relationship established in the Galilean text between *momentum* and *impetus*, which will have a profound and decisive influence on Hobbesian physics. As Galluzzi emphasized, *impetus* seems to physically express the consequences

---

[63] Galilei, *Le Mecaniche*, *OG*, II, 159.

[64] Ibid.: '... propensione ad andare al basso, cagionata non tanto dalla gravità del mobile, quanto dalla disposizione che abbino tra di loro diversi corpi gravi; mediante il qual momento si vedrà molte volte un corpo men grave contrapesare un altro di maggior gravità.'

[65] Ibid.

[66] Ibid.

[67] Ibid.: 'Centro della gravità si diffinisce essere in ogni corpo grave quel punto, intorno al quale consistono parti di eguali momenti: sì che, imaginandoci tale grave essere dal detto punto sospeso e sostenuto, le parti destre equilibreranno le sinistre, le anteriori e le posteriori, e quelle di sopra quelle di sotto; sì che il detto grave, così sostenuto, non inclinerà da parte alcuna, ma, collocato in qual si voglia sito e disposizione, purché sospeso dal detto centro, rimarrà saldo. E questo è quel punto, il quale andrebbe ad unirsi col centro universale delle cose gravi, cioè con quello della terra, quando in qualche mezzo libero potesse discendervi.'

and tangible effects of the variation in the efficacy of the weight, recorded abstractly by the term *momentum*. At the same time, the *impetus* also represents the dynamic effect of the *momentum*, which if not expressed in the *conatus* would otherwise continue to indicate a 'propensity' only, that is, an abstract condition.[68]

3.4   *Le Mecaniche* and Hobbes' Galilean Physics

Before directly exploring the influence of these reflections on Hobbesian physics, particularly the connection between *momentum* and *impetus*, it would be advisable to turn our attention to Mersenne's translation of *Le Mecaniche*. The French text is simplified compared to the original and the French friar's struggle to grasp some-times obscure Galilean concepts, particularly that of *momentum*, is apparent. On the other hand, it should be observed that the periphrases chosen by Mersenne to express the Galilean terms may sometimes lead to misinterpretations of Galileo's physical concepts.[69]

We should start by noting that the Italian term '*gravità*' is rendered in French with '*pesanteur*,'[70] although the clarification added by Mersenne later on is well articulated, accurately specifying that the notion of gravity is linked to mass and specific weight. Secondly, propensity is rendered with the French term *inclination*, a word that we also find in Descartes. Lastly, Mersenne, in his definition of '*moment*' as '*inclination ... composée de la pesanteur absoluë du corps, de l'éloignement du centre de la balance, ou de l'appuy du levier*,' subtly modifies the Galilean defini-tion in order to avoid further interpretation of issues determined by the original text, in which momentum is defined as 'that downward impetus, composed of gravity, position and *more*,' where it is not immediately apparent to what this 'more' refers.[71] Mersenne's use of *centre de pesanteur*, which obviously renders the Italian '*centro*

---

[68] See Galluzzi 1979a, b, 205.

[69] On this subject see Shea 1977.

[70] Mersenne 1634. Now in Mersenne 1985, 443 : 'La *pesanteur* d'un corps est l'inclination naturelle qu'il a pour se mouvoir, et se porter en bas vers le centre de la terre. Cette pesanteur se rencontre dans le corps pesans à raison de la quantité des parties materielles, dont ils sont composez; de sorte qu'ils sont d'autant plus pesans qu'ils ont une plus grande quantité desdites parties souz une mesme volume.'

[71] Ibid., 444: 'Le *moment est l'inclination du mesme corps*, lors qu'elle n'est pas seulement consid-erée dans ledit corps, mais coinjonctement avec la situation qu'il a sur le bras d'un levier, ou d'une balance; et cette situation fait qu'il contrepese souvent à un plus grand poids, à raison de sa plus grande distance d'avec le centre de la balance. Car cet éloignement estant joint à la propre pesan-teur du corps pesant, luy donne une plus forte inclination à descendre: de sorte que cette inclination est composée de la pesanteur absoluë du corps, de l'éloignement du centre de la balance, ou de l'appuy du levier. Nous appellerons donc toujours cette inclination composée, *moment*, qui répond au 'ροπή des Grecs.'

della gravità,' also simplifies the original text, because here the term *momentum* is actually suppressed.[72]

However, the semantic plurality of the term *momentum* already emerges in *Le Mecaniche*, where it is interpreted as a mechanical *momentum* with two interdependent meanings: it is both downward *impetus*—therefore connected to gravity—but it also features in the sense of the relationship between distance and weight in the scales or lever scales.

If we focus on Chapter 23 of *De corpore*, with its significant title: *De centro aequilibri eorum quae premunt secundum rectas parallelas,*[73] we can see that the discussion opens with the issue of the *scales* and, just like Galileo in *Le Mecaniche*, Hobbes provides a list of the definitions that he intends to illustrate, revealing significant analogies with Galileo's lexicon and speculations.

Firstly, like Galileo, Hobbes describes the scales, which consist of a 'strait line, whose middle point is immovable, all the rest of its point being at liberty; and that part of the scale, which reaches from the centre to either of the weights, is called the *beam.*'[74] By examining the scale, Hobbes sets out the physical concepts that Galileo had already outlined in *Le Mecaniche*: '*A Equiponderation* (Lat. *Æquilibrium*) is when the endeavour of one body, which presses one of the beams, resists the endeavour of another body pressing the other beam, so that neither of them is moved; and the bodies, when neither of them is moved, are said to be *equally posed.*'[75] The concept of *weight* (*pondus*) is very important. It is defined as 'the aggregate of all the endeavours, by which all the points of that body, which presses the beam, tend downwards in lines parallel to one another; and the body which presses is called the *ponderant.*'[76] All the other definitions reveal significant similarities to Galilean language, with evident interdependence between the concepts of *momentum, pondus* and, obviously, *conatus*:

IV. *Moment* (*Momentum*) is the power which the ponderant has to move the beam, by reason of a determined situation.

V. *The plane of equiponderation* (*Planum aequilibri*) is that by which the ponderant is so divided, that the moments on both sides remain equal.

VI. The *diameter of equiponderation* (*Diameter aequilibri*) is the common section of the two planes of equiponderation.[77]

---

[72] Ibid.: 'Le centre de pesanteur de chaque corps est le point autour duquel toutes les parties dudit corps son également balancées, ou équiponderantes: de sorte que si l'on s'imagine que le corps soit soustenu, ou suspendu par ledit point, les parties qui sont à main droite, contrepeseront à celles de la gauche, celles de derriere à celles de devant, et celles d'en haut à celles d'en bas, et se tiendront tellement en équilibre, que le corps ne s'inclinera d'un costé ni d'autre, quelque situation qu'on luy puisse donner, et qu'il demeurera tousjours en cet estat. Or le centre de pesanteur est le point du corps qui s'uniroit au centre des choses pesantes, c'est-à-dire au centre de la terre, s'il y pouvoit descendre.'

[73] *De corpore*, XXIII, 1, *OL*, I, 286.

[74] Ibid.; Eng. trans. *EW*, I, 351.

[75] Ibid.

[76] Ibid.

[77] The English Translation also adds 'and is in the strait line by which the weight is hanged.' Ibid., 352.

VII. The *center of equiponderation* (*Centrum aequilibri*) is the common point of the two diameters of equiponderation.[78]

The analogies with the text of *Le Meccaniche* are more than evident, although we should underscore a fundamental difference: Hobbes only uses the term *momentum* to indicate one of the two meanings that recur in the Galilean text. He only uses the word *momentum* (*moment* in the English edition) to indicate the relationship between weight and the length of the arm of the scales and not with the meaning of *impetus*.

Indeed, on the following page, Hobbes defines proportionality between *momenta* in the scales as follows:

The moment (*momentum*) of any ponderant applied to one point of the beam, to the moment of the same or an equal ponderant applied to any other point of the beam, is as the distance of the former point from the centre of the scale, to the distance of the latter point from the same centre.[79]

Likewise, it is interesting to note that Hobbes does not use the term *gravity*, but instead talks of *pondus* (*weight* in English), just as Mersenne used the term *pesanteur*, and uses this to indicate the 'aggregate of all the endeavours (*aggregatum omnium conatuum*), by which all the points of that body, which presses the beam, tend downwards in lines parallel to one another; and the body which presses is called the *ponderant*.'[80] What is more, the weight is indicated as the *conatus* (that recurs here with the meaning of *impetus*) of the individual points making up the heavy body. On the other hand, Hobbes seems to have omitted all explicit reference to gravity in relation to the mass of the body, which instead is present in Galileo's text and emphasized in Mersenne's translation.

Despite this, one of the meanings with which the term *conatus* recurs in *De corpore* has a strong kinetic connotation and is directly dependent upon the notion of *momentum velocitatis*, specific to Galilean physics.[81]

On the 'third day' of *Discourses and Mathematical Demonstrations Relating to Two New Sciences*, Galileo, when discussing uniformly accelerated motion, expresses the intent—through the words of Salviati—not to deal with the cause of the acceleration, which was still obscure and inaccessible, but instead to 'investigate and demonstrate some of the properties of accelerated motion.' Galileo maintains that 'the moments of its velocity go on increasing after departure from rest in simple

---

[78] Ibid. Compared to the translation, the Latin original offers a better understanding of the Galilean origin of the concepts expressed by Hobbes: 'IV. *Momentum* est ponderantis, pro certo situ, certa ad movendum radium potentia. V. *Planum aequilibrii* est quo ponderans dividitur, ita ut momenta utrinque sint aequalia. VI. *Diameter aequilibrii* est duorum aequilibrii planorum sectio communis. VII. *Centrum aequlibrii* est duarum aequilibrii diametrorum commune punctum.'

[79] Ibid., XXIII, 4, 288; Eng. Trans. 353.

[80] Ibid. 287: '*Pondus* est aggregatum omnium conatuum, quibus singula puncta corporis, quod radium premit, in rectis sibi invicem parallelis conantur; ipsum autem corpus premens ponderans nominatur.'

[81] See Leijenhorst 2004, 183.

proportionality to the time, which is the same as saying that in equal time intervals the body receives equal increments of velocity.'[82] Following on from Salviati's reflections, it is Sagredo, a little later on, who provides the precise formulation for the law of uniformly accelerated motion, in which the term *momentum* also recurs: '*Motum aequabiliter, seu uniformiter, acceleratum dicimus eum, qui, a quiete recedens, temporibus aequalibus aequalia celeritatis momenta sibi superaddit.*' (We define equally or uniformly accelerated motion, that which, starting from rest, increases its *momenta* of speed in equal times).[83]

The term *momentum* therefore expresses an infinitesimal degree of speed and recurs several times throughout the discussion of uniformly accelerated motion. It is obviously not possible to reproduce the entire Galilean argument here, but we can use the reconstruction offered by Galluzzi to sum up the problem of the meaning of *momentum velocitatis* as it recurs in the *Discourses*. The *momentum velocitatis* unfolds within an instant of time and can be considered a quantity of distance (constant or increasing) that is added to the previous ones. It is, therefore, a sort of 'hinge' between time and space. The very close connection between the *momentum velocitatis* and the distance covered during the same time refers, by analogy, to the *momentum gravitatis*. The latter concept has always been interpreted and illustrated by Galileo as the tendency, propensity, and inclination to cover a certain path (at uniform speed) over a given time. As a result, the *Discourses* present *momentum* as the tendency, propensity, and inclination for rapid movement (of motion that is still uniform) that is no longer constant, but variable according to the distance from rest.[84]

The analogies with Hobbesian *conatus*, defined as 'motion through some length, though not considered as length, but as a point'[85] are profound. Nevertheless, Hobbes does not refer to the notion of velocity in this instance. In fact, he expresses this concept in relation to the instant or the point, using the term *impetus*, a notion that is defined in Chapter 15 immediately after that of *conatus*: 'I define *impetus*, or quickness of motion, to be the swiftness or velocity of the body in the several points of that time in which it is moved. In which sense *impetus* is nothing else but the quantity or velocity of endeavour.'[86] Consequently, *impetus* is the velocity considered at any point of time, or the velocity of the conatus itself.[87]

---

[82] Galilei, *Discorsi e dimostrazioni intorno a due nuove scienze*, *OG*, VIII, 202 (my italics); Eng. trans. Finocchiaro, 340.

[83] Ibid., 205.

[84] See Galluzzi 1979a, b, 371.

[85] *De corpore*, XXII, 1, *OL*, I, 271; Eng. Trans. *EW*, I, 333.

[86] *De corpore*, XV, 2, *OL*, I, 178; Eng. trans. 206. In the English edition, the discussion is broader and more detailed. The fact that Hobbes was well aware of Galileo's speculations regarding uniformly accelerated motion has been widely demonstrated by Jesseph, who emphasized that the mathematical demonstrations present in chap. 16 of *De corpore* follow those of Galileo in the fourth day of his *Discourses and Demonstrations*. On the other hand, as Jesseph himself notes, John Wallis had observed the very close analogy between the geometric argument in § 9 of Chapter 16 of *De corpore* and Galileo's speculations. See Jesseph 2004, 204 ff.

[87] On the connection between *impetus/conatus* and the importance of the concept of *conatus* in Hobbesian dynamics, see also Gaukroger 2006, 287 ff.

The relation between *impetus* and velocity takes back to Galileo's *Discourses*, particularly to the Theorem II in the 'fouth day,' where Galileo uses the term *momentum* as a synonym of *impetus*: 'If a body falls from rest with a uniformly accelerated motion, then the spaces traversed are to each other as the squares of the time intervals employed in traversing them.'[88] The Italian scientist goes on to illustrate the Theorem, explaining that when two uniform motions cover two different distances in the same space of time, it can be said that their *momenta* are equal to their distances and *momentum* is primarily used here to indicate speed.[89] Galileo's text clearly reveals that Hobbes' definition of *impetus* is also strongly influenced by that of Galileo, just like his definition of *conatus*.

However, before returning to Hobbes' definitions of *impetus* and *conatus*, with regard to the range of different terminological meanings in which these two words recur, we must stress that in the second edition of the *Discourse on Bodies in Water* (*Discorso intorno alle cose che stanno in su l'acqua o che in quella si muovono*, 1612), where Galileo clarifies the meaning of the term *momentum* (in response to those who accused him of using the term without providing specific indications as to the nature of this concept), he insists on the connection between *force* and *impetus* in relation to gravity.[90]

The connection between the concepts of *force* and *impetus* also emerges in Hobbes' works. In *De motu, loco et tempore*, Hobbes provides a definition of *impetus* that is closely connected to force and inseparable from it: 'It must be known that motion and impetus are the same thing, and that it is yet called impetus and force (*vim*). In fact, impetus and force are synonyms, inasmuch as impetus produces other movement.'[91] In addition to the specific issue (which here concerned whether or not the continuous presence of a motor was necessary to preserve motion),[92] the terminology adopted by Hobbes suggests that they match the Galilean principles. Indeed,

---

[88] Galilei, *Discorsi e dimostrazioni intorno a due nuove scienze*, OG, VIII, 280; Eng. Trans. Finocchiaro, 346: 'Si aliquod mobile duplici motu aequabili moveatur, nempe horizontali et perpendiculari, impetus seu momentum lationis ex utroque motu compositae erit potentia aequalis ambobus momentis priorum motuum.'

[89] See Galluzzi 1979a, b, 374.

[90] Galilei, *Discorso intorno a le cose che stanno in su l'acqua e che in quella si muovono*, OG, IV, 68: 'Momento appresso i meccanici significa quella virtù, quella forza, quella efficacia, con la quale il motor muove e 'l mobile resiste; la qual virtù depende non solo dalla semplice gravità, ma dalla velocità del moto, dalle diverse inclinazioni degli spazii sopra i quali si fa il moto, perché più fa impeto un grave descendente in uno spazio molto declive che in uno meno. Ed in somma, qualunque si sia la cagione di tal virtù, ella tuttavia ritien nome di momento.' See Galluzzi 1979a, b, 241.

[91] *MLT*, XXI, 12, 260.

[92] In this passage, Hobbes once again argues against White, using the Galilean arguments from the 'second day' of the *Dialogue* (see Galilei, *Dialogo sopra i due massimi sistemi del mondo*, OG, VII, 182), where Galileo stated that a heavy body dropped by a man on horseback will not stop its motion immediately, but will be braked progressively by the friction of the earth with which it comes into contact.

the English philosopher asserts that impetus is a *motion* and maintains that the terms *impetus* and *force* are synonyms.

What is more, Hobbes adds an interesting clarification, where he writes that 'that which contributes to some extent to the size of the moving body, also contributes to its impetus. It is therefore necessary for there to be an impetus in every part of the moved body, just as Galileo believes.'[93] This passage tells us that Hobbes not only establishes a relationship between the mass of a certain body and its *impetus*, but also, quoting Galileo directly, he maintains that the impetus of the body is present in every part of the body itself (since it coincides with the *conatus* of its individual points).

Hobbes' conclusion leads to the identification of *impetus* and *motion*.[94] In support of his idea that the *conatus* is *always* and in every case a movement in action, he appeals to the authority of Galileo: 'Galileo denies that impetus is before motion, or that it is anything other than motion itself.'[95]

These passages from Hobbes reveal a stringent interpretation of the science of Galilean motion. Hobbes' mechanical philosophy contemplates every action solely in terms of matter and motion, meaning that each *impetus* or *conatus* must necessarily be expressed in kinetic terms. Galileo's position, adopted and upheld by Hobbes, is to express the *conatus* in strictly dynamic terms and rule out the existence of any *vis*, *quality* or *property*, which were present in some late medieval philosophers.[96] Galileo's development of the problem therefore enabled Hobbes to conceive this force or principle of movement in a different way even from Descartes (according to whom, as we know, inclination was not already a movement).

On the other hand, in *De corpore*, the English philosopher uses terminology that differentiates between *conatus* and *impetus* and the second term is used to express *'the quantity or velocity of endeavour.'*[97]

---

[93] *MLT*, XXI, 12, 260.

[94] Ibid., 261: 'Idem enim est impetus quod motus sive velocitas, non causa eius neque effectus.'

[95] Ibid.

[96] The concept of *impetus* as *virtus derelicta* was introduced in European Latin world during the Middle Ages by the Scotist Francesco de Marchia, but it is primarily in Jean Buridan that the term acquires fundamental importance in physics. Buridan claims, in contrast to the Aristotelian theory of *antiperistasis* (on which, see *infra*), the projectile has an *impetus*, a force, or a quality, proportional to its mass and motion (see Jean Buridan, *In De Caelo*, III, q.). Buridan also uses *impetus* to explain the phenomenon of the increased velocity of heavy bodies in free fall, claiming that *impetus* increases with the speed of the projectile (Ibid, II, 1. 12). See Crombie 1953, 250 ff. For an extensive and detailed discussion of *impetus* in medieval philosophy: Maier 1968, 113 ff. See also Grant 1996a, b, 93 ff., who underscores the Arabic origin of the notion of *impetus*, which is rooted in the concept of *mail*, advocated by Avempace. On the influence of Avempace on Galileo, the following classic study is always useful: Moody 1951. On the elements present among medieval physicists, also common to Galileo, see Clavelin 1968, 75–126. See also Hooper 1996, which presents two different interpretations of the word in Galileo's work: one, from his youth, closer to the interpretation of Francesco de Marchia and a second later one, from his maturity, closer to the meaning attributed to it by Buridan (ibid., 159).

[97] *De corpore*, XV, 2, *OL*, I, 178; Eng. Trans. *EW*, I, 207.

As we shall see, this linguistic differentiation is due to the need perceived by Hobbes to justify the multiplicity of meanings that Galileo expressed solely with the term *momentum*, but that instead required—according to Hobbes—greater clarification. Nevertheless, the analysis of *impetus* does not exhaust the problem of *conatus*, which, as we have observed, presents significant conceptual polysemy and is used, exactly like Galilean *momentum*, to define the concept of gravity.

3.5   'Gravity Is Nothing Other Than a *Conatus*'

The concept of *conatus* is extremely relevant in Hobbes' discussions on the subject of gravity, as emerges clearly in the *Tractatus Opticus II*. Here, Hobbes describes the phenomenon by seeking out the cause of gravity in the internal movements of the heavy body and in the alteration of these motions (which therefore also produces the hardness of the body).[98] He uses the well-known example of the sieve: when it is moved the grains inside it are concentrated '*in medium loco*.'[99] The same principle can be used 'for the explication of the nature of falling bodies: gravity is nothing other than a conatus of each body towards a place.'[100]

   The problem is taken up again more extensively in *De motu, loco et tempore*, where Hobbes, after having candidly admitted that he does not possess sufficient knowledge to solve the age-old problem of the nature of gravity, provides a definition of the phenomenon that looks back to his optical treatise, but is expressed in a more extensive and detailed manner:

> It is well known that gravity is nothing other than the conatus of some bodies towards the center of the Earth; in truth, that conatus is either the motion of the whole body, as when the entire body falls, or the motion of its parts of which pressure consists, a motion by which the parts are driven forward, although the whole body does not advance.[101]

The comparison between the Galilean law of falling bodies and Hobbes' discussion of gravity has already been studied by Cees Leijenhorst, who provided a very precise description of the way in which the theme evolved from *De Motu* to *De corpore*, and the influence exercised over Hobbes by his friend and colleague, Pierre Gassendi.[102] Nevertheless, it is advisable to look at some of Leijenhorst's observations, so as to expand the discussion of the problem and compare the speculations featured in *De motu, loco et tempore* and *De corpore* with those in later works, as the subject offers additional ideas for exploring the nature of *conatus*. What is more,

---

[98] See *supra*, chap. 1, § 4.

[99] *TO II*, chap. 1, § 11, f. 198*v*/152–153: '… notandum id quod experientia Certissimum est, in cribro sic moto grana, paleis et scrupis permixta ea agitatione segregari, et haec inter se in medium locum congregari.'

[100] Ibid.

[101] *MLT*, X, 11, 180 (my trans.).

[102] See Leijenhorst 2004, 167 ff.

a detailed examination of the theme suggests that Hobbes' reflections on the phenomenon of gravity are not influenced only by Gassendi but also, in part, by Marin Mersenne.

We have seen that, in both the *Tractatus Opticus II* and *De motu, loco et tempore*, Hobbes associates gravity with the conatus of some bodies towards the center of the earth. Despite this, it is important to emphasize that in the second work, he distinguishes between motion 'of the whole body, as when the entire body falls' and 'motion of its parts of which pressure consists, a motion by which the parts are driven forward, although the whole body does not advance.'[103] Gravity is conceived as the motion of the internal parts of the body, which present a *conatus*, without the entire body moving.

Yet, as we know, Hobbesian physics rules out any concept of self-movement and, as a result, Hobbes adds that the motion of a falling body must necessarily be produced by an external mover.[104] He also distinguishes between three types of action produced by a mover, so that 'every mover moves by either pushing or pulling, or by impeding the pre-existing movement of the thing to be moved, and diverting it as a result.'[105] Hobbes lingers on the third type of movement and recalls the model of the spinning top, which we encountered in the first chapters on the motions of the earth.

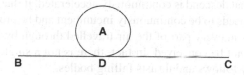

Taking some round body such as A (*see figure*), one supposes that it is at rest, while its parts are moved in keeping with the figure of the whole, for example circularly, or in another way. We then have the straight line BC that touches body A, along the small section represented by D, above which straight line another body is moved. It is clear that the motion of the body, which is moved along BC, will disturb the motion of the parts of body A. Therefore, it will happen that body A will either be split in D or, if it is too hard for this to take place, it will escape to conserve the motion of its parts: this withdrawal is the motion of which we speak. This movement presents itself to us most clearly in the case of spinning tops: these, rotating with a fast and whirling movement, rebound from every lateral contact with a direct movement.[106]

This type of reasoning is applied to justify the downward fall of falling bodies and in an attempt to explain the nature of gravity. Firstly, Hobbes maintains that the

---

[103] *MLT*, X, 11, 180.

[104] Ibid.: 'Qui quaerunt itaque quae sit causa gravitatis, hoc quaerunt quo motore saxum, vel aliud corpus descendit per aërem, vel premit manum vel aliam rem qua sustentatur. Primum igitur satis iam plerisque philosophis manifestum esse puto, ipsum saxum non movere seipsum; habet ergo motorem externum.'

[105] Ibid.

[106] Ibid, 180–181 (my trans.).

particles that make up the heavy bodies are subject to a movement that characterizes the *essence* of the heavy body.[107] This is the same argument that he had used to describe the property of hard bodies, both in his letters to Descartes,[108] and in *De motu, loco et tempore*, thereby establishing a sort of analogy between the hard body and the heavy body. This movement, of which the weight of the heavy body consists,

> ... is not a movement of the whole, but inside the whole, in the parts that do not emerge from the whole; therefore, the movement of the individual parts is a sort of rotation on itself, otherwise it would emerge from the whole with a movement in a straight line; that is to say, the whole would be corrupted, as occurs in the case of putrefaction.[109]

Hobbes deems that inside the heavy body (as inside the hard body and, generally speaking, in any body) there is a motion that characterizes the behavior of the particles within this body, subject to a whirling and circular movement. Following on from this conception of the matter that makes up heavy bodies, Hobbes formulates a theory on gravity and free falling. He starts by ruling out that gravity and, in particular, the acceleration of gravity should be attributed to the thrust of a motor because, if that were the case, 'the degree of velocity that the thrusting body imparts to the falling body from the start should be that with which the falling body always necessarily descends, or a lesser degree due to air resistance, unless a new impulse is added to the descending body.'[110] Instead, continues Hobbes, 'since the motion of the falling bodies that descend is continuously accelerated, if that motion is a thrust, the thrusting body needs to be continuously incumbent and be present with the same increase in velocity in every part of the air travelled through by the falling body—something that cannot be conceived. In fact, there is not a single body that thrusts, which continuously follows and thrusts falling bodies.'[111]

On the other hand, Hobbes also discards the theory of the traction of the falling body by a motor that drags it in the direction of the center of the earth, for one simple reason: this entails the existence of a force that adheres to the falling body, that is to say one pertaining to another tractor body, but this reasoning necessarily leads to a regression to the infinite.[112] The only remaining alternative, towards which Hobbes is inclined, is focused on what he refers to as the 'circular movement of the air:'

> It remains to be thought that it takes place in the third manner, that is to say that there is circular movement of the air everywhere. In which, if a body rises through the air due to the action of a force, the motion of its internal parts, of which its essence consists, is altered due

---

[107] Ibid, 181.

[108] Hobbes to Mersenne for Descartes, 7 February 1641, *AT*, III, 302: '...spiritus subtiles et liquidos, vehementia motus sui, posse constituere corpora dura, ut adamantem, et lentitudine, alia corpora mollia, ut aquam vel aërem.'

[109] *MLT*, X, 11, 181. In this passage, Hobbes seems to refer to the Baconian discussion of the theme of corruption of bodies, which is caused by the spirits leaving the body. See, for example: Bacon, *Historia Vitae et Mortis*, in Bacon 2007, 172.

[110] *MLT*, X, 11, 181 (my trans.).

[111] Ibid.

[112] Ibid.

to the discordance and diversity of the paths; so it happens that the body continuously recoils, always recuperating its position, so that it arrives at the center of the earth, unless obstructed by a force. It therefore seems to me to be impossible to describe those motions of the parts in the various kinds of bodies, that is to say to illustrate the nature of gravity entirely.[113]

Hobbes' argument is rather abstruse and, on the basis of what we can deduce from the cited passage, Hobbes himself realized the difficulties associated with his reasoning. However, before exploring the specific solution proposed by him, it is interesting to stress that (as Leijenhorst has already revealed)[114] the two alternatives examined and ruled out by Hobbes in his examination were both also contemplated in the *Epistolae de motu* by Pierre Gassendi.[115] As we know, Gassendi's speculations were already known to Hobbes prior to writing *De motu, loco et tempore* and are taken up by Hobbes in his work of 1643.

In reality, we have to observe that both the authors seek to tackle two connected but different problems at the same time: the *cause of gravity* and the *cause of the acceleration* of gravity. In this regard, Gassendi supposed that the accelerated fall of the falling object was produced by the combination of two concomitant forces: firstly, a *vis attrahens* is activated, exercised by the earth, which emits chains of magnetic particles able to 'harpoon' the objects, and a *vis impellens*, that is to say a pressure produced by the air that, driven out of its place by the falling body (already activated by the *vis attrahens*), occupies the spaces left free by the falling body and continuously presses upon the falling body itself, causing an increase in velocity.[116]

There is an evident echo of Aristotle in this idea. In fact, in order to justify the movement of projectiles, Aristotle had come up with the theory of ἀντιπερίστασις, which supposes that this movement is produced by the air shifted by the projectiles themselves, which then occupies the spaces vacated by the advancing object and propels it.[117]

Nevertheless, Gassendi was convinced that a single agent principle was not sufficient to justify both the fall and the acceleration, and so he added a second force (first in chronological order), which put the falling object into motion and caused the air to shift for the first time.

---

[113] Ibid. (my trans).

[114] See Leijenhorst 2004, 171 ff.

[115] See Palmerino 2004.

[116] See Gassendi, *Epistolae tres. De motu impresso a motore translato*, in Gassendi 1658, III, 478 ff. For a discussion of the relationship between the Galilean theory of falling heavy bodies and Gassendi's ideas, see Palmerino 2004, 153–157, and Palmerino 2008, 161 ff.

[117] See Aristotele, *Physics*, IV, 215a ff., see also ibid., VIII, 266b; *De Caelo*, III, 301b. As we know (see *supra*, footnote 96), this Aristotelian theory was criticized by Buridan, who demonstrated the contradictory implications of the explanation proposed by the Greek philosopher: firstly, that which offers resistence to motion (that is to say air) cannot also be its mover. Moreover, it is not explained how air, which is so easily divisible, 'supports a stone weighing ten pounds for any length of time'. Buridan, *In de Caelo*, III, 2, my translation. On this subject, see Crombie 1953, 250 ff.

The reasons for Hobbes' rejection of the explanation put forward by the friar from Digne can be sought in the nature of the *vis attrahens*: he evidently considered this part of Gassendi's argument to be specious and, by concentrating solely on the second part (relating to the *vis impellens*), he encountered some difficulties in justifying the acceleration of gravity, because the falling object logically ought to be held back by the friction of the air, and yet, as Galileo demonstrated in his *Discourses*, it continues to accelerate according to a precise law.

3.6   Research into the Cause of Gravity

The solution adopted by Hobbes in *De motu, loco et tempore* is, as we have seen, to seek the cause of the fall in the alteration of the circular motions to which the particles making up heavy bodies are subject.[118] Nevertheless, there is a unique element in Hobbes' theory that has not yet been highlighted by scholars and which should be underscored here. We have seen that Hobbes establishes a curious link between the *gravity* and *hardness* of the bodies. As we know from his correspondence with Descartes, Hobbes deemed the movement of the body's internal particles to be a bigger cause of resistance for some objects than others and this idea is used in *De motu, loco et tempore* not only to describe the phenomenon of terrestrial rotation,[119] but also to sketch out a possible explanation for gravity and the conatus associated with gravity itself.

> ... *conatus* consists of that, which is moved by a body that tends [to move]. Likewise, we say of heavy bodies that lie on the ground that they tend downwards, because when the impediment is removed, they descend; in fact, if they did not descend, it would not be said that they first tend [to descend]. Likewise, it is deemed that the taut bow tends to the restoration of the parts, so that once the impediment has been removed, the bow in fact returns [to its position], and once returned [to it] it no longer has conatus. Nevertheless, since the removal of the impediment is not an action, and yet an action is necessary in order for bodies at rest to move, [the only theory that] remains is that the driving principle of the fall of falling bodies and that of the return of the bow consist of the fact that there is some movement underway in the heavy body that tends to fall and in the same bow that tends to straighten.[120]

Here Hobbes makes an analogy between the force that leads to the restoration of the taut bow and the force of gravity to which the bodies are subject, using it again in subsequent works with the example of the crossbow.[121] The theory is based on the idea that there is a movement inside heavy bodies that involves their microparticles and, once the external impediment has been removed, these bodies are free to fall downwards or return to their initial position.

---

[118] *MLT*, X, 11, 181.

[119] See ibid., XXI, 10, 259; and XXIV, 10 ff., 298 ff.

[120] Ibid., XIII, 2, 195 (my trans.).

[121] See Baldin 2017.

A similar argument also featured in Gassendi and in the *Harmonie Universelle* by Mersenne, where the latter made an analogy between the resistance of stone to the pressure of the hand and the force underlying the phenomenon of the taut string.[122]

The comparison of the phenomenon of the bow and the fall of falling objects, together with Hobbes' use of the same principles to solve both the problems, returns, albeit with some changes (introduced by *De corpore*), in his later works too.[123] However, the explanation of gravity developed by Hobbes in *De motu, loco et tempore* remains problematic, and it is no coincidence that he proposes some adjustments to his theory in *De corpore*. In greater detail, the problems encountered by the Hobbesian model were primarily due to two things: firstly, he deemed the movement of the internal particles to be the cause of the resistance of the bodies and the cause of their weight and, on the basis of this definition, it is difficult to explain why hard but light bodies fall more slowly than heavier bodies. Secondly, Hobbesian reasoning suggests that there is already movement inside the bodies. This is difficult to reconcile with Hobbes' strict mechanism, which entails that the movement of a body is necessarily determined by the existence of a motor, that is to say an external and adjacent body.

The problem of the failure to consider the weight of the falling object in relation to the phenomenon of gravity is also to be found in the explanation formulated by Gassendi,[124] who proposed an alternative theory from 1646 onwards, exclusively considering the *vis attrahens* and excluding the thrust exercised by the air.[125] Nevertheless, despite Gassendi's change of perspective, Hobbes had to reconsider the old Gassendian solution, focused on the *vis impellens*, and this is demonstrated by the explanation of gravity given by Hobbes in Chapter 30 of *De corpore*, which diverges drastically from that given in *De motu, loco et tempore*.

It describes two alternative positions elaborated by earlier philosophers with regard to gravity: on the one hand, those who attribute the fall of falling bodies to

---

[122] Ibid. See Mersenne 1637, Livre III. *Des Mouuemens & du son des chordes*, 165 and Bernier 1684, II, 309–309. See also Gassendi to Jean Chapelain, in Gassendi 1658, III, 466b. As we know, Gassendi attributed atoms with an innate tendency to movement. On the concept of matter in the philosophy of Gassendi, see Bloch 1971, 210–229; Messeri 1985, 74–93; Osler 1994, 180 ff., and Gaukroger 2006, 262–276. Clericuzio focuses heavily on *materia actuosa* in Gassendi, a concept that—according to the author—removes Gassendi from a strictly mechanistic perspective. See Clericuzio 2000, 63 ff.

[123] In the *Dialogus physicus* he refers to a steel blade or a crossbow and explains that the return of the two bodies to their initial position is 'itself a local motion, but within an imperceptible space, and no less very fast, so as to cause a very fast movement' (*Dialogus physicus, OL*, IV, 248, my translation). Nevertheless, this recovery movement cannot take place in a straight line because motion of this type would necessarily cause the moving object to move in its entirety, 'therefore it is necessary for the movement to be circular, so that every point of the body that is dilated completes a small circle' (ibid., 249).

[124] See Palmerino 2004, 149.

[125] See Gassendi, 'Epistolae de proportione qua gravia decidnetia acceleratur'. In Gassendi 1658, III, 572 ff.

'some internal appetite,[126] by which, when they were cast upwards, they descended again, as moved by themselves, to such place as was agreable to their nature;'[127] on the other, those who believe that these same bodies are 'attracted by the earth.'[128]

The reasons adopted by Hobbes for challenging the first hypothesis should be sought in the principles of Hobbesian mechanical physics: 'There can be no beginning of motion, but from an external and moved body.'[129] Instead, he maintains that he agrees with the second alternative: 'To the latter, who attribute the descent of heavy bodies to the attraction of the earth, I assent.'[130] What is more, Hobbes maintains that the phenomenon of gravity has not yet been explained by anyone and therefore he proposes to formulate a correct solution for it.[131]

In reality, despite the premises, the explanation proposed by Hobbes in the fourth paragraph of Chapter 30 of *De corpore* also digresses considerably from that which Gassendi had developed in his *Epistolae de motu*. Hobbes' theory (*see figure*) is that a stone conducted in a straight line from point D (positioned on the surface of the earth at the equator) to point E, exerts a progressive thrust on parts of the air. Nevertheless, given that the earth also removes some air during its daily rotation and the particles of air are lighter than those making up the bodies, it therefore rises more quickly than the bodies themselves.

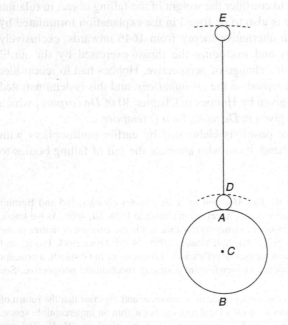

---

[126] *De corpore*, XXX, 2, *OL*, I, 414: '*Appetitum aliquem esse putarent internum*'.
[127] Ibid.
[128] Ibid.
[129] Ibid.
[130] Ibid.
[131] Ibid., 415.

Through the twofold action of the thrust exerted by the object and the movement of the earth, an empty space tends to form in the layer of air immediately around point E. As a result, in a world where no voids are permitted, we must suppose that the air from above comes down to occupy this space, pushing the falling body downward. What is more, the falling body receives successive thrusts from the air at every instant and in every point of the EA segment.

Hobbes cites Galileo in the next paragraph and adds that the latter should be credited with discovering the ratio between space and time in freefall motion, which is covered by Galileo's *Dialogi de motu*.[132] In fact, in the 'third day' of his *Discourses*, Galileo explored the problems of *De motu aequabili* and *De motu accelerato* and this latter part contains the famous law of falling objects,[133] which is taken up again by Hobbes in *De corpore*:

> And from hence it follows, by that which Galileus hath in his *Dialogues of Motion* demonstrated, that heavy bodies descend in the several times with such differences of transmitted spaces, as are equal to the differences of the square numbers that succeed one another from unity; which square numbers being 1, 4, 9, 16, &c. their differences are 3, 5, 7, &c.; that is to say, the odd numbers which succeed one another from unity.[134]

Ss Jesseph underscored,[135] the presence of Galileo and the influence of arguments developed by Galileo in his *Discourses* are also particularly evident in Chapter 26 of *De corpore*, where Hobbes discusses accelerated and uniform motion.[136] Here he defines *velocity* as 'that power by which a body can in a certain time pass through a certain lenghth,' while *impetus*, in its turn, is 'velocity taken in one point of time only.'[137] In the third paragraph, Hobbes takes up the Galilean law of uniformly accelerated motion,[138] while in the ninth he discusses the behavior of a moving body moved by two concomitant and concurrent movers:

> If a body be carried by two movents together, which meet in any given angle, and are moved, the one uniformly, the other with motion uniformly accelerated from rest, that is, that the proportion of their impetus be as that of their times, that is, that the proportion of their lengths be duplicate to that of the lines of their times, till the line of greatest impetus acquired by acceleration be equal to that of the line of time of the uniform motion; the line

---

[132] Ibid., XXX, 5, 417–418. The law of falling heavy objects makes its appearance in a letter from Galileo to Paolo Sarpi dated October 16, 1604, (*OG*, X, 115). However, here the increase in velocity is deemed to be proportional to space and not to time. It was in his *Dialogue* that Galileo presented the scientific community with the correct law of falling heavy bodies (See Galilei, *Dialogo sopra i due massimi sistemi del mondo*, *OG*, VII, 248).

[133] Galilei, *Discorsi e dimostrazioni intorno a due nuove scienze*, *OG*, VIII, 209–210. For a detailed analysis of the mathematical and geometrical discussion of naturally accelerated motion in Galileo's texts, see Clavelin 1968, 285 ff.

[134] *De corpore*, XXX, 5, *OL*, I, 417–418; Eng. Trans. *EW*, I, 514.

[135] See Jesseph 1999, 104–110; Jesseph 2016, 73–75.

[136] See *De corpore*, XVI, *OL*, I, 184 ff.

[137] Ibid., 184–185; Eng. Trans. *EW*, I, 218–219.

[138] Ibid., 186–187.

in which the body is carried will be the crooked line of a semiparabola, whose base is the impetus last acquired, and vertex the point of rest.[139]

Hobbes' argument exactly follows that present in the 'fourth day' of the *Discourses*,[140] and it is also interesting to highlight that those pages by Galileo also refer extensively to the notion of *momentum*,[141] which we examined earlier and whose conceptual and speculative significance are firmly grasped by Hobbes.

This additional reference to Galileo allows us to make an observation on the underlying philosophy shared by Galileo and Hobbes. Indeed, Galileo was primarily interested in the physical analysis of phenomena and did not believe his exploration of the behavior of falling bodies to be lacking, although he was unable to explain the *essence* of gravity.[142]

This is evident in his handling of the phenomenon during the 'second day' of the *Dialogue Concerning the Two Chief World Systems*, where Salviati, debating with Simplicio, observed that we know absolutely nothing about the nature of gravity, or rather about its *essence*.[143] The passage in question takes up some considerations formulated in the *Letters on Sunspots*, where Galileo maintained that 'attempting the essence, I consider to be an undertaking that is no less impossible and no less vain in terms of effort, as regards both nearby elementary substances and remote and celestial ones.'[144] The clear anti-Aristotelian implications of the passage outline a significant aspect of Galilean thinking and epistemology, which we also find in Mersenne and Hobbes and which is at the basis of modern philosophical and scientific thinking: the *philosopher* Galileo maintains that the hidden essence of the phenomena is impenetrable and that the aim of the researcher is to describe the behavior of natural entities, using the tools provided by mathematics.[145]

---

[139] Ibid., XVI, 9, 196; Eng. Trans. *EW*, I, 232.

[140] See Galilei, *Discorsi e dimostrazioni, OG*, VIII, 272 ff.

[141] In the *Theorema III, Propositio III* (*OG*, VIII, 282), Galileo writes: '*momenta autem celeritatis sunt inter se ut spatia, quae iuxta ipsa momenta eodem conficiuntur tempore*.'

[142] In an article on the physics that Descartes outlines in his exchange of letters with Mersenne (which differs from the essentially 'narrative' physics of the *Principia* due to its markedly mathematical dimension), Garber claims that *gravity* as a 'natural tendency of heavy bodies to fall towards the centre of the earth' is a fundamental element of Galileo's 'paradigm' (in the meaning coined by Kuhn 1962, esp. 30 ff.), but that it could not have been shared by Descartes: 'because they contain nothing over and above extension, bodies can have no innate tendencies at all'. See Garber 2000, esp. 122. Nevertheless, in the light of Galileo's criticism of Kepler with regard to the phenomenon of tides, it is difficult to believe that a concept of an innate tendency in bodies (which is therefore foreign to the principles of mechanism) was present in Galileo's epistemology. As we observed in the previous chapter and as we shall see even more clearly in the following chapter, Galileo made an ongoing effort to explain every physical phenomenon in mechanistic terms, including the structure of matter.

[143] See Galilei, *Dialogo sopra i due massimi sistemi del mondo, OG*, VII, 248.

[144] Galilei, *Istoria e dimostrazioni intorno alle macchie solari e loro accidenti*, Letter III, *OG*, V, 187.

[145] This should not lead us to believe that metaphysical and theological problems were alien to Galileo's thinking, as extensively demonstrated by the Copernican letters. On this subject see, for example, Redondi 1997.

Nevertheless, above and beyond every reference to Galileo in *De corpore*, it is advisable to return to the explanation of gravity drawn up by Hobbes. This actually proves to be no less problematic than the one in *De motu, loco et tempore*.[146] Firstly, the model is only able to explain the fall of a body raised to a certain height above the ground, thanks to the intervention of a force, and does not justify freefall starting from rest. Moreover, although Hobbes claims to share the idea of those philosophers who believe gravity to be produced by a force of attraction exercised by the earth, nevertheless, the cause of falling is in no way attributed to the attraction of the earth (it is, if anything, an indirect cause), but to a sort of *vis impellens* exercised by the air on the falling body.[147]

The reasons for the rejection of a theory based on earthly attraction, like the one put forward by Gassendi after 1646, should evidently be sought in Hobbes' opposition to any obscure force of attraction, which also led him—as we have seen—to reject Keplerian positions in Chapter 26 of the work.[148]

In any case, Hobbes himself must have noticed this anomaly. In his later scientific works, in which he proposes the same theory that we find in *De corpore*, without any major changes, the author does not provide any indications that would define this position as attractionist.[149] In the first of the *Seven Philosophical Problems* (1662), dedicated specifically to the problems of gravity, Hobbes once again uses his argument from 1655, to which he adds the example of the *basin* to the model of the sieve and attributes acceleration to the *impression* of the air that, thrust by the body, goes on to fill the space that had been occupied by the falling body.[150]

Meanwhile, a number of interesting aspects emerge from the more extensive argument in the *Decameron physiologicum* (1678). Although the theory of falling bodies is almost identical to that seen in earlier works, Hobbes adds two noteworthy reflections here. Indeed, *De motu, loco et tempore* includes a curious link that Hobbes seemed to make between heavy bodies and hard bodies: the hardness (or resistance) of the bodies was determined by the motions to which the internal particles that comprise them are subject, while the alteration of these motions, due to external movement, caused the object to fall. Hobbes denies this idea here, because he is backed up by empirical proof: hard bodies are not always heavier too.[151]

---

[146] See Leijenhorst 2004, 177–179.

[147] Hobbes adopts an identical model to explain the phenomenon of magnetism. See *De corpore*, XXX, 15, *OL*, I, 427 ff.

[148] See *supra*, chap. IV, § 4.

[149] See *Seven Philosophical Problems EW*, VII, 7–13; and also *Decameron physiologicum, EW*, VII, 147–151. In his *Dialogus physicus sive de natura aeris*, Hobbes simply observes that gravity 'is a conatus direct from every place towards the center of the Earth.' (*Dialogus physicus, OL*, IV, 250, my translation).

[150] See *Seven Philosophical Problems, EW*, VII, 7–9 (*Problemata Physica, OL*, IV, 306).

[151] *Decameron physiologicum, EW*, VII, 147: 'It is certain that when any two bodies meet, as the earth and any heavy body will, the motion that brings them to or towards one another, must be upon two contraries ways; and so also it is when two bodies press each other in order to make them hard;

Furthermore, having established that the link between heaviness and hardness is incorrect, on the next page, after referring again to Galileo's law according to which the velocity of falling bodies increases 'according to the odd numbers from unity,'[152] Hobbes maintains that this does not however give us any indication regarding 'the knowledge of why one body should have a great endeavour downward than another.'[153] Hobbes indicates a possible element that could determine the different acceleration of the falling body: the *figure*,[154] but in reality this was deemed to be problematic too.[155]

To conclude, even in the final work that Hobbes dedicates to natural philosophy, the English thinker openly states that he does not fully understand the *cause* of gravity: although the law of acceleration of a freefalling body is known thanks to Galileo, nevertheless we have no idea *why* one body falls with a greater *conatus* (*endeavour*) than another.

3.7   Towards a New Scientific Vocabulary

Above and beyond Hobbes' declared inability to provide reasons regarding the cause of gravity, it is important to underscore that the concept of *conatus* returns here too, at least once, in a markedly dynamic sense. At this point, we can therefore attempt to make a comparison between the different meanings of the term *conatus*, as it is presented in Hobbesian physics. Although in *De motu, loco et tempore*, the notion of *conatus* has a decidedly less technical meaning than the one it will acquire in *De corpore*, nevertheless certain aspects already feature that will go on to characterize the 'mature' conception of the notion.

Indeed, the definition in *De motu*, according to which the conatus is a motion of either the entire body, or of the particles that comprise it,[156] is used to indicate the principle of movement defined by Descartes as '*inclination à se mouvoir*,'[157] characterized, according to the French philosopher, by the absence of movement.

---

so that one contrariety of motion might cause both hard and heavy, but it doth not, for the hardest bodies are not always the heaviest; therefore I find no access that way to compare the causes of different endeavours of heavy bodies to descend.'

[152] *Decameron physiologicum*, *EW*, VII, 147: 'And further, the motion of the stone downward shall continually be accelerated according to the odd numbers from unity; as you know hath been demonstrated by Galileo. But we are nothing the nearer, by this, to the knowledge of why one body should have a great endeavour downward than another. You see the cause of gravity is compounded motion with exclusion of vacuum.'

[153] Ibid.

[154] Ibid. (my italics): 'It may be it is the *figure* that makes the difference. For though figure be not motion, yet it may facilitate motion, as you see commonly the breadth of a heavy retardeth the sinking of it.'

[155] Ibid., 148–149.

[156] See *MLT*, X, 11, 180.

[157] See Descartes, *La Dioptrique*, *AT*, VI, 88.

Hobbes is aware that, in the principle of motion or in the downward force exercised by a body suspended at a certain height, the body is still immobile as a whole, albeit for a fraction of a second. As a result, he uses the concept of *conatus* to express the notion of an impulse or principle of movement, despite staying within the boundaries of his dynamic conception of physics and his stringent mechanism. Hobbes therefore maintains that this impulse or *impetus* is already a movement in itself and must necessarily be expressed in dynamic terms.

Echoes of Galilean *momentum* can already be perceived in this connotation of the *conatus*, going on to be even more explicit in *De corpore*. As we have seen, the concept of *conatus* evolves more and more, in the wake of the different meanings of Galilean *momentum*, and will be used by Hobbes alongside other terms: *velocity, impetus, force*, and *momentum*. They appear in the work of 1655 and convey an attempt to master a series of associated, but different problems, contemplated in the light of the Galilean notion of *momentum*.

In Chapter 15 of *De corpore*, Hobbes explores the subject: *Of the nature, properties and divers considerations of motion and endeavour*, opening the treatise by defining the concepts that were present in Galilean physics and that will constitute the basic principles of his discussion of motion.

Firstly, when defining *velocity*, Hobbes maintains the Galilean interpretation that emerged in the theorems of Galileo's *Discourses*, indicating the relationship between space and time that characterizes the conception of velocity in modern physics. Nevertheless, he conceives it as 'motion considered as power, namely, that power by which a body moved may in a certain time transmit a certain length.'[158] This definition takes velocity itself back to *impetus* which, in fact, is defined like velocity itself, but considered at any point of time.[159] Once again, the determination of impetus, as *'quantitas sive velocitas ipsius conatus,'*[160] is fundamental: if impetus consists of velocity considered at any instant in time and *conatus* is *'motion made in less space and time can be given,'*[161] we can clearly see Hobbes' attempt to use terminological distinctions to clarify and explain two different concepts encompassed in the Galilean notion of *momentum*.

In the English edition of *De corpore*, the concept of *impetus* is discussed in greater depth, and the meaning that Hobbes attributes to the term indicates a very close connection with Galileo's *momentum velocitatis*.

---

[158] *De corpore*, XV, 1, *OL*, I, 176; Eng. trans. *EW*, I, 204–205: '...*velocitatem esse motum consideratum ut potentiam, qua mobile tempore certo certam potest transmittere longitudinem. Quod est brevius enunciari potest sic, velocitas est quantitas motus per tempus et lineam determinata.'*

[159] Ibid., XV, 2, 178: '...*impetum esse ipsam velocitatem, sed consideratam in puncto quolibet temporis in quo fit transitus.'*

[160] Ibid.

[161] Ibid., 177; Eng. Trans. 206. '...*conatum esse motum per spatium et tempus minus quam quod datur, id est, determinatur, sive expositione vel numero assignatur, id est, per punctum.'*

> *But considered with the whole time, it is the whole velocity of the body moved taken together*
> *throughout all the time, and equal to the product of a line representing the time, multiplied*
> *into a line representing the arithmetically mean* impetus *or quickness.*[162]

Compared to *De motu, loco et tempore*, Hobbes attributes a much more scientific connotation to the term *impetus*, making it translatable into a mathematical proportion, as Galileo did. As the treatise continues, it is evident that he is seeking to render the concept of increasing and decreasing *degrees of velocity*, which were inherent to the Galilean notion of *momentum*:

> And because as in equal times the ways that are passed are as the velocities, and the *impetus*
> is the velocity they go withal, reckoned in all the several points of the times, it followeth that
> during any time whatsoever, howsoever the *impetus* be increased or decreased, the length of
> the way passed over shall be increased or decreased in the same proportion; and the same
> line shall represent both the way of the body moved, and the several *impetus* or degrees of
> swiftness wherewith the way is passed over.[163]

Using the term *conatus*, which, significantly, is defined in Chapter 22 as 'motion through some length though not considered as length, but as a point,'[164] Hobbes attempts not only to provide a physical and dynamic definition for the notion of the principle of movement, but also to refer to the Galilean concept of *momentum velocitatis* that expresses the idea of the infinitesimal portion of time in relation to the space.

However, in the definition present in *De motu, loco et tempore*, which presented *conatus* as 'an actual movement, either of the entire body that tends [towards movement], or of its internal and invisible parts,'[165] the reference to the invisible particles that make up the bodies contained a clear nod to the principle of motion. In a body that begins to move because it is knocked by another object, all the internal parts react, and the principle of movement can be contemplated as the first internal motion of the particles that make up the body.

The added precision and conceptual clarity that distinguish *De corpore* from *De motu, loco et tempore* also emerge clearly with regard to the concept of *force*. In the 1643 text, *impetus* and *vis* were deemed to be synonyms,[166] while in 1655 Hobbes defines force as the '*impetus* [or quickness][167] of motion multiplied either into itself, or into the magnitude of the movent.'[168] The relationship between the *volume* of the

---

[162] *EW*, I, 207.

[163] Ibid.

[164] Ibid., 271; Eng. Trans. *EW*, I, 333: 'motum per longitudinem aliquam, consideratam autem non ut longitudinem, sed ut punctum'.

[165] *MLT*, XIII, 2, 195.

[166] *MLT*, XXI, 12, 260 (my italics): 'Sciendum autem est motum & impetum eandem rem esse, vocari autem *impetum*, atque etiam *vim*.'

[167] In the English edition Hobbes adds the concept of *quickness*. See *EW*, I, 212: 'I define Force *to be the* impetus *or quickness of motion multiplied either into itself, or into the magnitude of the movent, by means wherof the said movent works more or less upon the body that resists it.*'

[168] *De corpore*, XV, 2, *OL*, I, 179: '...*vim* definiemus *esse impetum multiplicatum sive in se, sive in magnitudinem moventis, qua movens plus vel minus agit in corpus quod resistit.*'

moving body and its *velocity*, which was only sketched out very briefly in *De motu*,[169] is now problematized and defined using a suitable term: the concept of *force*. Furthermore, Hobbes uses the words '*multiplicatum in se*' to refer specifically to the increased *impetus* in relation to the increased velocity, as emerges from the discussion of gravity.

Lastly, Hobbes also uses the concept of *momentum*, defined as the motion considered from the effect only which the movent works in the moved body,[170] and coincides with the '*excess of motion which the movent has above the motion or endeavour of the resisting body.*'[171] This definition refers directly to the meaning present in Chapter 23 of *De corpore*, where the thinker, discussing the scales, uses the term to indicate—just like Galileo[172]—the *compound force* of the lever: '*Moment* is the power which the ponderant has to move the beam, by reason of a determined position.'[173] In both cases, although we encounter two slightly different meanings of the term, they can still be traced back to Galilean definitions and linked to a physical situation of action and resistance between two forces. In the first case, the *momentum* indicates the force, or effect, of the mover on the resistant body, while in the second it expresses a condition of balance (or imbalance) between *forces*: the forces acting on the two arms of the scales.

To conclude, far from claiming that we have exhausted the issue of the relationships between Galilean physical concepts and the physics of Hobbes in this brief exploration, we can nevertheless undertake to make some overall observations regarding the problem of *conatus*. Through an intellectual path that begins with the assimilation of the vocabulary and the physical problems that emerged in the texts of Galileo, Hobbes attempts to clarify and explain some of the physical principles expressed by the Italian scientist through the word *momentum*, and does so using a more extensive and diversified terminology. In this attempt, he sometimes runs the risk of being rather vague and some concepts are only clear and comprehensible in the light of the comparison with Galilean *momentum*.

Nevertheless, Hobbes does not restrict himself to simply absorbing and reproducing certain Galilean physical concepts, but reworks them and, using a multitude of terms, attempts to translate a mechanical and static physics in terms of dynamic concepts.

As Lasswitz and Gargani have highlighted, Hobbes' substantial lack of mathematical ability prevented him from expressing his reflections on *conatus* in markedly scientific terms, but despite this it must be said that he succeeded in identifying the various problems associated with the Galilean notion of *momentum* and it is no

---

[169] *MLT*, XXI, 12, 260: 'Ex quo sequitur illud omne quod mobilis magnitudini aliquid contribuit, eiusdem etiam impetui contribuere.'

[170] *De corpore*, XV, 4, *OL*, I, 181; Eng. Trans. *EW*, I, 215.

[171] Ibid. Eng. 214–215: 'excessus motus corporis moventis super motum vel *conatus* corporis resistentis.'

[172] See Galilei, *Discorsi e dimostrazioni intorno a due nuove scienze*, *OG*, VIII, 154–155.

[173] *De corpore*, XXIII, 1, *OL*, I, 287; Eng. Trans. EW, I, 351: '*Momentum* est ponderantis, pro certo situ, certa ad movendum radium potentia.'

coincidence that the intellectual legacy left by Hobbes' *conatus* would be taken up again a few decades later by one of the fathers of infinitesimal calculus.[174]

References

Works of Thomas Hobbes:

Hobbes, Thomas. 1976. *Thomas White's De Mundo Examined*, ed. Harold Whitmore Jones. Bradford: Bradford University Press.

Other Works:

Bacon, Francis. 2007. *The Oxford Francis Bacon*, Vol. XII, ed. Graham Rees and Maria Wakely. Oxford: Oxford University Press.

Bernier, François. 1684. *Abrégé de la philosophie de M. Gassendi*, (2nd ed. Lyon. Reprint: *Corpus des œuvres philosophiques en langue française*, 6 Vols., ed. Sylvia Murr and Geneviève Stefani. Paris: Fayard, 1992).

Galilei, Galileo. 2001. *Dialogue Concerning the Two Chief World Systems*, Eng. Trans. Stillman Drake. New York: The Modern Library (1st ed. Los-Angeles-Berkeley: University of California Press, 1967).

Gassendi, Pierre. 1658. *Opera Omnia*, 6 Vols. Lyon: Anisson, & Devenet, 1658 (Stuttgart-Bad Cannstatt: Fromman Holzboog, 1994).

Mersenne, Marin. 1634. *Les Mechaniques de Galilee, Mathematicien et Ingenieur du Duc de Florence*. Paris: Henry Guenon (Reprint in Mersenne. 1985: 427–513).

———. 1637. *Harmonie Universelle*, 2nd Vol. Paris: Pierre Ballard.

———. 1985. *Questions Inouyes*, (Corpus des œuvres de philosophie en langue française) (Paris: Fayard, 1985).

Oresme, Nicole. 1968. *Le livre du Ciel et du Monde*, ed. Albert D. Menut and Alexander J. Denomy. Madison/Milwaukee/London: The University of Wisconsin Press.

Leibniz, Gottfried Wilhelm von. 1923—. *Sämtliche Schriften und Briefe*. Darmstadt-Leipzig-Berlin: Deutsche Akademie der Wissenschaften.

White, Thomas. 1642. *De mundo dialogi tres*. Paris: Moreau.

Secondary Sources:

Baldin, Gregorio. 2017. Archi, spiriti e *conatus*. Hobbes e Descartes sui principi della fisica. *Historia Philosophica* 15: 147–165.

---

[174] Howard Bernstein observed that Leibniz, who is unanimously considered one of the fathers of infinitesimal calculus, used the concept of *conatus* in his early writings with the same meaning coined by Hobbes. See Leibniz, 'Vorarbeiten zur theoria motus abstracti' in Leibniz 1923—, VI, 2, 171 'Omnis actio corporis est motus, omnis motus est in tempore. Motus autem in tempore minori quolibet dato, intra spatium minus quolibet dato est conatus'. See Bernstein 1980, esp. 26. The analogy between Hobbesian *conatus* and the ideas of Leibniz was also suggested by Watkins 1965, 123–132.

Barnouw, Jeffrey. 1990. Hobbes's Causal Account of Sensation. *Journal of the History of Philosophy* 18 (n. 2): 115–130.

———. 1992. Le vocabulaire du *conatus*. In *Hobbes et son vocabulaire*, ed. Yves-Charles Zarka, 103–124. Paris: Vrin.

Bernhardt, Jean. 1989b. L'apport de l'aristotelisme à la pensée de Hobbes. In *Thomas Hobbes. De la métaphysique à la politique*, ed. Martin Bertman and Michel Malherbe, 9–15. Paris: Vrin.

Bernstein, Howard R. 1980. *Conatus*, Hobbes, and the Young Leibniz. *Studies in History and Philosophy of Science* 11: 25–37.

Bertman, Martin. 2001. *Conatus* in Hobbes' De Corpore. *Hobbes Studies* 14: 25–39.

Bianchi, Luca. 1997. La struttura del cosmo. In *La filosofia nelle Università: secoli XIII-XIV*, ed. Luca Bianchi, 269–303. Florence: La Nuova Italia.

Bloch, Olivier René. 1971. *La philosophie de Gassendi: Nominalisme, matérialisme et métaphysique*. The Hague: Martinus Nijhoff.

Brandt, Frithiof. 1928. *Thomas Hobbes' Mechanical Conception of Nature*. Copenhagen/London: Levin & Mungsgaard-Librairie Hachette (or. Ed. 1921).

Bucciantini, Massimo. 2003. *Galileo e Keplero. Filosofia, cosmologia e teologia nell'Età della Controriforma*. Turin: Einaudi.

Buridan, Jean. 1942. *Iohannis Buridani Quaestiones super libris quattuor De caelo et mundo*, ed. Ernst A. Moody. Cambridge, MA: Medieval Academy of America.

Camerota, Michele. 2004. *Galileo Galilei e la cultura scientifica nell'età della Controriforma*. Rome: Salerno Ed.

Clavelin, Maurice. 1968. *La philosophie naturelle de Galilée*. Paris: Librairie Armand Colin.

Clericuzio, Antonio. 2000. *Elements, Principles and Corpuscles. A Study of Atomism and Chemistry in Seventeenth Century*. Dordrecht: Kluwer.

Collins, Jeffrey R. 2002. Thomas Hobbes and the Blackloist Conspiracy of 1649. *Historical Journal* 45: 305–331.

———. 2005. *The Allegiance of Thomas Hobbes*. Oxford: Oxford University Press.

Crombie, Alistair C. 1953. *Augustine to Galileo. The History of Science A.D. 400–1650*. Cambridge, MA: Harvard University Press.

Drake, Stillman. 1970. *Galileo Studies. Personality, Tradition, and Revolution*. Ann Arbor: The University of Michigan Press.

Festa, Egidio. 1999. Le galiléisme de Gassendi. In *Géométrie, atomisme et vide dans l'école guliléenne*, ed. Egidio Festa, Vincent Jullien, and Maurizio Torrini, 213–227. Fontenay St.-Cloud: ENS Éditions.

Finocchiaro, Maurice A. 2008. *The Essential Galileo*. Indianapolis/Cambridge: Hackett Publishing Company.

Funkenstein, Amos. 1986. *Theology and the Scientific Imagination from the Middle Ages to the Seventeenth Century*. Princeton: Princeton University Press.

Galluzzi, Paolo. 1979a. *Momento. Studi galileiani*. Rome: Edizioni dell'Ateneo & Bizzarri.

———. 1979b. Il tema dell'"ordine" in Galileo. In *Ordo. Atti del II colloquio internazionale del Lessico Intellettuale Europeo*, ed. Marta Fattori and Massimo L. Bianchi, 235–277. Rome: Ateneo & Bizzarri.

———. 1993. Gassendi e *l'affaire Galilée* delle leggi del moto. *Giornale critico della filosofia italiana* 72 (1): 86–119.

Garber, Daniel. 2000. A Different Descartes. Descartes and the Programme for a Mathematical Physics in His Correspondence. In *Descartes' Natural Philosophy*, ed. Stephen Gaukroger, John Schuster, and John Sutton, 113–130. London-New York: Routledge.

Gargani, Aldo G. 1971. *Hobbes e la scienza*. Einaudi: Turin.

Gaukroger, Stephen. 2002. *Descartes' System of Natural Philosophy*. Cambridge: Cambridge University Press.

———. 2006. *The Emergence of a Scientific Culture*. Oxford: Oxford University Press.

Geymonat, Lodovico. 1957. *Galileo Galilei*. Turin: Einaudi.

Grant, Edward. 1981. *Much Ado About Nothing: Theories of Space and Vacuum from Middle Ages to the Scientific Revolution*. Cambridge: Cambridge University Press.
———. 1996a. *The Foundation of Modern Science in the Middle Ages*. Cambridge: Cambridge University Press.
Grant, Hardy. 1996b. Hobbes and Mathematics. In *The Cambridge Companion to Hobbes*, ed. Tom Sorell, 108–128. Cambridge: Cambridge University Press.
Henry, John. 2016. Hobbes, Galileo and the Physics of Simple Circular Motion. *Hobbes Studies* 29: 9–38.
Hooper, Wallace. 1996. Inertial Problems in Galileo's Preinertial Framework. In *The Cambridge Companion to Galileo*, ed. Peter Machamer, 146–171. Cambridge: Cambridge University Press.
Jacquot, Jean, and Harold Whitmore Jones. 1973. Introduction. In *Thomas Hobbes, Critique du De Mundo de Thomas White*, 9–102. Paris: Vrin.
Jesseph, Douglas M. 1999. *Squaring the Circle. The War Between Hobbes and Wallis*. Chicago/London: University of Chicago Press.
———. 2004. Galileo, Hobbes and the Book of Nature. *Perspectives on Science* 12 (2): 191–211.
———. 2016. Hobbes on 'Conatus': A Study in the Foundations of Hobbesian Philosophy. *Hobbes Studies* 29 (1): 66–85.
Koyré, Alexandre. 1966. *Études galiléennes*. Paris: Hermann (or. Ed. 1939).
Kuhn, Thomas S. 1962. *The Structure of Scientific Revolutions*. Chicago: University of Chicago Press.
Lasswitz, Kurd. 1963. *Geschichte der Atomistik von Mittelalter bis Newton*, 2 Vols. Hildesheim: Georg Olms, (or. Ed. 1890).
Leijenhorst, Cees. 2002. *The Mechanisation of Aristotelianism. The Late Aristotelian Setting of Thomas Hobbes' Natural Philosophy*. Leiden/Boston/Köln: Brill.
———. 2004. Hobbes and the Galilean Law of Free Fall. In *The Reception of the Galilean Science of Motion in Seventeenth-Century Europe*, ed. Carla Rita Palmerino and J.M.M.H. Thijssen, 165–184. Dordrecht/Boston/London: Kluwer.
Lupoli, Agostino. 2001. Power (Conatus-Endeavour) in the 'Kinetic Actualism' and in the 'Inertial' Psychology of Thomas Hobbes. *Hobbes Studies* 14: 83–103.
Maier, Annaliese. 1968. *Zwei Grundprobleme der Scholastischen Naturphilosophie*. Rome: Edizioni di Storia e Letteratura.
Malherbe, Michel. 1984. *Thomas Hobbes ou l'oeuvre de la raison*. Paris: Vrin.
McLaughlin, Peter. 2000. Force, Determination and Impact. In *Descartes' Natural Philosophy*, ed. Stephen Gaukroger, John Schuster, and John Sutton, 81–112. London/New York: Routledge.
Messeri, Marco. 1985. *Causa e spiegazione: la fisica di Pierre Gassendi*. Milan: Franco Angeli.
Mintz, Samuel I. 1952. Galileo, Hobbes and the Circle of Perfection. *Isis* 43: 98–100.
Moody, Ernest A. 1951. Galileo and Avempace, The Dynamics of the Leaning Tower Experiment. *Journal of the History of Ideas* 12 (3): 375–422.
Paganini, Gianni. 1990. Hobbes, Gassendi e la psicologia del meccanicismo. In *Hobbes Oggi*, ed. Bernard Willms et al., 351–445. Milan: Franco Angeli.
———. 2008. Le néant et le vide: les parcours croisés de Gassendi et Hobbes. In *Gassendi et la modernité*, ed. Sylvie Taussig, 177–214. Turnhout: Brepols.
———. 2010. Introduzione. In *Moto, luogo e tempo*, ed. Thomas Hobbes, 9–104. Turin: Utet.
Palmerino, Carla Rita. 2004. Galileo's Theories of Free Fall and Projectile Motion as Interpreted by Pierre Gassendi. In *The Reception of the Galilean Science of Motion in Seventeenth-Century Europe*, ed. Carla Rita Palmerino and J.M.M.H. Thijssen, 137–164. Dordrecht/Boston/London: Kluwer.
———. 2008. Une force invisible à l'œuvre: le rôle de la *vis attrahens* dans la physique de Gassendi. In *Gassendi et la modernité*, ed. Sylvie Taussig, 141–176. Turnhout: Brepols.
Parodi, Massimo. 1981. *Tempo e spazio nel medioevo*. Turin: Loescher.

Pietarinen, Juhani. 2001. Conatus as Active Power in Hobbes. *Hobbes Studies* 14: 71–82.
Prins, Jan. 1996. Hobbes on Light and Vision. In *The Cambridge Companion to Hobbes*, ed. Tom Sorell, 129–156. Cambridge: Cambridge University Press.
Redondi, Pietro. 1997. I fondamenti metafisici della fisica di Galileo. *Nuncius* 12: 267–289.
Robinet, Alexandre. 1990. Hobbes: structure et nature du *conatus*. In *Thomas Hobbes: Philosophie première, théorie de la science et politique*, ed. Yves-Charles Zarka and Jean Bernhardt, 127–138. Paris: Presses Universitaires de France.
Roux, Sophie. 2004. Cartesian Mechanics. In *The Reception of Galilean Science of Motion in Seventeenth-Century Europe*, ed. Carla Rita Palmerino and J.M.M.H. Thijssen, 24–66. Dordrecht/Boston/London: Kluwer.
Sarasohn, Lisa T. 1985. Pierre Gassendi, Thomas Hobbes and the Mechanical World-View. *Journal of the History of Ideas* 46 (3): 363–379.
Schmitt, Charles B. 2001. *Filosofia e scienza nel Rinascimento*. Florence: La Nuova Italia.
Sergio, Emilio. 2001. *Contro il Leviatano. Hobbes e le controversie scientifiche 1650–1665*. Soveria Mannelli: Rubbettino.
Shea, William R. 1977. Marin Mersenne: Galileo's 'traduttore-traditore'. *Annali dell'Istituto e Museo di Storia della Scienza di Firenze* A. 2 (1): 55–70.
———. 1991. *The Magic of Numbers and Motion*. Canton: Science History Publication.
Southgate, Beverley C. 1993. *Covetous of Truth.' The Life and Work of Thomas White, 1593–1676*. Dordrecht/Boston/London: Kluwer.
Spragens, Thomas A. 1973. *The Politics of Motion. The World of Thomas Hobbes*. London: Croom Helm.
Watkins, John W.N. 1965. *Hobbes's System of Ideas*. London: Hutchinson.
Westfall, Richard S. 1971. *The Construction of Modern Science. Mechanisms and Mechanics*. Cambridge: Cambridge University Press.

Chapter 4
The Paradoxes of Matter

4.1 Solid Bodies, Fluids, and Very Fine Powder

In the first day of his *Discourses and Mathematical Demonstrations Relating to Two New Sciences*, Galileo illustrates the theme of the 'first new science,' which concerns the 'resistance which solid bodies offer to fracture.'[1] This is the most comprehensive and complex discussion left to us by the Italian scientist regarding his concept of matter,[2] in which the passages devoted to the paradoxes of the infinite are particularly noteworthy.[3] Galileo tackles the subject of the composition of the *continuum* by focusing the argument on a geometric problem[4]: in order to be divisible into infinitely divisible parts, the line and every *continuum* must necessarily be composed of an infinite number of indivisibles 'and if the parts are infinite in number, we must conclude that they are not finite in size, because an infinite number of finite quantities would give an infinite magnitude.'[5] Galileo writes:

---

[1] Galilei, *Discorsi e dimostrazioni matematiche intorno a due nuove scienze*, *OG*, VIII, 47.

[2] See, in this regard: Biener 2004, esp. 266 ff.

[3] Carla Rita Palmerino focused on the problem of the continuum and, especially, the solution developed by Galileo for the paradox of *rota aristotelis*. See Palmerino 2000, and Palmerino 2001.

[4] See Palmerino 2000, 290. The subject of the mathematical and physical composition of the continuum was explored extensively in the ancient and medieval world. See Murdoch 1982. William of Ockham took the Aristotelian theory up again, firmly denying the real existence of *mathematical points* (see, for example, Ockham, 'Expositionis in libros artis logicae Proemium et Expositio in librum Porphyrii de praedicabilibus.' In Ockham 1974–1988, II, 205 ff; McCord Adams 1987, I, 201 ff). Ockham's position was shared by Jean Buridan and Albert of Saxony, as well as by the Mertonians Thomas Bradwardine and William Heytesbury (see Murdoch 1982, 573–575), while the existence of *indivisibles* was upheld by Henry Harclay, Walter Chatton, Gerard Odon and Nicolas Bonet (ibid., 575–576). See also Murdoch 2009. On the problem of the *continuum* and the *minima naturalia* in the Middle Ages see Maier 1984, 271 ff.

[5] Galilei, *Discorsi e dimostrazioni matematiche intorno a due nuove scienze*, *OG*, VIII, 80; Eng. trans. Galilei 2003, 34.

© Springer Nature Switzerland AG 2020
G. Baldin, *Hobbes and Galileo: Method, Matter and the Science of Motion*,
International Archives of the History of Ideas Archives internationales
d'histoire des idées 230, https://doi.org/10.1007/978-3-030-41414-6_4

> A division and a subdivision which can be carried on indefinitely presupposes that the parts
> are infinite in number, otherwise the subdivision would reach an end; and if the parts are
> infinite in number, we must conclude that they are non-quanta, because an infinite number
> of quanta would give an infinite magnitude. And thus we have a continuous quantity built
> up of an infinite number of indivisibles.[6]

Naturally, the infinite 'subdivision' mentioned in the cited passage is a mathematical subdivision. Galileo supposes that the spatial entities that form the geometric shapes—for example the straight line—must be conceived as a *continuum* made
up of parts that can be mathematically divided indefinitely. These parts are *non-
quanta*, that is to say lacking in magnitude, and are infinite.

Nevertheless, there is a terminological and conceptual shift in Galileo's handling
of the topic because the scientist goes on to explore the nature of *fluids*, progressing
from divisibility conceived in purely mathematical – and, thus, hypothetical – terms
to a real subdivision present in the physical universe.[7] Galileo's argument is developed on the basis of a suggestion made by Salviati: he theorizes that, 'having broken
up a solid into many parts, having reduced it to the finest of powder, even so resolved
it into *its infinite atoms that are no more divisible*,' it could be conceived as a *fluid*.
Spurred on by the observation made by Sagredo, who asks whether fluids are such
because 'resolved into their first infinite indivisible components,'[8] the spokesman
for Galileo's opinions expounds upon an extremely interesting theory. Firstly,
Galileo maintains that, even when reducing any *hard* body, such as stone or metal,
into the 'most minute and impalpable powder,' the finest particles into which it is
reduced, 'although when taken one by one are, on account of their smallness, imperceptible to our sight and touch, are nevertheless *quanta, shaped, and capable of
being counted*.'[9] In other words, even supposing that any hard body can be smashed
into the smallest of pieces with a hammer or a mortar, it would still be composed of
finite particles that can be counted, that is to say corpuscles of a certain magnitude
and shape and, as a result, of finite number. Nevertheless, Galileo believes that
water and fluids have a different composition:

> But if we attempt to discover such properties in water we do not find them; for when once
> heaped up it immediately flattens out unless held up by some vessel or other external retain-
> ing body; when hollowed out it quickly rushes in to fill the cavity; and when disturbed it
> fluctuates for a long time and sends out its waves through great distances. Seeing that water
> has less firmness than the finest of powder, in fact has no consistence whatever, we may, it
> seems to me, very reasonably conclude that the *minima* into which it can be resolved
> (because the lesser consistence of whatsoever finest powder, that, in fact, has no consis-
> tence), they are *very different from minima and divisible quanta*; indeed the only difference
> I am able to discover is that the former are *indivisibles*.[10]

---

[6] Ibid.; Eng. Trans. 34, slightly modified.

[7] Le Grand 1978 has been the first to focus on this terminological and conceptual shift. See also
Baldini 1977a, esp. 66.

[8] Galilei, *Discorsi e dimostrazioni matematiche intorno a due nuove scienze*, *OG*, VIII, 85; Eng.
trans. Galilei 2003, 36–37, slightly modified.

[9] Ibid.; Eng. Trans. 37, slightly modified.

[10] Ibid., 85–86 (my italics); Eng. trans. 37, slightly modified.

Galileo maintains that *fluids* are characterized by a certain corpuscular composition, meaning that their *particles* are of a different nature to those of solid bodies. Despite this, the Italian scientist does not suppose the existence of two material entities that are entirely different from one another.[11] Indeed, as we shall see, he imagines that the liquefaction process also splits metals up into their ultimate components by virtue of the action of fire. Before examining his position in greater detail, it is worth noting that the subdivision of matter into *solid* and *fluid* bodies can also be found in a number of writings by Hobbes. In his *Tractatus Opticus II*, the lexicon of his argument seems to be drawn from Galileo, at least in part:

> The bodies in the universe are of two kinds, some whose parts are coeherent among them and that are not easily separated, and which are called *hard*, some more, some less. The other [bodies are those] whose parts spread because of every impress of motion, and separate from each other, and these [bodies] are called *fluids*. Whether a body is really so fluid that according to some reflection is divided in parts always divisible, and in the same way the same thing could always be separated in separable parts, could not be discussed here.[12]

Hobbes also seems to propose a division here between hard and fluid bodies. Nevertheless, we know that he believed hardness to be produced by the movement of the internal parts of the bodies: the faster and more whirling these internal movements, the harder and more resistant the bodies to external pressure. Yet in Chap. 26 of *De corpore*, Hobbes returned to the composition of fluids and his words noticeably echo the discussion outlined by Galileo on the first day of his *Discourses*:

> Wherefore a fluid body is always divisible into bodies equally fluid, as quantity into quantities; And though many men seem to conceive no other difference of *fluidity*, but such as ariseth from the different magnitudes of the parts, in which sense dust, though of diamonds, may be called fluid; yet I understand by *fluidity*, that which is made such by nature, equally in every part of the fluid body; not as dust is fluid, for so a house which is falling in pieces may be called fluid; but in such manner as water seems fluid, and to divide itself into parts perpetually fluid.[13]

The analogies with the passages cited from Galileo's *Discourses* are evident, starting with the example of powder found in both authors. Moreover, Hobbes, just like Galileo, seems to linger here on the idea that the nature of fluid bodies is different from that of hard or resistant bodies.[14]

---

[11] See Baldini 1977a, 20–21. On the subject of the structural isomorphism of the universe as an endemic feature of some seventeenth-century philosophies and on the importance of the concept in modern science, see Palmerino 2011. On Galileo: 309 ff. On corpuscularism and its importance in modern science (particularly in Italy), see Baldini 1977b.

[12] *TO II*, chap. I, § 13, fol. 200v/154 (my italics, my trans.): 'Corporum in universum duo sunt genera; unum quorum partes inter se ita cohaerent ut non facile separentur, qualia sunt quae vocantur Dura, alia magis, alia minus; alterum quorum partes ad omnem motum impressionem diffluunt et aliae ab aliis dirimuntur, quod vocant fluidum. Utrum vero sit corpus aliquod ita fluidum ut quaemadmodum cogitatione in semper divisibilia dividi, ita re ipsa in semper separabilia separari possit, id hoc loco non est disputandum.'

[13] *De corpore*, XXVI, 4, *OL*, I, 347; *EW*, I, 426.

[14] A number of analogies between the handling of fluids in Galileo and Hobbes have been observed by Thomas Holden. See Holden 2004, 18.

However, on the other hand, it is also worth underscoring an important difference: Hobbes openly distances himself from atomism in numerous works, but primarily in a letter to Samuel Sorbière dated February 1657, in which he reiterates the
differences between his conception of matter and that of Epicurus.[15] In an earlier
missive, Sorbière had shown himself to be dissatisfied with the reasons adopted by
Hobbes in the second paragraph of Chap. 26 of *De corpore* to challenge the existence
of the vacuum, asking his friend for further clarifications in relation to his concept
of matter.[16] Hobbes answered that he did not believe Epicurus' theory to be absurd
at all, but, simply, that the latter called a *vacuum* 'what Descartes calls "*subtle matter*" and what I call "*extremely pure ethereal substance*," of which no part is an
atom, and each part is divisible (as quantity is said to be) into further divisible
parts.'[17] Furthermore, going back to his claims in the third paragraph of Chap. 26 of
De corpore,[18] he clearly specified the nature of this fluid and the morphological and
structural characteristics that obviously distinguish its microparticulate composition
from that of hard bodies. After listing the various problems encountered when
exploring the issue of the vacuum and atomism, Hobbes further clarifies the nature
and internal composition of fluids:

> All these things are difficult to explain. But these difficulties are necessarily incurred by
> those who think that the fineness of bodies arises from the fineness of their atoms, as if
> water, air, and other fluids were like milled corn, and as if they were more or less fine
> depending on whether they were knocked and thrown together into thicker or smaller par-
> ticles. In my opinion not only pure air but also pure water and even pure quicksilver are
> infinitely divisible (*in partes divisibilia esse sentio semper divisibiles*).[19]

By their very nature, fluids are therefore homogenous and entirely different to
non-fluid bodies. Consequently, ground wheat cannot be reduced to a fluid body,
despite being crushed into the finest particles, almost imperceptible to the senses.
On the contrary, particles of pure air, water or mercury are, by their very definition
of fluids, divisible into infinitely divisible parts.

As in *De corpore*, here too Hobbes seems to refer back to Galileo's observations
on the internal composition of fluids, which he deemed to be infinitely divisible.
Nevertheless, Hobbes is far from considering the existence of ultimate components
of matter and even further from describing them as indivisible, as Galileo did.

---

[15] On this subject see primarily Giudice 1997. The question of the corpuscular composition of fluid
and the theme of atomism will be discussed in greater depth in the next paragraph.

[16] See Samuel Sorbière to Hobbes, from Paris, [23 January/] 2 February 1657, *CH*, I, 433–435.

[17] Hobbes to Samuel Sorbière, from London, 6 [/16 February] 1657, *CH*, I, 443, Eng. trans. Noel
Malcolm, 445.

[18] *De corpore*, XXVI, 3, *OL*, I, 340: 'Neque causa apparet major, quare materia mundana inter-
puncta debeat esse, ad admittendum motum, spatiis vacuis quam spatiis plenis, plenis inquam sed
fluidis'. What is more, regarding the nature of fluid, he added that those who uphold the vacuum,
'… dum naturam fluidi contemplantur, constare ilud imaginantur tanquam ex duri granulis, quomodo farina fluida fit e frumento molito; cum tamen fluidum concipere possibile sit tanquam
natura sua aeque homogeneum, ac est ipsa atomus vel ipsum vacuum.'

[19] Hobbes to Samuel Sorbière, from London, [February 6/16] 1657, *CH*, I, 443; Eng. trans. 446.

Hobbes returns to and develops the theme of the fluid in his *Dialogus physicus de natura aeris*,[20] where he compares his concept of matter with that of certain members of the Royal Society. Additional references to this issue feature in subsequent works,[21] but it is primarily here that Hobbes insists on the particular composition of fluids. He states that he is not yet able to subscribe in any way to the division between fluid bodies and non-fluid bodies on the basis of the dimensions of the internal particles that make up these bodies, as established by his adversaries. Hobbes uses the image of powder once again and claims that, according to the distinction proposed by his critics, the rubble of 'St Paul's Cathedral,'[22] could be described as fluid. He also pressed his interlocutors, who maintained that air had to be composed of non-fluid parts:

> Truly, you who cannot accept infinite divisibility, tell me what appears to you to be the reason why I should think it more difficult for almighty God to create a fluid body less than any given atom whose parts might actually flow, than to create the ocean. Therefore, you make me despair of fruit from your meeting by saying that they think that air, water, and other fluids consist of nonfluids: as if they were to call fluid a wall whose ruined stones fell around the place.[23]

Conversely, if we suppose that fluids are made up of non-fluid particles, we must necessarily conclude that the rubble of St Paul's cathedral, like that of any other solid body that crumbles, is fluid.

We have emphasized the profound analogies with Galileo's claims in his *Discourses*. Nevertheless, in order to understand Hobbes' concept of matter, it is opportune to analyze the discussion he devotes to Galileo's argument in greater detail.

4.2 Maximum Resolution, Indivisibles, and Atoms

The problem of the structure of matter is one of the crucial themes of Galilean thought and has been extensively and variously explored by philosophical historiography.[24] The most recent analyses have highlighted that this theme not only occupied

---

[20] On the subject of fluid in the *Dialogus physicus* see primarily Shapin and Schaffer 1985, 115 ff.

[21] See, for example, *Decameron physiologicum*, *EW*, VII, 108–109 and 138–139.

[22] *Dialogus physicus de natura aeris*, *OL*, IV, 244–245: 'Ego, contra, distinctionem non capio inter fluida et non fluida, quam sumitis a magnitudine partium; nam si caperem, ruina illa, sive rudera illa, quae jacent in ecclesia Paulina, mihi dicenda essent fluida'.

[23] Ibid. Eng. trans. Shapin and Schaffer 1985, 354.

[24] Lasswitz had already drawn attention to Galileo's atomism, see Lasswitz 1963, II, 37 ff. Further significant contributions on this topic include: Shea 1970; Le Grand 1978. Baldini 1977a broadly develops the issue, comparing the *Discourse on Bodies in Water* (1612) with the *Discourses and Mathematical Demonstrations* (1638). Baldini maintains that Galileo's 'corpuscularism' is key to the distinction between *primary and secondary qualities* (ibid., 16–17), although he affirms that the Galilean position differs clearly from the modern imitators of atomism such as Gassendi. Pietro Redondi considered three phases of the Galilean concept of matter: an initial moment linked to a personal interpretation of certain Aristotelian theories taught at the Collegio Romano; a phase

Galileo during his later years, but that the first signs of interest in this subject can actually be traced back to his Paduan period.[25] Interesting observations on the Galilean concept of matter also emerge from the *Discourse on Bodies in Water*[26] but even more so from his *Discourse on Comets* (1619), and *The Assayer* (1623). In the former work, Galileo supported a timidly atomistic position (receiving numerous accusations because of this, particularly from the Aristotelian Vincenzo Di Grazia).[27] Here, with regard to the nature of metals, Galileo maintained that they appear to be divided into 'last and least particles' following the smelting process. He claims that 'by employing the most subtle and acute instruments, such as the weakest parts of fire,' we shall '*resolve*' the metal '(perhaps) into its smallest particles, so that they shall no longer offer any resistance to fracture, nor will it be possible to divide them, especially using instruments larger than needles of fire.'[28]

In the same work, Galileo reflected on the difference between the nature of solid and fluid bodies, but here he seemed to hold a different position to the one he would adopt many years later in his *Discourses*. In fact, in his 1612 text, water and other fluids were considered to be made from *contiguous* rather than *continuous* parts:

> Two manners of penetration, therefore, offer themselves to us: one in bodies with continuous parts, where division seems necessary; the other in aggregates of non-continuous parts, which are merely contiguous only and where there is no need for division. I have not yet resolved where water and other fluids may be deemed to be made up of continuous or simply contiguous parts. I am inclined to believe they are contiguous.[29]

When reading the cited passage, we have the impression that Galileo was aware he had not reached a definitive solution regarding the microscopic nature of fluid

---

characterized by physical atomism, which reached its peak in 1623 with the publication of *The Assayer*, and, lastly, a third period, after 1634, when Galileo's atomism was characterized by a mathematical and geometrical concept of the continuum, which culminates in his *Discourses*. See esp. Redondi 1985. The question of atomism is also central to his well-known *Galileo's eretico*. See Redondi 2009, esp. 16–31. Redondi deems that the 'shift' from a physical interpretation of atomism to the geometrical one of the *Discourses* was also due to reasons of a 'theological' nature). Nonnoi 2000 claimed that in the first day of the *Discourses* Galileo did not develop the problem in a definitive and complete manner and insisted on the difference between the concept of *minima* and that of *atoms*. See, also Biener 2004, Gómez Lopez 2001, and Gómez Lopez 2008.

[25] See Galluzzi 2011, 36–37. As Galluzzi observed (taking up some observations already made by Redondi, see Redondi 1985, 537–539), some Galilean passages concerning the subject of matter and atomism present profound analogies with the observations sketched out by Paolo Sarpi in some of his *Pensieri*, which the Servite presumably composed in around the 1580s (Galluzzi 2011, 70 and note). We we will be returning to this subject later on.

[26] Baldini also focused on this. See Baldini 1977a, 6 ff. The author also claims that the *Discourse* is the first text in which Galileo criticizes not only individual aspects of Aristotelian physics, but disputes the very foundations of Aristotle's natural philosophy (ibid., 9).

[27] See Palmerino 2000, 292 ff.

[28] Galilei, *Discorso intorno alle cose che stanno in su l'acqua e che in quella si muovono*, *OG*, IV, 106.

[29] Ibid., 105–106 (my trans.). On this topic of the *Discourse* and on the development of the subject in Galileo's later works, see Clavelin 1968, 441 ff.; Shea 1970, 13–14; and, above all, Baldini 1977a, 11 and 17.

bodies,[30] but an important development can be observed in his famous work of 1623: *The Assayer.* Here, when discussing the distinction between the different *qualities*, Galileo expresses himself in these terms:

> Having shown that many sensations which are supposed to be qualities residing in external objects have no real existence save in us, and outside ourselves are mere names, I now say that I am inclined to believe heat to be of this character. Those materials which produce heat in us and make us feel warmth, which are known by the general name of 'fire,' would then be a multitude of minute particles having certain shapes and moving with certain velocities. Meeting with our bodies, they penetrate by means of their extreme subtlety, and their touch as felt by us when they pass through our substance is the sensation we call 'heat.'[31]

The Assayer is packed with references to *minute particles*, of a certain size and shape and whose movements, slow or swift, are the cause of our sensations.[32] However, Galileo allows for an exception that concerns the phenomenon of light: he alludes to its different microscopic structure and to its particular ontological constitution compared to sound and heat.[33] Although he maintained that his argument was too uncertain to be expressed in a printed text, he nevertheless reneged on this precautionary reticence and stated:

> ... when such attrition stops at or is confined to the smallest quanta, their motion is temporal and their action calorific only; but when their ultimate and highest resolution into *truly indivisible atoms* is arrived at, light is created. This may have an *instantaneous motion*, or rather an instantaneous expansion and diffusion, rendering it capable of occupying immense spaces by its—I know not whether to say its *subtlety*, its *rarity*, its *immateriality*, or some other property which differs from all these and is nameless.[34]

Galileo supposes that *light*—unlike sound and heat—is a substance, a matter that populates the physical universe.[35] However, it has unique characteristics that distin-

---

[30] Galluzzi maintains that this excerpt represents a further step forward by the scientist towards developing a 'binary' concept of atomism, which contemplates two levels (equivalent, symmetrical, and interchangeable) at the same time: physical and mathematical. See Galluzzi 2011, 6.

[31] Galilei, *Il saggiatore, OG*, VI, 350; Eng. trans. Drake 1957, 277.

[32] This idea had already been put forward by Galileo some years earlier, as recorded by the scientist's correspondence. In a letter from as early as 1619 he refers to *shape, size and motion* as characteristics typical of the *smallest parts* that make up bodies and are the fundamental components of the phenomenology of the sensation that Galileo would go on to develop in *The Assayer.* In a letter to Galileo dated 8 August 1619, Giovanni Battista Baliani expressed his doubts regarding the theory of matter that Galileo had set out through the pen of his pupil Guiducci in the *Discourse on Comets.* At the bottom of the page, Galileo had noted that the only real characteristics that can be attributed to bodies are *shape, size and motion.* See Giovanni Battista Baliani to Galileo, 8 August 1619, *OG*, XII, 475. See also Galluzzi 2011, 68 ff.

[33] Galilei, *Il saggiatore, OG*, VI, 350: 'E come ai quattro sensi considerati ànno relazione i quattro elementi, così credo che per la vista, senso sopra tutti eminentissimo, abbia relazione la luce, ma con quella proporzione d'eccellenza qual è tra 'l finito e l'infinito, tra 'l temporaneo e l'istantaneo, tra 'l quanto e l'indivisibile, tra la luce e le tenebre.'

[34] Ibid., 352; Eng. Trans. Drake 1957, 278.

[35] Baldini 1977a emphasized that in a letter to Fortunio Liceti, dated 25 August 1640, Galileo openly disputed the position that had been attributed to him by the Aristotelian philosopher La Galla, who accused him of claiming light to be a material and corporeal substance (see Galileo to

guish it from other bodies: it is 'resolved' into its least parts, that is to say into 'truly indivisible atoms' and, because of its 'subtlety,' 'rarity' or, even, 'immateriality,' is endowed with instantaneous movement, which enables it to 'occupy immense spaces.'

Light can therefore be conceived as an extremely fluid body, composed of infinite indivisible particles lacking in magnitude. Consequently, in *The Assayer* light is the physical entity that best illustrates the *continuum*, just as Galileo describes it in his *Discourses and Mathematical Demonstrations*.[36] In later years, the Italian scientist reworks this concept and sets it out more clearly, as can be seen in his correspondence. The idea emerges that light is an extremely rarefied substance, representing the archetype of maximum rarefaction, being composed of infinite indivisibles and producing all the other bodies that populate the physical world by condensing itself.[37]

In his *Dialogue Concerning the Two Chief World Systems*, when discussing the phenomenon of magnetism, Galileo includes some brief considerations that seem to anticipate the extensive discussion of the subject developed later in his *Discourses*. Indeed, he supposes that contact between two ferrous surfaces can be established between the '*infinity of points* on both surfaces.'[38] In other texts from the same period, Galileo reflects on the composition of the continuum and affirms that 'saying that the continuum consists of parts that are *always divisible*, along with saying that the continuum consists of *indivisibles* are the same thing.'[39] He maintains that the continuum is infinitely divisible precisely and solely because it is composed of indivisibles.[40]

Galileo's considerations reach a turning point, or completion, in the last major work of his mature years: his *Discourses*, where he presents the theory we encountered previously, contemplating the seamless passage of the continuum from the *geometrical* to the *physical* dimension.

The geometrical discussion develops out of the solution proposed by Galileo in reference to the mathematical paradox known as *rota aristotelis*.[41] Taking two concentric circles of different dimensions (*figure b*) and imagining that the larger one is rotated

---

Fortunio Liceti, 25 August 1640, *OG*, XVIII, 233–234). Nevertheless, he considered it to be composed of infinite corpuscles, so that it could be defined as something 'simultaneously' corporeal and incorporeal.

[36] However, we must observe that in the *Discourses* of 1638, Galileo would claim that light has no instantaneous velocity, but moves in time. See Galilei, *Discorsi e dimostrazioni matematiche intorno a due nuove scienze*, *OG*, VIII, 88–9.

[37] See Galluzzi 2011, 76–78.

[38] Galilei, *Dialogo sopra i due massimi sistemi*, *OG*, VII, 435–6; Eng. Trans. Galilei 2001, 409: '…il toccamento si fa di innumerabili minime particelle, se non forse degli infiniti punti di ambedue le superficie'. See also Galluzzi 2011, 82–84.

[39] Galilei, *Postille alle esercitazioni filosofiche di Antonio Rocco*, *OG*, VII, 745 (my italics).

[40] Ibid. See Galluzzi 2011, 87.

[41] See Palmerino 2000, 281 ff. In general, on the history of Aristotle's wheel paradox, which originates in a problem covered in the pseudo-Aristotelian *Questions of Mechanics*. See Drabkin 1950, and Costabel 1964.

on its own tangent, it necessarily drags the smaller one with it and the two travel the same distance. The problem that emerges concerns the difficulty of explaining how the smaller circle can draw a path which is identical to that of the bigger circle, when completing a single revolution and always remaining in contact with the tangent.[42]

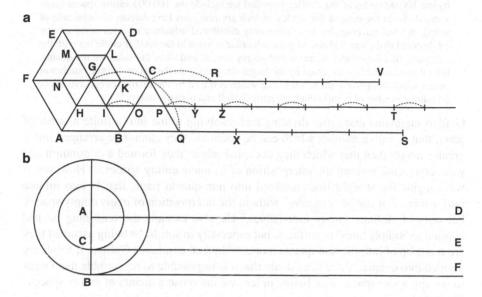

Figures **a** (above) and **b** (below)

The strange solution proposed by Galileo is based on the idea that the circle is a polygon with infinite sides and that the two segments described by concentric circles (*figure b*) are only seemingly equal,[43] but are actually different due to the presence of a much higher number of non-quanta empty spaces than those inside the segment marked out by the circumscribed circle.[44] As we have observed, Galileo's

---

[42] On the discussion of the *rota aristotelis* in the *Discourses* and the importance of the geometrical problem in reference to the concept of the continuum developed by Galileo, see Dijksterhuis 1961, 421–423.

[43] Galilei, *Discorsi e dimostrazioni matematiche intorno a due nuove scienze*, OG, VIII, 95–96: '… gl'infiniti lati indivisibili del maggior cerchio con gl'infiniti indivisibili ritiramenti loro, fatti nell'infinite istantanee dimore de gl'infiniti termini de gl'infiniti lati del minor cerchio, e con i loro infiniti progressi, eguali a gl'infiniti lati di minor esso cerchio, compongono e disegnano una linea eguale alla descritta dal minor cerchio, contentente in sé infinite sovrapposizioni non quante, che fanno una costipazione e condensazione senza veruna penetrazione di parti quante, quale non si può intendere nella linea divisa in parti quante, quale è il perimetro di qualsivoglia poligono, il quale, disteso in linea retta, non si può ridurre in minor lunghezza se non col far che i lati si sovrapponghino e penetrino l'un l'altro.'

[44] As observed by Palmerino 2000, 286–287, in the preface to *Méchaniques de Galilée*, Mersenne rejects the solution to the *rota aristotelis* based on the expedient of condensation and rarefaction. See Mersenne 1634a in Mersenne 1985, 433.

argument is driven by the idea that circles can be considered as polygons with infinite sides (*figure a*), meaning that:

> In the case of polygons with 100,000 sides, the line traversed by the perimeter of the greater, i.e., the line laid down by its 100,000 sides one after another, is equal to the line traced out by the 100,000 sides of the smaller, provided we include the 100,000 vacant spaces interspersed. So in the case of the circles, (which are polygons have having an infinitude of sides), the line traversed by the continuously distributed infinitude of sides is the greater circle equal to the line laid down by the infinitude of sides in the smaller circle but with the exception that these latter alternate with empty spaces; and since the sides are not quanta, but infinite, so the line traversed by the larger circle consists then of an infinite number of points which completely fill it; while that which is traced by the smaller circle consists of an infinite number of points which are partially full and partially empty.[45]

Galileo maintains that 'after dividing and resolving a line into a finite number of parts, that is, into a number which can be counted,' they cannot be arranged into a greater length than that which they occupied when 'they formed a continuum and were connected without the interposition of as many empty spaces.'[46] However, if we imagine the straight line 'resolved into non-quanta parts, that is into infinite indivisibles,'[47] it can be '*extended*' without the interposition of many empty spaces, but only of 'infinite empty indivisibles.' He also extends the reasoning he had applied to 'simple lines' to surfaces, but especially to solids, 'it being assumed they are made up of infinite, non-quanta atoms.'[48] In fact, continues Galileo, if solids are divided into quanta, 'there is no doubt that it is impossible to reassemble them so as to occupy wider spaces than before unless we interpose a quanta of empty spaces.' Nevertheless:

> ... if we consider the highest and last resolution into the primary components, non-quanta and infinite in number, then we shall be able to think of them as indefinitely extended in space, without the interposition of quanta, but of infinite non-quanta spaces.[49]

Galileo does not limit his theory to geometry, but also extends it to physics. This enables him to explain the phenomena of *rarefaction* and *condensation*:

> This contraction of non-quanta but infinite parts, without the interpenetration of (finite) quanta parts and the previously mentioned expansion of *infinite indivisibles by the interposition of indivisible vacua* is, in my opinion, the most that can be said concerning the *contraction and rarefaction of bodies*, unless we give up the impenetrability of matter and introduce empty quanta spaces.[50]

---

[45] Galilei, *Discorsi e dimostrazioni matematiche intorno a due nuove scienze*, *OG*, VIII, 71; Eng. trans. Galilei 2003, 28, slightly modified.

[46] Ibid., 72; Eng. Trans. 25.

[47] Ibid.: '... risoluta in parti non quante, cioè nei suoi infiniti indivisibili.'

[48] Ibid.

[49] Ibid. (my italics).

[50] Ibid., p. 96. As underscored by Galluzzi, the first day of the *Discorsi* marks the completion of the process undertaken by Galileo since the years of the Pisan *De motu,* in which he considers the mathematical and physical interpretation of the natural world to be perfectly equivalent and comparable. See Galluzzi 2011, 84.

Consequently, not even an 'ounce of gold might be rarefied and expanded until its size would exceed that of the earth,'[51] without admitting empty spaces of any magnitude. The only vacuum admitted by Galileo is, so to speak, a particular form of *vacuum interspersum*, which contemplates the existence of empty indivisibles mixed with the material indivisibles, that is to say the atoms. Galileo also extends this reasoning to the analysis of fluids and considers the atomic structure of *water*. He supposes that its *minima* have a different structure than all the other smallest parts: they are *indivisible atoms*.[52] Nevertheless, the Galilean theory of the continuum does not only apply to water: it is interesting to note what Galileo writes afterwards, when he alludes to the liquefaction of metals through smelting, through the action of the burning glass (that is to say one of those reflective parabolic surfaces, discussed by his pupil Bonaventura Cavalieri).[53] These metals, as a result of the action of the Sun's rays, are melted and become fluid, and are dissolved 'into their ultimate components, infinite, indivisibles.'[54]

4.3   A Subtle Fluid

Galileo's concept of matter did not fail to arouse perplexity, even among the circle of his closest followers, such as Bonaventura Cavalieri.[55] In France, in a letter to Mersenne written in autumn 1638, in which he examined Galileo's *Discourses* in detail, Descartes highlighted certain issues linked to Galileo's theory of matter and his solution of the *rota aristotelis* problem. Descartes stated that Galileo's proposal of considering solid bodies, having once become fluid upon smelting, to be divided into an infinite number of indivisibles, was nothing more than 'an imagination that was easy to refute and for which no proof was given.'[56] Similar criticism is directed

---

[51] *OG*, VIII, 96; Eng. Trans. Galilei 2003, 51.

[52] Ibid., 86.

[53] See Cavalieri 1632.

[54] Galilei, *Discorsi e dimostrazioni matematiche intorno a due nuove scienze*, *OG*, VIII, 86 (my italics).

[55] In a letter dated 2 October 1634, Cavalieri wrote to Galileo that his position did not force him to uphold the continuum composed of indivisibles (*OG*, XVI, 136–138, esp. 138). Subsequently, the two scholars discussed the composition of the continuum and indivisibles. See Cavalieri 1989, 725–769. On the discussion between the two and Cavalieri's contribution to the development of the Galilean theory, see Giusti 1980, 40 ff.; Galluzzi 2011, 109–111; Jullien 2015, and Palmerino 2010. See also Brunschvicg 1913, 162–167; Baroncelli 1992; Festa 1992; Radelet-De Grave 2015.

[56] Descartes to Mersenne, 11 October 1638, *AT*, II, 383. Upon examining the *Discourses and Mathematical Demonstrations,* Descartes deemed the Galilean plan to 'examine matters of physics with mathematical reasons' to be laudable and declared himself to be in total agreement with Galileo regarding this matter (ibid., *AT*, II, 380). Despite this, he made some detailed criticisms of Galileo's work. See Shea 1978, especially as regards the problem of the continuum and *indivisibles*: 152 ff.

at the mathematical aspect of Galileo's treatise: the opinion according to which the straight line drawn by the circle is composed of infinite 'actual' (that is, real) points is also described as 'pure imagination.'[57] Descartes' criticism was also directed to what he refers to as the 'glue,' which Galileo 'would combine ... with the vacuum in order to [explain] the link between the parts and the bodies.' This glue would be attributed, in Galileo's opinion, 'to other small empty spaces that can in no way be imagined.'[58] Descartes' criticism is very interesting, because it refers to Galileo's explanation of the cohesion of solid bodies, which the Italian scientist attributed to the presence of very small vacuums within the microscopic structure of the solid bodies. It is precisely the penetration of these very small vacuums by the 'extremely fine particles of fire' that permits the metals to liquefy during the smelting process, because the particles of fire free them 'from the attraction which these same vacua exert upon them and which prevents their separation.'[59]

Descartes, on the other hand, identified a different reason for the cohesion of solid bodies, referring to corpuscular links of another nature: a sort of ramified structure that he considered to be specific to the particles that make up the bodies.[60]

Hobbes' position on the subject of the composition of bodies and their cohesion differs from both that of Galileo and Descartes, but it is appropriate to consider Hobbes' first formulation of his theory in a letter to Mersenne regarding his dispute with Descartes. Hobbes maintains that the cause of the hardness of bodies lies in the movement or, to be precise, in the 'violence' of the motion of the spirits.[61] What is more, he maintains that his opinion is 'no less suitable for explaining the phenomenon of hardness than M. Descartes' hypothesis, which posits certain knots and entanglements of his atoms, which are meant to make the parts of hard bodies stick together.'[62] Hobbes indeed supposes the existence of an internal spirit inside the

---

[57] Ibid., *AT*, II, 384 (my trans.)

[58] Ibid., *AT*, II, 382 (my trans.).

[59] Galilei, *Discorsi e dimostrazioni matematiche intorno a due nuove scienze*, *OG*, VIII, 67; Eng. Trans. Galilei 2003, 19: '...le sottilissime particole del fuoco, penetrando per gli angusti pori del metallo (tra i quali, per la lor ristrettezza, non potessero passare i minimi dell'aria né di molti altri fluidi), col riempiere i minimi vacui tra esse fraposti liberassero le minime particole di quello dalla violenza con la quale i medesimi vacui l'una contro l'altra attraggono, proibendogli la separazione; e così, potendosi liberamente muovere, la lor massa ne divenisse fluida, e tale restasse sin che gl'ignicoli tra esse dimorassero; partendosi poi quelli e lasciando i pristini vacui, tornasse la lor solita attrazione, ed in conseguenza l'attaccamento delle parti.'

[60] Descartes illustrates his conception of the microscopic structure of solids, distinguishing it from that of water, in Descartes, *Les Météores*, *AT*, VI, 233; Eng. Trans. Garber 2009, 115: 'I assume that the small particles of which water is composed are long, smooth, and slippery like little eels, which are such that however they join and interlace, they are never thereby so knotted or hooked together that they cannot easily be separated; and on the other hand, I assume that nearly all particles of earth, as well as of air and most other bodies, have very irregular and rough shapes, so that they need be only slightly intertwined in order to become hooked and bound to each other, as are the various branches of bushes that grow together in a hedgerow.' On the structure of matter in Descartes see: Gaukroger 2002, 130–134.

[61] Hobbes to Mersenne for Descartes, 7 February 1641, *AT*, III, 302.

[62] Ibid., *AT*, III, 302–303.

bodies, also described as a 'subtle and fluid body,' which he considers to be similar to Descartes' *subtle matter*, whose movement causes the hardness of bodies. In fact, according to Hobbes, it should not be attributed to the greater cohesion of the internal parts, but instead to the rapid motion of this fluid or internal spirit.[63]

Unlike Galileo, Hobbes does not consider the penetration of particles inside the body to be necessary in order to split the atomic or corpuscular bonds, or to strengthen them. In order to explain the changes in state, he deems the change of speed of the subtle fluid present in the body to be sufficient. However, Hobbes' theory is not without its difficulties. Among other things, it is necessary to consider the microscopic or corpuscular composition of this subtle fluid. In his later works, Hobbes makes substantial changes to his theory of hard bodies, and in *De corpore* he suggests that various phenomena can cause bodies to harden, including the movement of the internal particles.[64] And yet the theme of fluid returns not only in *De corpore*, but in almost all his later scientific works, starting with the aforementioned *Dialogus physicus*.[65] Despite this, Hobbes' natural philosophy presents a significant shortcoming as regards the microscopic composition of matter: until 1648 the English philosopher adopts a position that we can describe as timidly vacuist.[66] He admitted the existence of a *vacuum disseminatum*,[67] only opposing the idea of a *vacuum separatum*, like Galileo.[68] It is fundamental to stress, however, that Hobbes was never a firm supporter of the existence of the vacuum and his recourse to the concept of the *vacuum disseminatum* or *interspersum* (within the light source) is merely functional to his theory of the vision and propagation of light.[69] In *De*

---

[63] In response to the objections made by Descartes, who did not agree that Hobbes' subtle fluid was comparable to his *subtle matter*, Hobbes set out his theory more clearly. See *AT*, III, 302, *CH*, I, 63: 'Neque enim ego dixi durescere corpora per ingressum spirituum, neque mollescere per exitum eorundem; Sed spiritus subtiles & liquidos, vehementiâ motûs sui, posse constituere corpora dura, ut adamantem; & lentitudine, alia corpora mollia, ut aquam vel aërem.'

[64] In *De corpore*, XXVIII, *OL*, I, 384–385, Hobbes considers various causes for bodies becoming hard: freezing; the cohesion of atoms within a coherent whole (ibid., 386–7); the 'exhalation' of fluid particles from the body, thanks to which the remaining hard particles become compacted (ibid., 388) and, lastly, Hobbes considers his old explanation, which contemplates the fast, whirling motion of particles (ibid.).

[65] The idea that the hardness of bodies can be attributed solely to internal motion is also found in *TO II*, 154, and in *MLT*, XXIV, 11, 299).

[66] See Kargon 1966, 57; Pacchi 1965, 240; Pacchi 1978; Bernhardt 1993; Giudice 1997, esp. 481 ff.

[67] As we know, until February 1648 Hobbes had admitted the possibility of a *vacuum interspersum* (see Hobbes to Marin Mersenne, from Saint Germain, 7[17] February 1648, *CH*, I, 165) and his attitude towards the vacuum changed, starting in May of the same year (see Hobbes to Marin Mersenne, from Saint Germain, 15 [25] May 1648, *CH*, I, 172).

[68] The terminology used to distinguish between *vacuum disseminatum* or *interspersum* and *vacuum separatum* comes from Gassendi. See Gassendi, *Syntagma*, In Gassendi 1658, I, 186. The issue is discussed in depth over the following pages. See also Grant 1981, 70–71 and 206–213 and Osler 1994, 183.

[69] Hobbes was forced to admit the existence of these small empty spaces due to his model of light propagation. This emerges clearly in the *Tractatus Opticus II* and returns in *De motu, loco et tempore* (*MLT*, IX, 2, 161) where Hobbes confirms that the movement of the light source is not admis-

corpore, Hobbes adopts a clearly anti-vacuist position, dissociating himself from the exponents of classic atomism and their modern followers.[70] In Chap. 7 of this work, he claims that 'whatsoever is divided, is divided into such parts as may again be divided' or, according to a different formulation, that 'the least divisible thing is not to be given.'[71] Likewise, in Chap. 26, when arguing against the doctrines of Lucretius and Epicurus, he proposes his concept of the perennially divisible fluid that we have already encountered.[72]

In his *Dialogus physicus*, where the microscopic composition of bodies is central to the debate between the two speakers, Hobbes returns to the idea that air is a fluid, that is to say that it is 'easily divisible into parts that are always still fluid and still air, such that all divisible quantities are there in any quantity.'[73]

As we have seen, his position echoes Galileo's treatise on the microparticulate composition of water in his *Discourses*, although Hobbes is far from considering it to be composed of *infinite indivisibles*.

In the continuation of the dialogue, Speaker B—who represents the position attributed by Hobbes to members of the Royal Society—indicates that the magnitude of the parts (that is to say the size of the corpuscles that compose them) marks the difference between fluid and non-fluid bodies, drastically opposing the idea of infinite division. To that, the Hobbesian character replies that 'infinite division cannot be conceived, but infinite divisibility can easily be,'[74] and he clearly rejects the definition of fluid on the basis of the size of the parts that make it up. The problem

sible, 'sine vacuo, vel spatiolis vacuis inter partes interiectis; sed quia neque impossibile est vacuum immaginari, neque possibile probare quòd omne spatium sit corpore aliquo repletum, nihil impedit quin partes solis motum talem habere possint'. In the *First Draught* Hobbes tackles the objection that motion within the light source cannot take place without the existence of the vacuum. See *FD*, fol. 6–7/96: 'I suppose, that there is a vacuity made by such dilatation, but find no impossibility, nor absurdity, nor so much as an improbability in admitting vacuity, for no probable argument hath ever beene produced to the contrarie, unlesse wee should take a space or extension for a body or thing extended and thence conclude because space is everywhere imaginable, Therefore bodie is in every space, For who knowes not that Extension is one thing and the thing extended another …'

[70] Hobbes' relationship with atomism has been explored since the nineteenth century, with the important study by Lasswitz (see Lasswitz 1963, II, 224 ff.) in which the author maintained that Hobbes' theory of fluids was what made it so difficult to place the author in the history of atomism. The issue was also taken up again by Pacchi in particular, who had already underscored the irreconcilability of atomism and antivacuism (see Pacchi 1965, 238–242, and primarily Hobbes 1978). See also Kargon 1966, 54–62. The difficulty involved in reconciling Hobbesian antivacuism with atomism was reiterated by Giudice (Giudice 1997). On the other hand, Agostino Lupoli, making particular reference to a passage from the *Decameron physiologicum*, in which Hobbes seems to suggest the existence of atoms that are created hard by an eternal cause (*EW*, VII, 134), maintains that Hobbes' concept of atoms is compatible with his fluidism, if we suppose a sort of physical indivisible. See Lupoli 1999, esp. 597 and Lupoli 2006, 539–554.

[71] *De corpore*, VII, 13, *OL*, I, 89; Eng. trans. *EW*, I, 100.

[72] Ibid., XXII, 17 ff., *OL*, I, 283 ff.

[73] *Dialogus physicus de natura aeris*, *OL*, IV, 244; Eng. Trans, Shapin and Schaffer 1985, 353.

[74] Ibid.; Eng. Trans, Shapin and Schaffer 1985, 354.

is taken up again in the final pages of the work, which explore the *cohesion* and *hardness* of bodies. Here, character B indicates three plausible causes for the phenomenon, which are: 'first, the magnitude of the parts; second, that the parts of the surface mutually touch; third, the entangled position of the parts.'[75]

Roughly speaking, the three causes of the above can respectively be associated with: (1) the opinion of certain members of the Royal Society (at least according to Hobbes' interpretation); (2) Galileo's theory, and, lastly, (3) the position adopted by Descartes, who identified a sort of bond between the corpuscles. Speaker A admits that all three causes can be contemplated as possible. Nevertheless, he upholds the need to indicate the cause of what is defined as the primary *hard body*, continuing:

> But if they (*that is the scientists from the Royal Society*) say hard bodies are made out of primary hard bodies, why do they not also think that fluid bodies are made out of primary fluid bodies? Could the greatest fluids, such as the aether, be created, and the least not be? He that first made a body hard or fluid could, if he had pleased, have made it greater or less than any other given body. If a fluid is made from nonfluid parts, as you have said, and such a hard body from hard parts, does it not follow that neither fluids nor hard bodies may be made from primary fluids?[76]

In the cited passage, it is claimed that if atoms or corpuscles are created hard *ab aeterno*, then as a result there is no reason to doubt that fluids must also always have been fluids and, therefore, by their very nature, divisible into infinitely divisible parts.

Meanwhile, as regards the cause of hardness, Hobbes again puts forward the same theory he had set out 20 years earlier in his letters to Descartes and in *De motu, loco et tempore*, focusing on the notion of motion. The cause of the fluid body is solely rest, while that of the hard body is 'some motion fit to produce that effect'[77] that is to say, the movement that Hobbes defines as *simple circular motion*. As a result, every body that is resistant to pressure can be described as hard and, therefore, air compressed by a piston can also be considered a hard body.

Hobbes also proposes a theory regarding the cause of the so-called 'primary hard body.' He starts by establishing that 'the matter that is contained in the nerves,' which is a 'very tenuous spirit,' can be formed into flesh as a result of compression, thereby acquiring a solid state.[78] 'And such indeed could be the efficient cause of *primary hardness*. However, the cause of *secondary hardness*, that is, of hardness

---

[75] Ibid., 283; Eng. Trans, Shapin and Schaffer 1985, 387.

[76] Ibid.; Eng. Trans, Shapin and Schaffer 1985, 388.

[77] Ibid, 284.

[78] Ibid, 285. The phenomenon was explained in a different manner in the first chapter of *De homine*, where Hobbes maintained that the generation of flesh in the human body is produced in different stages: firstly, with the digestion of food, then with circulation that carries the transformed matter to the nerves and, lastly, this matter, filtered through the nerves, becomes flesh. See *De homine*, I, 2, *OL*, II, 2–5. See also Médina 2013b, 160–162. The idea, derived from Galen, that animal spirits present in the nerves are like an air or a very subtle wind 'qui, venant des chambres ou concavités qui sont dans le cerveau, s'escoule par ces mesmes tuyaux dans les muscles', is present, in different terms, not only in Descartes' *Dioptrics* (*AT*, VI, 110), but also in Campanella 2007, 47 ff. See Ernst 2010, 26 and 108 ff.

from the cohesion of primary hard bodies, could be that same simple circular motion together with their superficial or else entangled contact.'[79] In other words, Hobbes supposes—although it must be stressed that this is purely a hypothesis—that the primary hard body can be produced by compressing the fluid.[80]

4.4 Infinite Indivisibles and Undivided Divisibles

The arguments developed by Hobbes in his *Dialogus physicus* are fundamental for resolving the difficulties associated with the theme of fluid and, more generally, with the structure of matter. There can be no doubt about the fact that Hobbes inherited the terminology and concepts of Galileo's speculations, particularly from the *Discourses and Mathematical Demonstrations Relating to Two New Sciences*. Hobbes presents a concept of the fluid that is extremely similar to the one developed by Galileo, but he never states that a fluid is 'resolved' into in its 'least' parts. Furthermore, while Galileo indicated water as the paradigm of the fluid 'resolved' into its smallest indivisible components, that is to say into an infinite number of tiny atoms, on the other hand his argument regarding the smelting of metals suggested that the action of the finest parts of fire could also reduce these solid bodies to their smallest parts, that is to say into infinite atoms.[81]

In his *Dialogus physicus*, but especially in his *De corpore*, Hobbes always talks about *potential infinite divisibility* and never precisely indicates the concrete possibility of an 'actual' division. Nevertheless, he develops an interesting argument in this regard in Chap. 7 of his work. Here he writes:

> ... that which is commonly said, that space and time may be divided infinitely, is not to be so understood, as if there might be any infinite or eternal division; but rather to be taken in this sense, *whatsoever is divided, is divided into such parts as may again be divided;* or thus, *the least divisible thing is not to be given;* or, as geometricians have it, *no quantity is so small, but a less may be taken.*[82]

In his later mathematical works, such as the *Examinatio et Emendatio Mathematicae Hodiernae* (1660), Hobbes also goes on to reiterate that 'division is the work of the

---

[79] *Dialogus physicus de natura aeris*, *OL*, IV, 245; Eng. trans, Shapin and Schaffer 1985, 389–390.

[80] According to Lupoli, this reveals the not only conceptual, but real and ontological priority of the fluid over the prime hard body. See, in this regard Lupoli 2006, 549 and 553–534. Lupoli maintains that Hobbes thought of the existence of a primordial fluid matter, which would in some way pre-date 'creation' and was entirely lacking in motion, which it would acquire through the intervention (in a sort of creative act) of a primary corporeal mover, which would cause atoms to come together. In reality, I believe that Hobbes' hypothesis presents fewer 'metaphysical' connotations, as we shall see below.

[81] See Galilei, *Discorsi e dimostrazioni matematiche intorno a due nuove scienze*, 86–87; see Galluzzi 2011, 96.

[82] *De corpore*, VII, 13, *OL*, I, 89; *EW*, I, 100.

intellect,'[83] and as a result the divisibility of which he speaks is always a potential and conceptual divisibility.[84]

Nevertheless, the passage cited from *De corpore* is interesting because it suggests the impossibility of identifying a last indivisible element of the matter, like the atom, because any given spatial or temporal quantity is always potentially divisible. In Chap. 26, as we know, Hobbes maintains that particles of matter within the microscopic structure of bodies are interspersed not by empty spaces, as Galileo claimed with his theory of the interspersed vacuum, but by fluids (as instead was set out by Descartes' theory of *subtle matter*),[85] adding that so-called hard atoms can be 'congregated' thanks to the motion of 'an intermingled fluid matter ... into compounded bodies of such bigness as we see.'[86] As he continues, Hobbes denies that fluids can be composed of hard matter, 'in such manner as meal is fluid, made so by grinding of the corn,'[87] while on the following pages he proposes a three-way division of the matter present in the universe, dividing it into three categories of bodies: *fluid, consistent,* and *a mixture of both.* The definition of the fluid body is similar to the one Hobbes later puts forward in his *Dialogus physicus* and matches Galileo's position:

> Wherefore a fluid body is always divisible into bodies equally fluid, as quantity into quantities; and soft bodies, of whatsoever degree of softness, into soft bodies of the same degree. And though many men seem to conceive no other difference of *fluidity*, but such as ariseth from the different magnitudes of the parts, in which sense dust, though of diamonds, may be called fluid; yet I understand by *fluidity*, that which is made such by nature equally in every part of the fluid body; not as dust is fluid, for so a house which is falling in pieces may be called fluid; but in such manner as water seem fluid, and to divide itself into parts perpetually fluid.

The roots of Hobbes' fluid theory lie in the concept of 'subtle spirit,' or 'subtle fluid,' which Hobbes had already outlined during his polemical correspondence with Descartes, indicating that the English philosopher had already come up with his concept on matter in the early 1640s. Some confirmation of this is provided by *De motu, loco et tempore*, where Hobbes criticizes the theory of the *rare* and the *dense* proposed by Thomas White in the *nodus* III of the first dialogue of his *De mundo*.[88] White's argument, which Hobbes takes up almost word for word in Chap. 5, is based on the statement that air is more divisible than water and water more divisible than earth. Therefore, 'it is undeniable that the proportion of divisibility, or

---

[83] *Examinatio et emendatio mathematicae hodiernae, OL*, IV, 56: 'Divisio est opus intellectus'.

[84] On this subject see Pacchi 1965, 235 ff. and, in more strictly mathematical terms, see Jesseph 1999, 82 ff.

[85] Descartes, *La dioptrique, AT*, VI, 86–87: 'y ayant plusieurs pores en tous les corps que nous apercevons autour de nous, ainsy que l'expérience peut monstrer fort clairement; il est necessaire que ces pores soyent remplis de quelque matière fort subtile & fort fluide, qui s'estende sans interruption depuis les Astres jusques a nous.'

[86] *De corpore*, XXVI, 3, *OL*, I, 340; *EW*, I, 417.

[87] Ibid.

[88] See White 1642, 31 ff.

of the quantity of air compared to air is greater than the proportion of the divisibility of water compared to water, and in the same way of water compared to earth.'[89] White seeks to maintain that the rarity or density of a body is linked to a greater or less divisibility of the same and Hobbes objects that he mistakes *softness* with *divisibility*. In fact, 'soft things are cut by a knife and they can easily suffer a break in continuity.'[90]

In the previous paragraph, Hobbes openly admitted that he did not know the cause of the rare and the dense, therefore stating: 'I prefer not to know than to make a mistake,'[91] adding that if someone were able to explain the phenomenon he would have 'revealed the innermost aspects of physics.'[92] In the same chapter, Hobbes considered the phenomenon of gunpowder, which occupies a much bigger space after deflagration. This presented him with a problem: 'It is unclear how it can happen that the same place now contains more matter, now less, as the place is equivalent to the body that is located there.'[93] In other words, in *De motu, loco et tempore*, Hobbes' theory of matter presented serious difficulties when it came to expressing the concept of density, in terms of the quantity of matter in relation to the volume of the body.

Nevertheless, as regards the nature of fluid, in *De motu, loco et tempore* Hobbes limits himself to stating that 'everyone usually calls fluid that substance whose parts are easily separated the one from the other,'[94] seeming to refer expressly to the notion of divisibility present in his later works, where fluids are defined as those bodies whose parts are always divisible into equally fluid parts, although he does not set out his ideas as clearly here as he does later in *De corpore* and *Dialogus physicus*.

It should be observed that this division can only be theorized by conceiving the subtle fluid as a *continuum*, comprised of parts that are always fluid and always divisible, in keeping with the indications put forward by Galileo in his works, particularly his *Discourses and Mathematical Demonstrations Relating to Two New Sciences*. And yet, Hobbes' concept of matter also features substantial differences compared to that of Galileo: Hobbes denies the existence of the notion of the indivisible and, because of this, his interpretation of the mathematical continuum differs from that of Galileo.[95] In fact, in his *Six Lessons* (1656) he opposes the definition of

---

[89] *MLT*, III, 12, 124.

[90] Ibid.

[91] Ibid., III, 9, 123.

[92] Ibid.

[93] Ibid.

[94] Ibid., XI, 8, 185.

[95] On the interpretation of indivisibles in Hobbes and his criticism of Torricelli's demonstration of the construction of the 'acute hyperbolic solid' through the application of the indivisible method, see Giorello 1990, 237–239; Mancosu and Vailati 1991, 67–68; Jesseph 1993. Jesseph demonstrates that Hobbes did not reject the indivisible method outright and, indeed, that he reflected extensively upon Cavalieri's work, drawing some observations from it (contained in a manuscript now conserved in Chatsworth House: Hobbes C.1.5). What is more, in chap. 26 of *De corpore*,

the point as *indivisible* with the definition of the point as *undivided*.[96] Hobbes sees division as a purely mental operation and the consideration of the point as undivided is solely a process of abstraction from a spatial *continuum*. This idea is reiterated in the *Examinatio et Emendatio Mathematicae Hodiernae* (1660),[97] as it is in other later geometrical works, where Hobbes strongly opposes the idea of conceiving an *indivisible non-quantum* because, in his opinion, *non-quantum* coincides with 'nothing'[98] and nothing cannot form a noticeable quantity.

On the other hand, Hobbes—like Galileo before him—implements a conceptual shift away from physics to geometry, transposing the arguments and reflections developed in the field of physical science to that of mathematics and vice versa.

It is interesting to observe that Hobbes also makes frequent use of the notion of the *atom*, both in *De corpore* and in *Dialogus physicus*. In Chap. 26 of *De corpore*, Hobbes claims that 'the immense space which we call the world is the aggregate of all bodies which are consistent and visible, as the earth and the stars; or invisible, as the small atoms' and, lastly: 'that most *fluid ether*, which so fills all the rest of the universe, as that it leaves in it no empty place at all.'[99]

The problem of atoms is always connected to Hobbes' theory of hardness, which also returns in the *Seven Philosophical Problems*.[100] In his *Decameron physiologicum*, Hobbes claims that the only factor able to make a body hard is movement,[101] but, on the other hand, he states that 'whatsoever perfectly homogeneous is hard consisteth of smaller parts, or, as some call them, *atoms*, that were made hard in the beginning, and consequently by an eternal cause.'[102] Despite this, Hobbes claims, in the same chapter, that not only air, but also 'all the other bodies,' are divisible 'into parts divisible. For no substance can be divided into nothings.'[103] In other words, although he was convinced of the existence of corpuscles or hard atoms ever since the existence of the world, Hobbes seems to reject the notion of physical indivisibility.

Hobbes proposed a demonstration taken from *Propositio 23* of Cavalieri's *Exercitationes Geometricae Sex* (ibid., 183–186). See also Jesseph 1999, 185–189; Holden 2004, 96–99. On Torricelli's indivisibles, see De Gandt 1989, 1992; Bortolotti 1989; Bascelli 2015.

On the subject of *indivisibles* and the transition from the concept of *indivisible* to the Leibnizian concept of *infinitesimal*: Malet 1996, esp. 11–50. On the 'prehistory' of the concept of *indivisibile*, see Celeyrette 2015.

[96] *Six Lessons*, EW, VII, 201.

[97] See *Examinatio et emendatio mathematicae hodiernae*, OL, IV, 55–56.

[98] See *De Principiis et Ratiocinatione Geometrarum*, OL, IV, 391–392.

[99] *De corpore*, XXVI, 5, OL, I, 347–348; Eng. Trans. EW, I, 426.

[100] See *Seven Philosophical Problems*, EW, VII, 38: 'But for one general cause of hardness it can be no other than such an internal motion of parts as I have already described, whatsoever may be the cause of the several concomitant qualities of their hardness in particular.' See, in general, Chap. 5, which discusses 'hard and soft'. Ibid., 32–38.

[101] *Decameron physiologicum*, EW; VII, 133: 'But yet this is most certain, that nothing can make a hard body of a soft, but by some motion of its parts.'

[102] Ibid., 133–134.

[103] Ibid., 129.

His *Decameron* includes another interesting passage, where Hobbes develops a curious argument regarding the cause of melting metals:

> For all motion compounded is an endeavor to dissipate, as I have said before, the parts of the body to be moved by it. If therefore hardness consists only in the pressing contact of the least parts, this motion will make the same parts slide off from one another, and the whole to take such a figure as the weight of the parts shall dispose them to, as in lead, iron, gold, and other things melted with heat. But if the small parts have such a figure as they cannot exactly touch, but must leave space between them filled with air or other fluids, then this motion of the fire, will dissipate those parts some one way, some another, the hard part still hard, as in the burning of wood or stone into ashes or lime.[104]

Hobbes indicates that the hardness of solid bodies consists of the contact of their least particles and maintains that the melting process is determined by the movement of fire, which, heating these particles, makes them slide off from one another. Lastly, if the metal corpuscles are not perfectly contiguous, but arranged in such a way as to permit the presence of small spaces filled by air or another fluid, the intervention of fire works in such a way as to separate these micro-particles.

His discussion of melting presents considerable analogies with the theory developed by Galileo on the first day of his *Discourses*. Here, Galileo wrote that fire, 'winds its way in between the most minute particles of this or that metal ... tears them apart and separates them' and provided this explanation of the phenomenon:

> I have thought that the explanation might lie in the fact that the extremely fine particles of fire, penetrating the slender pores of the metal (too small to admit even the finest particles of air or of many other fluids), would fill the small intervening vacua (*minima vacua*) and would set free these small particles from the violence which these same vacua exert upon them and which prevents their separation. Thus the particles are able to move freely so that the mass [*massa*] becomes fluid.[105]

In truth, there are also substantial differences between the arguments set forth by Galileo and Hobbes. First and foremost, Galileo allows for the existence of a *vacuum interspersum*. What is more, he deems that not even the smallest particles of air, or of other fluids, can pass through the slender pores of metal, while Hobbes maintains that these spaces within the solid bodies can be filled by air or other fluids.

Nevertheless, where Hobbes' theory diverges from Galileo's argument, there are also clear convergences with the position adopted by another author well known to Hobbes. In his *Les Météores*, Descartes supposed that the pores that compose the microscopic structure of bodies flow with subtle matter and attributed the greater or lesser solidity of the bodies to the greater or lesser movement of this matter. He also added that when it does not have sufficient strength to bend and agitate the particles of water, 'they stop confusedly conjoined, one upon the other, and thereby compose a hard body, that is to say ice.'[106] Descartes identified the same difference between water and ice that there is between 'a pile of small eels both living and dead, floating

---

[104] Ibid., 134.
[105] Galilei, *Discorsi e dimostrazioni matematiche intorno a due nuove scienze*, *OG*, VIII, 66–67; Eng. Trans. Galilei 2003, 19, slightly modified.
[106] Descartes, *Les Météores*, *AT*, VI, 237.

in a fishing boat full of holes, through which the water of a river passes and moves them, and a pile of the same eels on the bank, fully dried and stiffened by the cold.'[107]

However, even in the *Decameron* Hobbes states that he is still convinced that the difference between hard bodies and fluid bodies always ultimately lies in movement: 'For the parts of the hardest body in the world can be no closer together than to touch; and so close are the parts of air and water, and consequently they should be equally hard, if their smallest parts had not different motions.'[108] Hobbes therefore states that even particles of fluid bodies cannot be deemed 'closer together than to touch' and the reason for this specification should be traced back to his concept of place, as suggested by a passage from *An Answer to Arch-bishop Bramhall's Book Called the Catching of the Leviathan* (1668)[109]:

> I have seen, and so have many others, two waters, one of the river and the other mineral water, so that no man could discern one from the other from his sight; yet when they are both put together the whole substance could not be distinguished from milk. Yet we know the one was not mixed with the other, so as every part of the one to be in every part of the other, for that is impossible, unless *two bodies can be in the same place*.[110]

As already indicated in *De motu, loco et tempore*,[111] and reiterated in *De corpore*,[112] the place is no other than the space occupied by the body and, as a result, it is not possible for two particles of different fluids to occupy the same place. These corpuscles of non-homogeneous fluids cannot even 'merge' and produce a new substance, as Aristotle would have it,[113] because Hobbes does not conceive of any

---

[107] Ibid. The image of particles of water resembling small eels was also cited critically in the second treatise on optics. See *TO II*, chap. I, § 21, fol. 202*v*/158.

[108] *Decameron physiologicum*, *EW*, VII, 133: 'For the parts of the hardest body in the world can be no closer together than to touch; and so close are the parts of air and and water, and consequently they should be equally hard, if their smallest parts had not different motions.'

[109] The work was published posthumously in 1682, but written in 1668, as Hobbes writes in the preface, where he indicates that he had composed it 10 years prior to Bramhall's text *The Catching of Leviathan*, published in 1658. See *An Answer to a book published by Dr. Bramhall, EW*, IV, 282.

[110] *An Answer, EW*, IV, 309–310.

[111] In *MLT*, III, 3, 118, the place is defined in these terms: 'Quoties autem corporis alicuius spatium reale coincidet cum parte aliqua spatii imaginarii, illam partem, qua cum coincidit, vocamus corporis illius, *locum*. Quod si cum spatio aliquo imaginario, corporis nullius spatium reale coincidat, tum vocamus spatium illud imaginarium *vacuum*.'

[112] In chap. 7 of *De corpore*, entitled *Place and Time*, Hobbes does not provide a definition of place (see *De corpore, OL*, I, 81 ff.), since it was already present in the previous chapter (*De corpore*, VI, 6, 62–63): 'locus est spatium quod a corpore adaequate impletur vel occupatur, ignorare non potest; et qui *motum* concipit, nescire non potest quod motus est loci unius privatio et alterius acquisitio.' In the aforementioned manuscript *De Principiis*, however, the definition was more in keeping with the language used in *De motu*. See *De Principiis*, (National Library of Wales, Ms 5297), *MLT*, Appendix II, 453: 'Space (which I always understand imaginary) which is coincident with the magnitude of any body is called the *place* of that body, and then the body is said to be placed.' Regarding the distinction between real space and imaginary space and the importance of these concepts in Hobbesian philosophy, see Schuhmann 1992.

[113] See Aristotle, *De generatione et corruptione*, I, 5, 328a. On the problem of substantial change or mixture see *below*, section 4.

'generation' of new matter. From this we can deduce that every substance is composed by infinitely divisible particles and that these particles must always be extended and occupy a specific place.

Nevertheless, this does not necessarily imply acceptance of the classic atomist theory, nor does it conflict with the idea of the divisibility of bodies. On the other hand, Hobbes had an illustrious predecessor in this: in *Les Météores*, Descartes stated: 'I do not conceive the particles of terrestrial bodies as atoms or small indivisible parts, but … considering them all the same matter, I believe that each could be yet divisible in an infinity of ways and that they differ from one another solely as stones of different shapes that would yet all have been detached from the same rock.'[114]

4.5 'Transmutations' and 'Transubstantiations' of Matter

Since fluid bodies are perfectly divisible into parts that are always fluid and always perfectly divisible, they represent the entity closest to the concept of the *continuum* that is perfectly divisible into parts. Despite this, in *De corpore* Hobbes seems to compare *ether*, or *subtle fluid*, with *materia prima*, since 'the parts of the pure *æther*, as if it were the *first matter* (*materia prima*), have no motion at all, but what they receive from bodies which float in them, and are themselves fluid.'[115] Hobbes returns several times to the notion of *materia prima* and explores this concept, causing a number of different interesting elements to emerge. As we know, Hobbes considered the universe to be a reality populated exclusively by moving bodies and the concept of the body is linked to the notion of *prime matter* (*materia prima*), defined both in *De motu, loco et tempore* and in *De corpore* as the matter common to all things. The discussion is more extensive in the first work, which contains some invaluable indications for correctly interpreting the later developments in *De corpore*. Firstly, we know that the concepts of *body*, *matter* and *material* are perfectly

---

[114] Descartes, *Les Météores*, AT, VI, 238–239. On the differences between the Cartesian concept and the atomistic theory of matter, see De Buzon 2013, 243 ff.

[115] *De corpore*, XXVII, 1, *OL*, I, 364; *EW*, I, 448 (my italics). Lupoli drew attention to this passage and believes that the Hobbesian concept of fluid has a 'metaphysical' dimension. It could be identified, in some way, with a sort of 'primordial' prime matter. See Lupoli 1999, 588; Lupoli 2006, 540–541. However, in my opinion Leijenhorst quite rightly objected that in the cited passage Hobbes does not identify the two terms at all, but simply limits himself to comparing them (see Leijenhorst 2004, 90 ff.). He also maintains that Hobbes is referring here to the traditional Aristotelian concept of prime matter, which is considered inert (ibid., p. 91), but Hobbes, on the contrary, proposes a radically different conception of prime matter, which he identifies with corporeality in general. What is more, in reference to the identification of divinity with the corporeal spirit or fluid proposed by Lupoli, Leijenhorst (ibid., 81–88) cites some passages from Hobbesian works that are difficult to reconcile with this idea. For the differences between the Hobbesian concept of prime matter and the one specific to the Aristotelian tradition see Leijenhorst 2002, 150–155, where the author also makes an interesting comparison with Zabarella's idea of prime matter.

identical according to Hobbes: they indicate 'everything that occupies the space that we imagine.'[116] Nevertheless:

Considered by itself [*simpliciter*] this thing is called body, matter instead is when it is compared with that from which it was produced. *Simpliciter*, wood is said to be *body*, but the same wood, inasmuch as a bench is made from it, is said to be the *matter* of the bench. ... It is therefore clear that the prime matter (*materia prima*) is nothing more than the body in general, and that it does not exist in any way more than all the other universals exist, which are words ... not real and proper things. It is also clear why Aristotle correctly removes every quality and action from prime matter, attributing it only with potential, that is to say with mobility and quantity, meaning the corporality or materiality, which pertain by their essence to all bodies, inasmuch as they cannot be separated with the intellect [from the body]; if, therefore, we consider prime matter as one thing, whatever reality that matter is and it thereby exists as a thing; but then it must not be called the matter of that thing, that is to say of itself, nor prime matter, since this latter, existing as a common, cannot be called singular.[117]

Hobbes states that matter has the same ontological status as *universals*: it is a pure name. However, that is only partly true, because he identifies *body* with *matter*: *prime matter* can therefore be conceived as the mere corporality or materiality of the body. It is not an entity that exists *per se*, nor is it any entity that populates the physical world but, instead, is the result of a process of abstraction by the human faculties that, by isolating the reality of the body as such, only succeed in grasping its materiality.[118]

The coinciding nature of *matter* and *body* is reiterated again in Chap. 26 of the work, where Hobbes returns to the same idea:

Body, however, and *matter*, are names of the same thing, although considered from different points of view; the same thing considered inasmuch as it exists is simply called body, and when considered as being susceptible to a new form or a new figure it is called *matter*.[119]

In the manuscript known as *De Principiis*, which presents many of the fundamental definitions that come together in *De corpore* and therefore represents a sort of *trait d'union* between the two main works, matter is defined as a *pure name*. However, it is of capital importance because it is what cannot be born, nor perish:

The common matter of all things which the philosophers call *materia prima* is not a distinct body from all other bodies nor one of them but a name only, signifying a body to be considered without considering any form or any accident except only *magnitude* or *extension* and aptitude to receive *form and accident*. So as *materia prima* is corpus universale i.e. a body considered universally whereof it cannot be said that there is no form or no accident, but in which it may be said no form or accident besides quantity and aptitude to receive form or accident is considered i.e. brought into argument or account.[120]

---

[116] *MLT*, VII, 3, 146.

[117] Ibid., 147 (my trans.).

[118] See *supra*, chap. 2, § 2.

[119] *MLT*, XXVII, 1, 312 (my trans.).

[120] Id., *De Principiis* (National Library of Wales, Ms 5297), *MLT*, Appendix II, 457.

As we saw in Chap. 2, physical changes to bodies lead to the generation or destruction of *accidents* and, following these changes, the body appears to be changed externally. Accidental changes are what is perceived by the sensory organs and, as we know, some of these accidents are *essential* to the very concept of *body*, such as *size* and *magnitude*. Hobbes, states that there is an *eternal matter*, which remains despite every alteration or destruction to which the body is subject. In keeping with the Aristotelian philosophical tradition, this matter is known as *prime matter*.[121] It is, however, fundamental to emphasize that Hobbesian *prime matter* has no autonomous subsistence outside the bodies and is in fact a mere name,[122] although Hobbes also insists, in *De corpore*, that it cannot be considered 'a name of vain use' because:

> ... since all singular things have their forms accidents certain, *materia prima*, therefore, is body in general, that is body considered universally, not as having neither form nor any accident, but in which no form nor any other accident but quantity are at all considered, that is they are not drawn into argumentation.[123]

Hobbes' considerations on matter become even more interesting if compared with a theme outlined by Galileo in the 'first day' of his *Dialogue*. Here, when disputing the Aristotelian idea that he calls *transmutation of substance* (*transmutazione sustanziale*)—which supposes that generation is produced *ex novo* through the action of opposites—Galileo suggests an alternative with echoes of Democritus and Lucretius,[124] conceiving every change solely as an accidental modification of a matter that always remains the same:

> ... I never was fully convinced of any transmutation of substance (always confining ourselves to strictly natural phenomena) according to which matter becomes transformed in such a way that it is utterly destroyed, so that nothing remains of its original being, and another quite different body is produced in its place. If I fancy to myself a body under one aspect, and then under another quite different, I do not think it impossible for transformation to occur by a simple transposition of parts, without any corruption or the generation of anything new, for we see similar metamorphoses daily.[125]

The above passage reveals a concept of matter with Democritean connotations, which Galileo shared with Paolo Sarpi.[126] According to these authors, nothing is

---

[121] See Aristotle, *De generatione et corruptione*, I, 5, 320 b; *Metaphysics*, 1029 b.

[122] See *De corpore*, VIII, 24, *OL*, I, 105; *EW*, I, 118–119, slightly modified.

[123] Ibid.

[124] On the atomistic motifs present in the philosophy of Galileo, please make special reference to the cited studies by Shea, Redondi and Galluzzi. For a detailed and accurate comparison with the philosophy of Lucretius and the *De rerum natura*, see the interesting contribution by Camerota 2008.

[125] Galilei, *Dialogo sopra i due massimi sistemi*, *OG*, VII, 64–65, Eng. trans. Galilei 2001, 40. See also Galluzzi 2011, 82.

[126] Campanella maintains that Galileo and Sarpi were both 'Democritean' in a letter to Peiresc. See Campanella to Nicolas-Claude Fabri de Peiresc, in Campanella 2010, 454: 'io son certissimo ch'il signor Galileo in molte cose, massime nei principii, è con Democrito, e dal discorrer c'ha fatto meco in Roma, e da quel che scrive nell'opuscolo *De natantibus* e nel *Saggiatore*, el padre Castelli

created *ex nihilo* and nothing is reduced to nothing, but every transient physical entity is the result of modifications, that is to say of continuous and different compositions of a matter that remains, despite any generation and corruption.[127]

In his *Decameron physiologicum*, when objecting to the idea that ice can be transformed into crystal, Hobbes supports the same idea upheld in Galileo's *Dialogue*: he rules out any 'transubstantiation of bodies by mixture,'[128] or compound, because 'mixture is no transubstantiation.'[129] Hobbes uses the term 'transubstantiation' to emphasize his opposition to an Aristotelian idea that—according to his interpretation—had 'mystical' rather than physical connotations,[130] and he expresses the same position proposed by Galileo in his *Dialogue*, although he used the term 'transubstantiation' of substance.[131]

---

[e]t monsignor [C]iampoli e condiscepoli così [per] tal lo difendeno. El fra Paolo *ab antiquo* si sa essere stato democritico, perché Giovan Battista Porta suo amico quando stava in Napoli fra Paolo, e col quale han fatto molte operazioni chimiche, me l'ha narrato. El signor Galileo conversò con lui, quando eravamo in Padua nel 1593.' See also Ernst and Canone 1994, esp. 363–364. See also the interesting observations made by Favino 1997, where the author also shows Ciampoli's warm support for the atomistic doctrines of his master Galileo. On Ciampoli see Favino 2015. There are numerous interesting analogies between the thinking of Ciampoli and the texts by Gassendi. As regards the link between Galileo and Sarpi, it is superfluous to underline that during his period in Padua, Galileo had the opportunity to work in collaboration with the Servite friar, who presumably contributed to the discoveries that Galileo published in his *Sidereus Nuncius*. In this regard see Cozzi 1979, 164 ff.; Sosio 1995, and Bucciantini, Camerota and Giudice, 24–43. Favaro's classic study is also useful. See Favaro 1966, II, 69 ff. See also Frajese 1994, 63 ff. Wootton maintains that Sarpi was a materialist, but not an atomist, because 'he believed that matter could be changed: it was not made up of unalterable unitary atoms, but would be compressed, expanded and so on.' See Wootton 1983, 15. However, see the following note.

[127] Galluzzi 2011, 82 emphasizes the profound analogy between this passage from Galileo's *Discourses* and the idea that features in one of Paolo Sarpi's *Pensieri*. See the 'Pensiero 111' in Sarpi 1996, 130: 'La materia delle cose naturali è corpo, perché, facendosi le trasmutazioni, il corpo è quello che sempre resta e non trasmutasi mai, e li suoi termini sono superficie, linea e punto, co' quali terminato acquista figura.' In my opinion, an equally significant mention is made in another *pensiero*, in which Sarpi criticises the Aristotelian notion of transmutation, just as Galileo would do. See the 'Pensiero 332', ibid., 267: 'la nutrizione si può far, senza alcuna trasmutazione, solo per congragazion e separazione.' Sarpi's concept of matter is extremely similar to that of Hobbes, with whom he shares the same 'Galilean root,' as I have emphasized in Baldin 2013, esp. 105–111. On the similarities between Sarpi and Hobbes on the political aspects, see also Baldin 2016a, and Baldin 2015.

[128] *Decameron physiologicum*, *EW*, VII, 132: 'Transubstantiation of bodies by mixture'.

[129] Ibid.: 'Mixture is no transubstantiation'.

[130] Hobbes considered *transubstantiation* a mere absurdity and, in fact, in Chap. 8 of his *Leviathan* he places the concept of *transubstantiation* in the category of *insignificant speech* (see Hobbes 2012, 122). Negative references to this dogma also appear elsewhere, as in *Historia Ecclesiastica*, *OL*, V, 404.

[131] It is interesting to observe that Hobbes' statements seem to 'confirm' Orazio Grassi's accusations made against Galileo, according to whom Galileo's 'Democritean' concept of matter is incompatible with the dogma of transubstantiation. Sarsi 1626, *OG*, VI, 486–487: 'Galilaeus vero diserte asserit, calorem, colorem, saporemque ac reliqua huiusmodi, extra sentientem, ac proinde in pane ac vino, pura esse nomina: ergo, abscedente panis ac vini substantia, pura tantum qualitatum nomina remanebunt. Quid ergo perpetuo opus miraculo est, puris tantus nominibus sustentan-

The passage from Hobbes' *Decameron*, as well as the excerpt cited from his *Answer* to Bramhall's text in which Hobbes denies that there can be a perfect fusion of two different waters such as to generate a new matter, refer to an idea found in Aristotle's *De Generatione et corruptione*. Here, Aristotle, criticizing the position adopted by Democritus, characterized *mixture* (μίξις) as a process in which 'any part of one of the two components must be found in any part of the other component.'[132] On the contrary, Galileo and Hobbes advocate a Democritean concept of matter and consider every change to be an aggregation or recomposition of a matter that always remains the same. In fact, matter cannot be transformed in its entirety. Instead, accidental changes generate a myriad of transient bodies that populate the natural world and can be experienced by the senses.

Nevertheless, according to Hobbes, matter evades the senses and, as a result, our investigation too, remaining entirely lacking in determination.[133] This does not mean that this reality is absolutely nothing, because it is identified with corporality in general and is what remains above and beyond all accidental mutations.

The concept outlined by Galileo is much clearer in Hobbes' thinking, where it is expressed openly and taken all the way through to extreme consequences. Hobbes maintains that 'all the accidents can be generated or perish excepted the body' (*Generari et perire possunt accidentia omnia praeter corpus*). Similarly, 'the bodies and the accidents through which the bodies appear variously differ in this way, that the bodies are entities not generated, while accidents are in fact generated things, but not entities' (*Corpora et accidentia sub quibus varie apparent ita differunt ut corpora sint res non genitae, accidentia vero genita non res*).[134]

Matter, which coincides with the pure materiality of the body, is not generated out of nothing, nor is it reduced to nothing and, therefore, it must necessarily be eternal. In *De motu, loco et tempore* it is no coincidence that Hobbes strongly criticizes Thomas White's assertion that he can demonstrate that the world is not eternal and that it is, indeed, the result of creation.[135] On the contrary, Hobbes states that the question is impossible to resolve through philosophy and natural reason, and recommends trusting in faith.[136]

---

dis? Videat ergo hic, quam longe ab iis distet, qui tanto studio harum specierum veritatem ac durationem firmare conati sunt, ut etiam divinam huic operi potentiam impenderint. Scio equidem lubricis ac versutis ingeniis videri posse, patere hinc etiam effugium aliquod, si fas sit sanctissimorum fidei praesidium dicta ad libitum interpretari, eaque a vero et communi sensu alio detorquere.' See also Shea 1970, 21, but above all Redondi 2009, 244 ff., who claims that this was the real reason for Galileo's condemnation.

[132] Aristotle, *De generatione et corruptione*, I (A), 10, 328a.

[133] Indeed, in *De corpore*, I, 8, *OL*, I, 9, Hobbes claims that 'Subjectum Philosophiae, sive materia circa quam versatur, est corpus omne cujus generatio aliqua concipi, et cujus comparatio secundum ullam ejus considerationem institui potest.'

[134] *MLT*, Appendix II, 457 (my trans.).

[135] See White 1642, 329.

[136] See *MLT*, XXXIII, 7, 379.

Without going into theological issues here, it is worth noting that the same sceptical approach to the origin of the universe is also expressed a decade later in *De corpore*, where Hobbes writes: 'I cannot therefore commend those that boast they have demonstrated, by reason drawn from natural things, that the world had a beginning.'[137]

References

Works of Thomas Hobbes:

Hobbes, Thomas. 2012. *Leviathan*, ed. Noel Malcolm, 3 Vols. Oxford, Clarendon Press.

Other Works:

Campanella, Tommaso. 2007. *Del senso delle cose e della magia,* ed. Germana Ernst. Rome-Bari: Laterza.
———. 2010. *Lettere*, ed. Germana Ernst. Florence: Olschki.
Cavalieri, Bonaventura. 1989. *Geometria degli Indivisibili*, ed. Lucio Lombardo Radice. Turin: Utet (1st ed. 1966).
Galilei, Galileo. 2001. *Dialogue Concerning the Two Chief World Systems*, Eng. Trans. Stillman Drake. New York: The Modern Library (1st ed. Los-Angeles-Berkeley: University of California Press, 1967).
———. 2003. *Dialogues Concerning Two New Sciences*, Eng. Trans. Henry Crew and Alfonso de Salvio. Mineola: Dover Publications (1st ed.: New York: The Macmillan Company, 1914).

---

[137] *De corpore*, XXVI, 1, *OL*, I, 336; Eng. Trans. *EW*, I, 412–413. See Paganini 2008, and, above all, Paganini 2015. I disagree with Arrigo Pacchi, who in numerous articles, maintained that a philosophical concept of God as the prime mover represents the necessary foundation for the existence of the universe in Hobbes' system. See Pacchi 1984 (now also in Pacchi 1998, 53–65), and Pacchi 1988. The question is also taken up again in another contribution by Pacchi, in which he compares the different interpretations of Hobbesian theology that emerged during the twentieth century with a number of philosophical and theological reflections made by Hobbes. See Pacchi 1990. I do not find Lupoli to be convincing either, who—starting with the claim that Hobbes' statements against the creation of the world made in *De motu, loco et tempore* and *De corpore* are limited to upholding the impossibility of philosophical *argument* in favour of creation, without reaching an all-out denial of it—maintains that it is possible to state that the eternity of the world is 'undoubtedly denied by Hobbes' (Lupoli 2006, 554 and 566–574). It is true that Hobbes incidentally states in 'Appendix ad Leviathan' (in Hobbes 2012, 1147; *OL*, III, 513), that the world was made 'without doubt from nothing', but he also indicates that this is deemed to be the truth because it is written in the Holy Scriptures and not on the basis of reasoning or philosophical exploration. In fact, it is necessary to bear in mind that Hobbesian statements on matters of faith are sometimes incoherent or even openly contradict his own philosophical arguments. For example, in *De motu, loco et tempore*, Hobbes, so as not to attribute materiality to God, maintains that the existence of *incorporeal substances* is a dogma of faith (*MLT*, IV, 3, 127), something that is openly denied in his later works, where the very concept of incorporeal substance is held to be a mere absurdity.

Gassendi, Pierre. 1658. *Opera Omnia*, 6 Vols. Lyon: Anisson, & Devenet, 1658 (Stuttgart-Bad Cannstatt: Fromman Holzboog, 1994).
Mersenne, Marin. 1634a. *Les Mechaniques de Galilee, Mathematicien et Ingenieur du Duc de Florence*. Paris: Henry Guenon (Reprint in Mersenne. 1985: 427–513).
——. 1985. *Questions Inouyes*, (Corpus des œuvres de philosophie en langue française) (Paris: Fayard, 1985).
Ockham, William of. 1974–1988. *Opera Philosophica et Theologica*, 17 Vols. New York: Cura Instituti Franciscani Universitatis S. Bonaventurae.
Sarpi, Paolo. 1996. *Pensieri, naturali, metafisici e matematici*, ed. Luisa Cozzi and Libero Sosio. Milan-Naples: Ricciardi.
Sarsi, Lotario. 1626. (*alias* Orazio Grassi). *Ratio ponderum librae et simbellae*. In *OG*, VI: 373–500.
White, Thomas. 1642. *De mundo dialogi tres*. Paris: Moreau.

Secondary Sources:

Baldin, Gregorio. 2013. Hobbes and Sarpi: Method, Matter and Natural Philosophy. *Galilaeana* 10: 85–118.
——. 2015. Thomas Hobbes e la Repubblica di Venezia. *Rivista di Storia della Filosofia* (4): 717–741.
——. 2016a. Hobbes, Sarpi and the Interdict of Venice. *Storia del pensiero politico* (2): 261–280.
——. 2016b. 'La reflexion de l'arc' et le *conatus*: aux origines de la physique de Hobbes. *Philosophical Enquiries. Revue des philosophies anglophones* 7: 15–42.
Baldini, Ugo. 1977a. La struttura della materia nel pensiero di Galileo. *De Homine* 56–58 (excerpt): 1–74.
——. 1977b. Il corpuscolarismo italiano del Seicento. Problemi di metodo e prospettive di ricerca. In *Ricerche sull'atomismo del Seicento. Atti del Convegno di Studio (14–16 ottobre 1976)*, ed. Ugo Baldini, Giancarlo Zanier, Paolo Farina, and Francesco Trevisani, 1–76. Florence: La Nuova Italia.
Baroncelli, Giovanna. 1992. Bonaventura Cavalieri tra matematica e fisica. In *Geometria e atomismo nella scuola galileiana (Atti del convegno)*, ed. Massimo Bucciantini and Maurizio Torrini, 67–101. Florence: Olschki.
Bascelli, Tiziana. 2015. Torricelli's Indivisibles. In *Seventeenth-Century Indivisibles Revisited*, ed. Vincent Jullien, 105–136. Dordrecht: Springer.
Bernhardt, Jean. 1993. La question du vide chez Hobbes. *Revue d'histoire des sciences* 46: 225–232.
Biener, Zvi. 2004. Galileo's First New Science: The Science of Matter. *Perspectives on Science* 12 (3): 262–287.
Bortolotti, Emilio. 1989. L'oeuvre géométrique d'Évangeliste Torricelli. In *L'œuvre de Torricelli: science galiléenne et nouvelle géométrie*, ed. François De Gandt, 111–146. Paris: Les Belles Lettres.
Brunschvicg, Léon. 1913. *Les étapes de la philosophie mathématique*. Paris: Blanchard.
Camerota, Michele. 2008. Galileo, Lucrezio e l'atomismo. In *Lucrezio: la natura e la scienza*, ed. Marco Beretta and Francesco Citti, 141–175. Olschki: Florence.
Cavalieri, Bonaventura. 1632. *Lo specchio ustorio, overo Trattato delle Settioni Coniche, et alcvni loro mirabili effetti*. Bologna: Clemente Ferroni.
Celeyrette, Jean. 2015. From Aristotle to the Classical Age, the Debate Around Indivisibles. In *Seventeenth-Century Indivisibles Revisited*, ed. Vincent Jullien, 19–30. Dordrecht: Springer.
Clavelin, Maurice. 1968. *La philosophie naturelle de Galilée*. Paris: Librairie Armand Colin.
Costabel, Pierre. 1964. La roue d'Aristote et les critiques françaises à l'argument de Galilée. *Revue d'histoire des sciences* 17: 385–396.

Cozzi, Gaetano. 1979. *Paolo Sarpi tra Venezia e l'Europa*. Turin: Einaudi.

de Buzon, Frédéric. 2013. *La Science cartésienne et son objet. Mathesis et phénomène*. Paris: Honoré Champion.

De Gandt, François. 1989. Les indivisibles de Torricelli. In *L'œuvre de Torricelli: science galiléenne et nouvelle géométrie*, ed. François De Gandt, 147–206. Paris: Les Belles Lettres.

———. 1992. L'evolution de la théorie des indivisibles et l'apport de Torricelli. In *Geometria e atomismo nella scuola galileiana (Atti del convegno)*, ed. Massimo Bucciantini and Maurizio Torrini, 103–118. Florence: Olschki.

Dijksterhuis, Eduard J. 1961. *The Mechanization of the World Picture: From Pythagoras to Newton*. Oxford: Clarendon Press.

Drabkin, Israel E. 1950. Aristotle's Wheel: Notes on the History of a Paradox. *Osiris* 9: 161–198.

Drake, Stillman. 1957. *Discoveries and Opinions of Galileo*. New York: Doubleday & Co..

Ernst, Germana. 2010. *Tommaso Campanella. Il libro e il corpo della natura*. Rome/Bari: Laterza.

Ernst, Germana, and Eugenio Canone. 1994. Una lettera ritrovata: Campanella a Periesc, 19 giugno 1636. *Rivista di Storia della Filosofia* (2): 353–366.

Favaro, Antonio. 1966. *Galileo Galilei e lo studio di Padova*, 2 Vols. Padua: Antenore (or. Ed. 1883).

Favino, Federica. 1997. A proposito dell'atomismo di Galileo: da una lettera di Tommaso Campanella a uno scritto di Giovanni Ciampoli. *Bruniana & Campanelliana* 3: 265–282.

———. 2015. *La filosofia naturale di Giovanni Ciampoli*. Florence: Olschki.

Festa, Egidio. 1992. Quelques aspects de la controverse sur les indivisibles. In *Geometria e atomismo nella scuola galileiana (Atti del convegno)*, ed. Massimo Bucciantini and Maurizio Torrini, 193–206. Florence: Olschki.

Frajese, Vittorio. 1994. *Sarpi scettico*. Bologna: Il Mulino.

Galluzzi, Paolo. 2011. *Tra atomi e indivisibili. La materia ambigua di Galileo*. Florence: Olschki.

Garber, Daniel. 2009. *Descartes Embodied*. Cambridge: Cambridge University Press.

Gaukroger, Stephen. 2002. *Descartes' System of Natural Philosophy*. Cambridge: Cambridge University Press.

Giorello, Giulio. 1990. Pratica geometrica e immagine della matematica in Thomas Hobbes. In *Hobbes Oggi*, ed. Bernard Willms et al., 215–255. Milan: Franco Angeli.

Giudice, Franco. 1997. Thomas Hobbes and Atomism: A Reappraisal. *Nuncius* 12: 471–485.

Giusti, Enrico. 1980. *Bonaventura Cavalieri and the Theory of Indivisibles*. Bologna: Edizioni Cremonese.

Gómez Lopez, Susana. 2001. Galileo y la naturaleza de la luz. In *Largo campo di filosofare*, ed. José Montesinos and Carlos Solis, 403–418. La Orotava: Fundaciòn Canaria Orotava de Historia de la Ciencia.

———. 2008. The Mechanization of Light in Galilean Science. *Galilaeana* 5: 207–244.

Grant, Edward. 1981. *Much Ado About Nothing: Theories of Space and Vacuum from Middle Ages to the Scientific Revolution*. Cambridge: Cambridge University Press.

Holden, Thomas. 2004. *The Architecture of Matter. Galileo to Kant*. Oxford: Oxford University Press.

Jesseph, Douglas M. 1993. Of Analytics and Indivisibles: Hobbes on the Modern Mathematics. *Revue d'histoire des sciences* 46: 153–193.

———. 1999. *Squaring the Circle. The War Between Hobbes and Wallis*. Chicago/London: University of Chicago Press.

Jullien, Vincent. 2015. Indivisibles in the Work of Galileo. In *Seventeenth-Century Indivisibles Revisited*, ed. Vincent Jullien, 87–103. Dordrecht: Springer.

Kargon, Robert H. 1966. *Atomism in England from Hariot to Newton*. Oxford: Clarendon Press.

Lasswitz, Kurd. 1963. *Geschichte der Atomistik von Mittelalter bis Newton*, 2 Vols. Hildesheim: Georg Olms, (or. Ed. 1890).

Le Grand, Homer E. 1978. Galileo's Matter Theory. In *New Perspectives on Galileo*, ed. Robert E. Butts and Joseph C. Pitt, 197–208. Dordrecht/Boston: Reidel Publishing Company.

Leijenhorst, Cees. 2002. *The Mechanisation of Aristotelianism. The Late Aristotelian Setting of Thomas Hobbes' Natural Philosophy*. Leiden/Boston/Köln: Brill.

———. 2004. Hobbes' Corporeal Deity. In *Nuove prospettive critiche sul Leviatano di Hobbes*, ed. Luc Foisneau and George Wright, 73–95. Milan: Franco Angeli.

Lupoli, Agostino. 1999. 'Fluidismo' e *corporeal deity* nella filosofia naturale di Thomas Hobbes: A proposito dell'hobbesiano 'Dio delle cause'. *Rivista di Storia della filosofia* (4): 573–609.

———. 2006. *Nei limiti della materia. Hobbes e Boyle: materialismo epistemologico, filosofia corpuscolare e Dio corporeo*. Milan: Baldini, Castoldi, Dalai.

Maier, Annaliese. 1984. *Scienza e filosofia nel Medioevo. Saggi sui secoli XIII e XIV*. Milan: Jaca Book.

Malet, Antoni. 1996. *From Indivisibles to Infinitesimals*. Bellaterra: Universitad Autònoma de Barcelona.

Mancosu, Paolo, and Ezio Vailati. 1991. Torricelli's Infinitely Long Solid and Its Philosophical Receptions in the Seventeenth Century. *Isis* 82: 52–70.

McCord Adams, Marilyn. 1987. *William Ockham*, 2 Vols. Notre Dame: Notre Dame University Press.

Médina, José. 2013b. Physiologie mécaniste et mouvement cardiaque: Hobbes, Harvey et Descartes. In *Lectures de Hobbes*, ed. Jauffrey Berthier, Nicolas Dubos, Arnaud Milanese, and Jean Terrel, 133–162. Paris: Ellipses.

Murdoch, John E. 1982. Infinity and Continuity. In *The Cambridge History of Later Medieval Philosophy*, ed. Norman Kretzmann, Anthony Kenny, and Jan Pinborg, 564–591. Cambridge: Cambridge University Press.

———. 2009. Beyond Aristotle: Indivisibles and Infinite Divisibility in the Later Middle Ages. In *Atomism in Late Medieval Philosophy and Theology*, ed. Christophe Grellard and Aurélien Robert, 15–38. Leiden/Boston: Brill.

Nonnoi, Giancarlo. 2000. Galileo Galilei, Quale atomismo? In *Atomismo e continuo nel XVII secolo*, ed. Egidio Festa and Romano Gatto, 275–319. Naples: Vivarium.

Osler, Margaret J. 1994. *Divine Will and Mechanical Philosophy. Gassendi and Descartes on Contingency and Necessity in the Created World*. Cambridge: Cambridge University Press.

Pacchi, Arrigo. 1965. *Convenzione e ipotesi nella filosofia naturale di Thomas Hobbes*. Florence: La Nuova Italia.

———. 1978. Hobbes e l'epicureismo. *Rivista di storia della filosofia* (1): 54–71.

———. 1984. Hobbes e il Dio delle cause. In *La storia della filosofia come sapere critico. Studi offerti a Mario Dal Pra*, ed. Nicola Badaloni et al., 295–307. Milan: Franco Angeli.

———. 1988. Hobbes and the Problem of God. In *Perspectives on Thomas Hobbes*, ed. G.A.J. Rogers and Alan Ryan, 171–189. Oxford: Clarendon Press.

———. 1990. Hobbes e la teologia. In *Hobbes Oggi*, ed. Bernard Willms et al., 101–121. Milan: Franco Angeli.

———. 1998. *Scritti hobbesiani (1978–1990)*, ed. Agostino Lupoli. Milan: Franco Angeli.

Paganini, Gianni. 2008. Hobbes alla ricerca del primo motore. Il De motu, loco et tempore. *Rinascimento* 48: 527–541.

———. 2015. Hobbes's Galilean Project. Its Philosophical and Theological Implications. *Oxford Studies in Early Modern Philosophy* 7: 1–46.

Palmerino, Carla Rita. 2000. Una nuova scienza della materia per la *Scienza Nova* del moto. La discussione dei paradossi dell'infinito nella prima giornata dei *Discorsi* galileiani. In *Atomismo e continuo nel XVII secolo*, ed. Egidio Festa and Romano Gatto, 275–319. Naples: Vivarium.

———. 2001. Galileo's and Gassendi's Solutions to the *Rota Aristotelis* Paradox: A Bridge Between Matter and Motion Theories. In *Late Medieval and Early Modern Corpuscolar Matter Theories*, ed. Cristopher Lüthy, John E. Murdoch, and William R. Newman, 381–422. Leiden/Boston/Köln: Brill.

———. 2010. The Geometrization of Motion: Galileo's Triangle of Speed and Its Various Transformations. *Early Science and Medicine* 15: 410–447.

———. 2011. The Isomorphism of Space, Time and Matter in Seventeenth-Century Natural Philosophy. *Early Science and Medicine* 16: 296–330.

Radelet-De Grave, Patricia. 2015. Kepler, Cavalieri, Guldin. Polemics with the Departed. In *Seventeenth-Century Indivisibles Revisited*, ed. Vincent Jullien, 57–86. Dordrecht: Springer.

Redondi, Pietro. 1985. Atomi, indivisibili e dogma. *Quaderni storici* 59 (Iss. 2): 529–571.

———. 2009. *Galileo eretico*. Rome/Bari: Laterza (or. Ed. 1983).

Schuhmann, Karl. 1992. Le vocabulaire de l'espace. In *Hobbes et son vocabulaire*, ed. Yves-Charles Zarka, 61–82. Paris: Vrin.

Shapin, Steven, and Simon Schaffer. 1985. *Leviathan and the Air-Pump*. Princeton: Princeton University Press.

Shea, William R. 1970. Galileo's Atomic Hypothesis. *Ambix* 17: 13–27.

———. 1978. *Descartes as Critic of Galileo*. In *New Perspectives on Galileo*, ed. Robert E. Butts and Joseph C. Pitt, 139–159. Dordrecht/Boston: Reidel Publishing Company.

Sosio, Libero. 1995. Galileo Galilei e Paolo Sarpi. In *Galileo Galilei e la cultura veneziana. Atti del convegno di studio*, 269–311. Venice: Istituto Veneto di Lettere ed Arti.

Wootton, David. 1983. *Paolo Sarpi. Between Renaissance and Enlightment*. Cambridge: Cambridge University Press.

Conclusion

It is well known that the most revolutionary element Galileo's *Assayer* consists of the 'reduction' of the so-called secondary qualities to primary qualities. Galileo only admitted the existence of real properties that were quantifiable and could be expressed in mathematical terms. As we have seen in this chapter and the ones preceding it, Hobbes shared Galileo's position and developed his natural philosophy upon this basis.

Nevertheless, it is interesting to linger a little more on some of the problems encountered by Galileo with regard to the sense that he considered to be 'eminent above all others,' that is to say vision. Galileo believed that vision, like all the other senses, had a particular relationship with its object, but the link that he established between *vision* and *light* was unique, expressing itself—to cite Galileo's own words—'in the proportion of the finite to the infinite, the temporal to the instantaneous, the *quantitative* to the *indivisible*, the illuminated to the obscure.'[1]

As we have seen, the phenomenon of light is paradigmatic when it comes to understanding Galileo's concept of matter and the continuum.[2] In his *Assayer*, Galileo establishes a difference between light and the other objects of the senses. In all the other phenomena contemplated, such as heat, he examines the 'smallest parts' of the bodies, which are still 'quanta' and 'numbered' and their motion is 'temporary.' On the contrary, 'when their ultimate and highest resolution into truly indivisible atoms is arrived at, light is created,' which is deemed to have 'an instantaneous motion, or rather an instantaneous expansion and diffusion.' Due to its '*subtlety*, its *rarity*, its *immateriality*, or *some other property which differs from all these and is nameless*,' light is able to occupy 'immense spaces.'[3] In *The Assayer*,

---

[1] Galilei, *Il saggiatore*, *OG*, VI, 350; Eng. trans. Drake 1957, 275.

[2] On this subject see Gómez Lopez 2008, 207–244.

[3] Galilei, *Il saggiatore*, *OG*, VI, 351–352; Eng. trans. Drake 1957, 278. In his *Copernican letters*, Galileo, taking up concepts and language specific to the late mediaeval tradition and linked to the metaphysics of light, spoke of a substance that was '*spiritosissima tenuissima e velocissima*,'

© Springer Nature Switzerland AG 2020
G. Baldin, *Hobbes and Galileo: Method, Matter and the Science of Motion*,
International Archives of the History of Ideas Archives internationales
d'histoire des idées 230, https://doi.org/10.1007/978-3-030-41414-6

light is considered a sort of universal primordial substance, resolved into its ultimate components. Upon being condensed, it gives life to all the other denser compositions of matter found in nature.[4] As a result, it also represents the paradigm of the maximum rarefaction of matter, being 'resolved' into its ultimate components, which are *truly indivisible* atoms.

Hobbes' concept of vision differs radically from that of Galileo on a fundamental matter: the English philosopher considers light to be nothing more than a movement propagated through the medium, while Galileo had ascribed it with an autonomous nature, considering it to be an extremely rarefied body, resolved into its indivisible atoms.[5] However, Hobbes' theory regarding the transmission of light is very similar to that of Descartes, who believed that light was 'the action or tendency to move of a certain very subtle matter that fills the pores of transparent bodies.'[6] Indeed, although he openly disputed Descartes' notion of tendency or 'inclination' to motion, Hobbes extracted the particular and exclusive dimension that Galileo had attributed to the phenomenon of light, assigning it—like Descartes—to the motion that is transmitted in the medium.[7] In this, Hobbes and Descartes both seem more 'Aristotelian'[8] than Galileo,[9] who, loyal to his 'Democritean' position, had described light as a real substance resolved into its ultimate indivisible atoms.[10]

Nevertheless, as we saw in Chapter 2,[11] Hobbes' explanation of light in the *Elements* was clearly inspired by the speculations in *The Assayer* and this is also apparent in his later works. For example, in *Seven Philosophical Problems*, Hobbes

---

which, however, was distinct from light and had calorific and, in some way, life-giving properties (*OG*, V, 301). On the similarities and differences between Galileo's position and that of the Renaissance Neo-Platonists, see Gómez Lopez 2008, 214–225. On the relationship between light and matter see also Galluzzi 2011, 61. For an analysis of the possible influence of the metaphysics of light and, especially, of the ideas of Robert Grosseteste on the optics and natural philosophy of Hobbes, see Alessio 1962, and Gargani 1971, 111 ff., who, however, focuses on the *Short Tract on First Principles*.

[4] See Galluzzi 2011, 75–78.

[5] See Galilei, Letter to Pietro Dini, *OG*, V, 301. This should not, however, lead us to believe that Galileo attributed an autonomous and qualitatively different ontological reality to light as compared to the matter of which the other bodies are comprised. On this subject and on its difference from the position of Kepler, see Bucciantini 2003, 234–235.

[6] Descartes, *La Dioptrique*, AT, VI, 197. Hobbes also supposes that light is a *conatus* transmitted through the medium. However, unlike his French colleague who speaks of 'tendency,' he considers this *conatus* to be a real movement in action, transmitted continuously through the medium.

[7] In 'Appendix to Leviathan' (Hobbes 2012, 1149; *OL*, III, 514), Hobbes writes: 'Lumen, enim, ut mihi videtur, Phantasma est, non res existens.'

[8] In fact, Aristotle considered light to be 'diaphanous in *entelecheia*' (Aristotle, *De Anima*, II (B), 7, 419a), claiming that its propagation and colors were determined by movement within the medium (ibid. 418b-419b).

[9] On the differences between the Galilean and Aristotelian concepts of light, see Redondi 1982, who also compares them with the theories of Giulio Cesare Lagalla and Fortunio Liceti.

[10] See Gómez Lopez 2008, 214–225 and 238 ff.

[11] See *supra*, chap. II, § 1.

states once again that our senses are nothing but 'fancy,' although the cause lies in a real body.[12]

As we know, this idea was also present in Descartes. In the sixth discourse of the *Dioptrics*, he claimed that 'the nature of our soul is such that the force of the movements that occur in that part of the brain from which the narrow fibers of the optic nerves originate causes it to have a sensation of light, and the way in which such movements occur causes it to have a sensation of color.'[13] For Descartes too, light and color are nothing more than movement, a movement that, once recognized by the brain, is 'perceived' by the soul.

Hobbes himself underscores Descartes' acumen, indicating how their ideas on optics converge. Indeed, he openly praises his 'colleague' in the *First Draught of the Optiques* for having been the first to associate light solely with movement. According to Hobbes, it is no surprise that all those who previously explored the nature of light encountered problems in identifying the correct placing of the light sources. This is because:

> ... they have all of them, (except onely Monsieur desCartes, now of late) supposed light and colour, that is to say the appearance of objects which is nothing butt our fancy, to bee some accident in the object it selfe, and so they sought the place of that which hath no place, for nothing hath a place butt bodie, and if improperlie wee assigne place to accidents, wee cannot assigne them any other then the place of the body whose accidents they are, They ought therefore to have enquired not of the place of the Image, but of the apparent or seeming place of the object itselfe. And seeing Mons. desCartes who only hath sett forthe the true principle of this doctrine, namely that the Images of objects are in the Fancie, and that they fly not through the aire, under the empty name of Species intentionales but are made in the braine by the operation of the objects themselves.[14]

Descartes was therefore the first to realize that light and colors consist solely of sensation and do not reside in any way in the perceived object. And yet, despite this reference, we know that Descartes' *Dioptrique* cannot be considered the source of Hobbes' theory of vision, because the English philosopher formulated the idea that light and other sensations consist solely of movement before he had ever read Descartes' *Essais*.[15] Nevertheless, the affinity of the two authors with regard to this subject is extremely significant because it highlights their general philosophical outlook and enables us to place their respective philosophies correctly within the great philosophical and scientific current of the early seventeenth century: *mechanism*.

---

[12] *Seven Philosophical Problems*, EW, VII, 28: 'All sense is fancy, though the cause be always in a real body.' Hobbes also offers a significant clarification further on: 'I see by this that those things which the learned call the accidents of bodies, are indeed nothing else but diversity of fancy, and are inherent in the sentient, and not in the objects, except *motion* and *quantity*.' Ibid.

[13] Descartes, *La Dioptrique*, AT, VI, 130–131.

[14] *FD*, 74 *v* and *r*/335–336. See Minerbi Belgrado 1993, 41; Médina 2015a, 76 ff. The subject of the theory of light transmission is also associated with the interpretation of *primary and secondary qualities*. On this subject, in the thinking of Descartes, see Wilson 1993.

[15] Hobbes also claims in a letter dated October 1636 that light is nothing other than movement transmitted through the medium and in the brain (see *CH*, I, 37–38), while Kenelm Digby only sent him Descartes's work early in October of the following year (see *CH*, I, 51).

Descartes and Hobbes are often cited as key exponents of the new scientific paradigm, and yet Hobbes is the first author in whom an awareness of the revolutionary
scope of Galileo's thinking emerges. Galileo's science of motion and his theory of
matter are at the basis of the distinction between the so-called primary and secondary qualities and, therefore, of mechanism. In Hobbes, Galilean reflections acquire
the dimension of an epochal turning point, not only with regard to the Aristotelian
and scholastic tradition, but—thanks to the mathematization of nature—also as
regards the natural philosophies of the early modern *novatores*, such as Telesio and
Campanella. Galileo sought in every way, both insistently and stubbornly, to see
himself acknowledged as a *natural philosopher* and did not accept that his description of nature was a mere mathematical hypothesis. On the contrary, he believed that
mathematics was the tool and vehicle for achieving a *true* and *real* description of the
universe. Hobbes was one of the first to realize the importance of the work completed by Galileo, rightly identifying the Italian scientist as a philosopher, the greatest 'not only of our own century, but of all time.'[16]

---

[16] *MLT*, X, 9, 178; Eng. Trans. Hobbes 1976, 123.

Appendix

Robert Payne and the *Short Tract on First Principles*

In 1879, Ferdinand Tönnies discovered an anonymous manuscript (Harley MS 6796, fols. 297–308) in the Harley collection at the British Library. Untitled and undated, it was composed of around ten folios, which he considered to be a youthful draft by Hobbes, dating to around 1630.[1] The text was published as an appendix to the first edition of *Elements of Law*, edited by Tönnies in 1889, with the title of *Short Tract on First Principles*.[2]

Several scholars were perplexed by the attribution of the manuscript to Hobbes, although this was never openly contested until recently. Nevertheless, as early as 1886, George Croom Robertson expressed his doubts, despite considering the text to reflect Hobbes' general philosophical orientation.[3] While raising a number of questions about it, Frithiof Brandt used the *Short Tract* extensively in 1921,[4] as did nearly all the other scholars over the following years.[5]

The issue was reconsidered by Arrigo Pacchi in 1971. He admitted that the *Short Tract* had 'the fundamental principles of Hobbes' future speculation' and this is the reason why 'the manuscript was unquestionably attributed to Hobbes.' However, Pacchi had several reservations, maintaining that it could best be considered as a

---

[1] See Tönnies 1879, esp. 463–464. For a detailed historical and critical analysis of the text, see Napoli 1990.

[2] *A Short Tract on First Principles*, Appendix I, in *EL*, 193–210.

[3] Robertson 1886, 35, note: 'This Tract, No. 26, doubtless in Hobbes' handwriting, but otherwise giving no account of itself…'. Moreover, the English scholar lingered on the fact that the *Short Tract* presents an 'emanantistic' concept of light that is rejected by Hobbes in his printed works.

[4] See Brandt 1928, 10 ff.

[5] Like Gargani 1971 too, who devotes a chapter to the comparison between the theory of light proposed in the *Short Tract* and the speculations of Robert Grosseteste and Roger Bacon. See also Giudice 1996, and Giudice 1999, 17 ff.

© Springer Nature Switzerland AG 2020
G. Baldin, *Hobbes and Galileo: Method, Matter and the Science of Motion*,
International Archives of the History of Ideas Archives internationales
d'histoire des idées 230, https://doi.org/10.1007/978-3-030-41414-6

collective work of the so-called 'Newcastle Circle,' the main drivers of which were William Cavendish, Earl and then Duke of Newcastle, and his brother Charles, and which included the intellectuals Walter Warner and Robert Payne, the playwright Ben Jonson, the author Joseph Webbe, and Hobbes.[6]

Despite Pacchi's considerations, the *Short Tract* continued to be attributed mainly to Hobbes, and in 1988 Jean Bernhardt edited an edition of the work with a facing-page French translation, and an extensive commentary in the appendix.[7]

That same year, however, Richard Tuck wrote two articles on other aspects of Hobbes' ideas,[8] disputing the authenticity of the text and suggesting that in all likelihood it could be attributed to the clergyman Robert Payne, chaplain at Welbeck Abbey. Tuck's theory was contested by Perez Zagorin,[9] as well as by Karl Schuhmann, who pointed out the analogies between the *Short Tract* and Hobbes' later works.[10]

Nevertheless, in 2001, Timothy Raylor went through a precise and meticulous analysis of styles and handwriting to demonstrate that the hand in the *Short Tract* is that of Robert Payne almost beyond a shadow of a doubt.[11] Furthermore, he identified certain similarities between the text in question and the translations, written in 1635 by Payne, in the second part of a work by Benedetto Castelli: *Della misura dell'acque correnti*.[12] Regarding the doubts advanced by Zagorin about Payne's originality, Raylor specified that the information available to us presents him as an *intellectual factotum*,[13] versed in every kind of literary and scientific study. Payne also translated one of Galileo's works—*Le Mecaniche*[14]—from the Italian for Sir Charles Cavendish and this translation was attached to the *Short Tract* and the version of Castelli's work.[15] Raylor also highlighted the activity of the 'Newcastle Circle,' which gave rise to the theatrical work *Wit's triumvirate, or the Philosopher*,

---

[6] See Pacchi 1971, 15–16.

[7] See Bernhardt 1988, 59–274. Bernhardt's extensive critical commentary is, in any case, an excellent essay on Hobbesian natural philosophy.

[8] See Tuck 1988a, b.

[9] See Zagorin 1993.

[10] See Schuhmann 1995b, esp. up to 27, the final pages are dedicated to the comparison with Suarez.

[11] See Raylor 2001.

[12] See Castelli 1628. As noted by Raylor (Raylor 2001, 44), Payne translated the second part, entitled *Demostrazioni geometriche della misura dell'acque correnti*, calling it: 'Geometricall demonstrations of the measure of running-waters'. In British Library, MS Harley 6796, fols. 309–316 (the pagination of the translation in this manuscript follows that of the *Short Tract*).

[13] See Raylor 2001, 47 ff. On Payne, see also: Jacquot 1952b, 21; Feingold 1985, Malcolm 2002, 85 ff.

[14] Raylor refers to the work with the title under which it was published posthumously by Luca Danesi in 1649: *Della scienza mecanica* (Ravenna: Stamperia Camerali, 1649) (ibid. 33); however, he is referring to *Le mecaniche*, a text written by Galileo in around 1593, during his time in Padua (See Camerota 2004, 82 ff.) and that was translated, as we know, by Mersenne in 1634.

[15] The text by Galileo bears the title: *Of the Profitt wch is drawn from the Art Mechanics & it's Instruments. A Tract of Sig.^r Galileo Galilei, Florentine*, is dated November 11, 1636, and is contained in the aforementioned MS Harley 6796, fols. 317–337.

in which Galileo's presence and Campanella's concept of universal animation are evident.[16]

Raylor's considerations were followed by those of Noel Malcolm (published in 2002, in a collection of essays on Hobbes),[17] who made interesting observations on a philological, historical, and philosophical level. First of all, the text of the *Short Tract* contains a number of corrections in the same hand and ink as the principal text, which would appear to rule out the intervention of a copyist and instead suggests it is the work of the author and writer.[18] Furthermore, Malcolm claimed that many of the analogies identified by Schuhmann between the *Short Tract* and Hobbes' texts, are, upon more detailed analysis, less significant and far from showing any close match. Lastly, regarding strictly philosophical matters, Malcolm thinks that the text does not reflect the aspects of Hobbes' materialist mechanism that can be found in his early letters.[19]

Regarding documentary, historical, and philological observations, we can certainly agree with Malcolm. However, regarding certain more strictly philosophical aspects, we can make a number of observations. For example, in the *Short Tract* every action is explained in terms of local motion,[20] just as in the *Tractatus Opticus I*. The concept of *phantasma*, understood as a visual or mental representation, evokes the one present in the *Elements*;[21] the act of understanding is explained in merely physical terms,[22] and the author states that light, color, and heat are not qualities inherent to the species.[23] In any case, be it an individual creation of Payne's or, as Pacchi supposed a collective product of the 'Newcastle Circle,' there is no reason to completely ignore the *Short Tract*, as many Hobbesian scholars have done recently.

Nevertheless, we must admit that the affinities between this work and Hobbesian texts can be explained by turning to several common sources, not only to medieval optical texts, but also several works by Galileo, which were accessible to both Hobbes and Payne. Payne's work as a translator of Galileo, like that of Joseph Webbe, who translated the *Dialogue Concerning the Two Chief World Systems*,[24] testify to the keen interest of the 'Newcastle Circle' in Galileo's texts, an interest that was clearly shared not only by Payne, but also by Hobbes, who would often refer to Galilean speculations in his letters between 1635 and 1641.

---

[16] See Raylor 2001, 48–50. In this regard, see also the observations made by Sergio 2006b, 2007.

[17] See Malcolm 2002, esp. 104 ff.

[18] Ibid., 108–109.

[19] Ibid., 118 ff.

[20] Hobbes 1988a, 13

[21] Ibid. 40 and 45–46.

[22] Ibid., 49.

[23] Ibid., 45.

[24] Hobbes demonstrates that he is aware of Webbe's initiative to translate the dialogue (see Hobbes to William Cavendish, Earl of Newcastle, 26 January [/5 February] 1634, *CH*, I, 19), and the British Library conserves the manuscript of this unpublished translation (Harley 6320).

Payne's translations and the *Short Tract on First Principles* are of key importance regarding the first circulation of Galileo's ideas in England;[25] but they require
specific analysis that does not purely involve Hobbes' debt to the natural philosophy
of Mersenne and Galileo. Indeed, the correspondence testifies to a special focus on
Galileo's works during the Grand Tour and Hobbes probably started to reflect deeply
on these texts in Paris, in the company of Mersenne, who was very familiar with
Galileo's works.[26]

Consequently, I felt it was best not to take the *Short Tract on First Principles*
directly into consideration and to reflect on the works of Hobbes and his
correspondence.

---

[25] Regarding Galileo's initial reception in England, see: Feingold 1984a, 411–420; Giudice 2011.

[26] In all likelihood, Payne received the original Italian copy of *Le Mecaniche* via Hobbes and
William Cavendish, 3rd Earl of Devonshire, who received it from Mersenne in Paris. A note at the
foot of Payne's translation states that he worked directly from the Italian original and not from
Mersenne's French version: 'Raptim ex Italico in Anglicum sermonem transfusam. Novemb. 11°
1636, By Mr Robert Payen.' British Library, MS Harley 6796, fol. 337r.

Bibliography

Manuscripts:

British Library:

Ms Harley 3360, ff. VI + 193 r et v (First Draught of the Optiques), available online at: http://www.bl.uk/manuscripts/FullDisplay.aspx?ref=Harley_MS_3360
Ms Harley 6320, (Engl. Transl. by Joseph Webbe of Galileo's *Dialogo sopra i due massimi sistemi*).
Ms Harley 6796, ff. 193–266 r et v *(Tractatus Opticus II)*.

Chatsworth House, Derbyshire:

Ms Hobbes E.1 A (List of books kepts in the Hardwick Library, compiled by Hobbes).

Works of Thomas Hobbes:

Hobbes, Thomas. 1660. *Examinatio et Emendatio Mathematicae Hodiernae*. London: Crooke.
———. 1839a. *Thomae Hobbes Malmesburiensis Opera Philosophica quae Latine Scripsit Omnia, in unum corpus nunc primum collecta studio et labore Gulielmi Molesworth*, 5 Vols. London: Johannem Bohn.
———. 1839b. *The Collected English Works of Thomas Hobbes*, ed. William Molesworth, 11 Vols. London: Bohn, Longman, Brown, Green, and Longmans.
———. 1963. *Tractatus opticus II*, first complete edition by Franco Alessio. *Rivista critica di storia della filosofia* 18 (2): 147–228.
———. 1968. *Elementi di legge naturale e politica*, ed. Arrigo Pacchi. Florence: La Nuova Italia.
———. 1976. Thomas White's *De Mundo Examined*, ed. Harold Whitmore Jones. Bradford: Bradford University Press.

© Springer Nature Switzerland AG 2020
G. Baldin, *Hobbes and Galileo: Method, Matter and the Science of Motion*,
International Archives of the History of Ideas Archives internationales
d'histoire des idées 230, https://doi.org/10.1007/978-3-030-41414-6

————. 1983a. *Thomas Hobbes' A Minute or First Draught of the Optiques*, ed. Elaine C. Stroud, PhD Dissertation. Madison: University of Wisconsin–Madison.

————. 1983b. *De cive. The Latin Version*, ed. Howard Warrender. Oxford: Clarendon Press.

————. 1984: *The Elements of Law Natural and Politic*, ed. Ferdinand Tönnies. London: Frank Cass (1st ed. 1889).

————. 1988a. (probably by Robert Payne), *Court traité des premiers principes*, ed. Jean Bernhardt. Paris: Presses Universitaires de France.

————. 1988b. *Of Passions* (British Library: Ms. Harl. 6083), ed. Anna Minerbi Belgrado. *Rivista di Storia della Filosofia* (4): 729–738.

————. 1994. *The Correspondence of Thomas Hobbes*, ed. Noel Malcolm, 2 Vols. Oxford: Clarendon Press.

————. 2010. *Moto, luogo e tempo*, ed. Gianni Paganini, Utet, Torino.

————. 2012. *Leviathan*, ed. Noel Malcolm, 3 Vols. Oxford, Clarendon Press.

————. 2015. *De l'Homme*, ed. Jean Terrel et al. Paris: Vrin.

Hobbes Thomas. 1973. *Critique du* De Mundo *de Thomas White*, ed. Jean Jacquot and Harold Whitmore Jones. Paris: Vrin.

Other Works:

Aquinas, Thomas. 1964. *Quaestiones Disputatae*, 2 Vols. Turin-Rome: Marietti.

Aquinas, Thomas. 1989. *Summa Theologiae*. Bologna: Edizioni Studio Domenicano.

Aristotle. 1938—. *Works*, 23 Vols. (Loeb Classical Library) Cambridge, Mass.: Harvard University Press.

————. 1986. *De Caelo*, Eng. Trans. W.K.C. Guthrie. Cambridge, MA: Harvard University Press.

————. 1989. *The Metaphysics,* Eng. Trans. Hugh Tredennick. London: William Heinemann.

Aubrey, John. 1962. *Aubrey's Brief Lives*, ed. Oliver Lawson Dick. Bristol: Penguin Book.

Bacon, Francis. 1858–1874. *The Works of Francis Bacon*, 14 Vols, ed. James Spedding, Robert Leslie Ellis and Douglas Denon Heath. London: Longman, Green and Roberts.

Bacon, Roger. 1897–1900. *Opus Maius,* ed. John Henry Bridges, 2 Vols. Oxford: Clarendon Press. (Reprint: Frankfut am Main: Minerva, 1964).

Bacon, Francis. 2007. *The Oxford Francis Bacon*, Vol. XII, ed. Graham Rees and Maria Wakely. Oxford: Oxford University Press.

Bernier, François. 1684. *Abrégé de la philosophie de M. Gassendi*, (2nd ed. Lyon. Reprint: *Corpus des œuvres philosophiques en langue française,* 6 Vols., ed. Sylvia Murr and Geneviève Stefani. Paris: Fayard, 1992).

Boyle, Robert. 1999–2000. *The Works of Robert Boyle*, 14 Vols, ed. Michael Hunter and Edward B. Davis. London: Pickering and Chatto.

Buridan, Jean. 1942. *Iohannis Buridani Quaestiones super libris quattuor De caelo et mundo*, ed. Ernst A. Moody. Cambridge, MA: Medieval Academy of America.

Campanella, Tommaso. 2007. *Del senso delle cose e della magia,* ed. Germana Ernst. Rome-Bari: Laterza.

————. 2010. *Lettere*, ed. Germana Ernst. Florence: Olschki.

Castelli, Benedetto. 1628. *Della misura dell'acque correnti*. Rome: Stamperia Camerale.

Cavalieri, Bonaventura. 1632. *Lo specchio ustorio, overo Trattato delle Settioni Coniche, et alcvni loro mirabili* effetti. Bologna: Clemente Ferroni.

————. 1635. *Geometria indivisibilibus continuorum noua quadam ratione promota*.... Bologna: Clemente Ferroni.

————. 1989. *Geometria degli Indivisibili*, ed. Lucio Lombardo Radice. Turin: Utet (1st ed. 1966).

Clavius, Christoph. 1591. *Euclidis elementorum libri XV.* Köln: Ciotti.

Copernicus, Nicolaus. 2015. *De Revolutionibus Orbium Coelestium*, ed. Michel-Pierre Lerner, Alain-Philippe Segonds and Jean-Pierre Verdet, 3 Vols. Paris: Les Belles Lettres.

Descartes, René. 1982–1992. *Œuvres*, ed. Charles Adam and Paul Tannery, 11 Vols., 13 Tomes. Paris: Vrin (1st ed. 1887–1913).

Galilei, Galileo. 1960. *Le Mecaniche*, Eng. Trans. Stillman Drake. Madison: University of Wisconsin-Madison Press.

———. 1968. *Edizione Nazionale delle Opere*, ed. Antonio Favaro, 20 Vols. Rome: Barbera (1st ed. 1890–1909).

———. 1988. *Tractatio de praecognitionibus et praecognitis* and *Tractatio de demonstratione*, ed. William F. Edwards and William A. Wallace. Padua: Antenore.

———. 2001. *Dialogue Concerning the Two Chief World Systems*, Eng. Trans. Stillman Drake. New York: The Modern Library (1st ed. Los-Angeles-Berkeley: University of California Press, 1967).

———. 2003. *Dialogues Concerning Two New Sciences*, Eng. Trans. Henry Crew and Alfonso de Salvio. Mineola: Dover Publications (1st ed.: New York: The Macmillan Company, 1914).

Gassendi, Pierre. 1658. *Opera Omnia*, 6 Vols. Lyon: Anisson, & Devenet, 1658 (Stuttgart-Bad Cannstatt: Fromman Holzboog, 1994).

Gilbert, William. 1600. *De Magnete*. London: Peter Short.

Kepler, Johannes. 1937—. *Gesammelte Werke*, ed. Walther von Dick, Max Caspar and Franz Hammer. München: Beck.

Mersenne, Marin. 1623. *Quaestiones Celeberrimae in Genesim...* Paris: Cramoisy.

———. 1625. *La Verité des Sciences. Contre les Sceptiques ou Pyrrhonines*. Paris: Toussainct du Bray. (Reprint: Stuttgart-Bad Cannstatt: Friedrich Frommann Verlag, 1969).

———. 1634a. *Les Mechaniques de Galilee, Mathematicien et Ingenieur du Duc de Florence*. Paris: Henry Guenon (Reprint in Mersenne. 1985: 427–513).

———. 1634b. *Questions Inouyes*. Paris: Jacques Villery (Reprint in Mersenne. 1985: 5–103).

———. 1634c. *Questions Harmoniques*. Paris: Jacques Villery (Reprint in Mersenne. 1985: 105–98).

———. 1634d. *Les Questions Theologiques, Physiques, Morales et Mathematiques*. Paris: Henry Guenon (Reprint in Mersenne. 1985: 199–425).

———. 1634e. *Les Preludes de l'harmonie universelle ou questions curieuses*. Paris: Henry Guenon (Reprint in Mersenne. 1985: 515–660).

———. 1634f. *Traité des Mouvemens, et de la chute des corps pesans, & de la proportion de leurs differentes vitesses. Dans lequel l'on verra plusierurs experiences tres-exactes*. Paris: Jacques Villery (Reprint in *Corpus*, 2/1986: 25–58).

———. 1636a. *Harmonie Universelle*, 1st Vol. Paris: Sebastien Cramoisy.

———. 1636b. *Harmonicorum libri, in qvibus agitur de sonorvm natvra, cavsis, & effectibus...* Paris: Gvillelmi Bavdry.

———. 1637. *Harmonie Universelle*, 2nd Vol. Paris: Pierre Ballard.

———. 1639. *Les nouvelles pensees de Galileé, Mathematicien et Ingenieur du Duc de Florence*. Paris: Pierre Rocolet.

———. 1644a. *Cogitata physico matematica. In quibusdam naturae quàm artis effectus admirandi artissimis demonstrationibus explicantur*. Paris: Bertier.

———. 1644b. *Universae Geometriae Mixtaeque Mathematicae Synopsis*. Paris: Bertier.

———. 1932–1986. *Correspondance du P. Marin Mersenne*, ed. Paul Tannery, Cornelis de Waard, René Pintard, Bernard Rochot, 16 Vols. Paris: Éditions Universitaires de France et Centre National de la Recherche Scientifique.

———. 1985. *Questions Inouyes*, (Corpus des œuvres de philosophie en langue française) (Paris: Fayard, 1985).

———. 2003. *Traité de l'Harmonie Universelle*, ed. Claudio Buccolini. Paris: Fayard (or. Ed. 1627).

Ockham, William of. 1974–1988. *Opera Philosophica et Theologica*, 17 Vols. New York: Cura Instituti Franciscani Universitatis S. Bonaventurae.

Oresme, Nicole. 1968. *Le livre du Ciel et du Monde*, ed. Albert D. Menut and Alexander J. Denomy. Madison/Milwaukee/London: The University of Wisconsin Press.

Plato. 1914—. *Works* (Loeb Classical Library). Cambridge, MA: Harvard University Press.

Sarpi, Paolo. 1996. *Pensieri, naturali, metafisici e matematici*, ed. Luisa Cozzi and Libero Sosio. Milan-Naples: Ricciardi.

Sarsi, Lotario. 1619. (*alias* Orazio Grassi). *Libra Astronomica ac Philosophica*. In *OG*, VI: 109–179.

———. 1626. (*alias* Orazio Grassi). *Ratio ponderum librae et simbellae*. In *OG*, VI: 373–500.

Leibniz, Gottfried Wilhelm von. 1923—. *Sämtliche Schriften und Briefe*. Darmstadt-Leipzig-Berlin: Deutsche Akademie der Wissenschaften.

White, Thomas. 1642. *De mundo dialogi tres*. Paris: Moreau.

Secondary Sources:

Alessio, Franco. 1962. De Homine e A Minute of First Draught of the Optiques, di Thomas Hobbes. *Rivista critica di storia della filosofia* 17: 393–410.

———. 1985. *Introduzione a Ruggero Bacone*. Rome/Bari: Laterza.

Altieri Biagi, Maria Luisa. 1995. L'incipit del *Dialogo sopra i due massimi sistemi*. In *Galileo e la cultura veneziana*, 351–361. Venice: Istituto Veneto di Lettere ed Arti.

Anstey, Peter. 2000. *The Philosophy of Robert Boyle*. London/New York: Routledge.

———. 2013. The Theory of Material Qualities. In *The Oxford Handbook of British Philosophy in the Seventeenth Century*, ed. Peter Anstey, 240–260. Oxford: Oxford University Press.

Applebaum, Wilbur, and Renzo Baldasso. 2001. Galileo and Kepler on the Sun as Planetray Mover. In *Largo campo di filosofare*, ed. José Montesinos and Carlos Solís, 381–390. La Orotava: Fundaciòn Canaria Orotava de Historia de la Ciencia.

Armogathe, Jean-Robert. 1992. Le groupe de Mersenne et la vie académique parisienne. *Dix-septième siècle* 175: 131–139.

Auger, Léon. 1948. Le R. P. Mersenne et la physique. *Revue d'histoire des sciences* 2: 33–52.

Bailhache, Patrice. 1994. L'harmonie universelle: la musique entre les mathématiques, la physique, la métaphysique et la religion. In *Études sur Marin Mersenne*, Special issue of *Les Études Philosophiques* 68 (Iss. 1–2), ed. Jean-Robert Armogathe and Michel Blay, 13–24.

Baldin, Gregorio. 2013. Hobbes and Sarpi: Method, Matter and Natural Philosophy. *Galilaeana* 10: 85–118.

———. 2015. Thomas Hobbes e la Repubblica di Venezia. *Rivista di Storia della Filosofia* (4): 717–741.

———. 2016a. Hobbes, Sarpi and the Interdict of Venice. *Storia del pensiero politico* (2): 261–280.

———. 2016b. 'La reflexion de l'arc' et le *conatus*: aux origines de la physique de Hobbes. *Philosophical Enquiries. Revue des philosophies anglophones* 7: 15–42.

———. 2017. Archi, spiriti e *conatus*. Hobbes e Descartes sui principi della fisica. *Historia Philosophica* 15: 147–165.

Baldini, Ugo. 1977a. La struttura della materia nel pensiero di Galileo. *De Homine* 56–58 (excerpt): 1–74.

———. 1977b. Il corpuscolarismo italiano del Seicento. Problemi di metodo e prospettive di ricerca. In *Ricerche sull'atomismo del Seicento. Atti del Convegno di Studio (14–16 otto-bre 1976)*, ed. Ugo Baldini, Giancarlo Zanier, Paolo Farina, and Francesco Trevisani, 1–76. Florence: La Nuova Italia.

Banfi, Antonio. 1964. *Galileo Galilei*. Milan: Il Saggiatore (1st ed. 1949).

Barnouw, Jeffrey. 1989. *Respice Finem!* The Importance of Purpose in Hobbes's Psychology. In *Thomas Hobbes: de la métaphysique à la politique*, ed. Martin Bertman and Michel Malherbe, 47–59. Paris: Vrin.

———. 1990. Hobbes's Causal Account of Sensation. *Journal of the History of Philosophy* 18 (n. 2): 115–130.

———. 1992. Le vocabulaire du *conatus*. In *Hobbes et son vocabulaire*, ed. Yves-Charles Zarka, 103–124. Paris: Vrin.

Baroncelli, Giovanna. 1992. Bonaventura Cavalieri tra matematica e fisica. In *Geometria e atomismo nella scuola galileiana (Atti del convegno)*, ed. Massimo Bucciantini and Maurizio Torrini, 67–101. Florence: Olschki.

Bascelli, Tiziana. 2015. Torricelli's Indivisibles. In *Seventeenth-Century Indivisibles Revisited*, ed. Vincent Jullien, 105–136. Dordrecht: Springer.

Baumgold, Deborah. 2004. The Composition of Hobbes's *Elements of Law*. *History of Political Thought* 25 (Iss. 1): 16–43.

Beaulieu, Armand. 1982a. Découverte d'un livre de Mersenne. *Revue d'histoire des sciences* 25: 55–56.

———. 1982b. Lumière et matière chez Mersenne. *Dix-Septième Siècle* 136 (3): 311–316.

———. 1984. Les réactions des savants français au début du XVIIe siècle devant l'heliocentrisme de Galilée. In *Novità celesti e crisi del sapere*, ed. Paolo Galluzzi, 373–381. Florence: Giunti-Barbera.

———. 1986. Mersenne et l'Italie. In *La France et l'Italie au temps de Mazarin*, ed. Jean Serroy, 69–77. Grenoble: Presses Universitaires de Grenoble.

———. 1989. Torricelli et Mersenne. In *L'œuvre de Torricelli: science galiléenne et nouvelle géométrie*, ed. François De Gandt, 39–51. Paris: Les Belles Lettres.

———. 1990. Les relations de Hobbes et de Mersenne. In *Thomas Hobbes. Philosophie première, théorie de la science et politique*, ed. Yves-Charles Zarka and Jean Bernhardt, 81–90. Paris: Presses Universitaires de France.

———. 1992. Le group de Mersenne. Ce que l'Italie lui a donné—Ce qu'il a donné à l'Italie. In *Geometria e atomismo nella scuola galileiana (Atti del convegno)*, ed. Massimo Bucciantini and Maurizio Torrini, 17–34. Florence: Olschki.

———. 1993. L'attitude nuancée de Mersenne vers la chimie. In *Alchimie et philosophie à la Renaissance*, ed. Jean-Claude Margolin and Sylvain Matton, 395–403. Paris: Vrin.

———. 1995. *Mersenne. Le grand minime*. Bruxelles: Fondation Nicolas-Claude de Peiresc.

Bernhardt, Jean. 1977. Hobbes et le mouvement de la lumière. *Revue d'histoire des sciences* 30: 3–24.

———. 1979. La polémique de Hobbes contre la *Dioptrique* de Descartes dans le *Tractatus Opticus II* (1644). *Revue Internationale de Philosophie* 33: 432–442.

———. 1985a. Nominalisme et mécanisme chez Hobbes. *Archives de philosophie* 48: 235–249.

———. 1985b. Sur le passage de F. Bacon à Th. Hobbes. *Les Études Philosophiques* (4): 449–457.

———. 1986. Témoignage direct de Hobbes sur son 'illumination euclidéenne'. *Revue philosophique de la France et de l'Etranger* 176 (n. 2): 281–282.

———. 1988. Essai de commentaire. In *Thomas Hobbes, Court traité des premiers principes*, 61–274. France: Presses Universitaires de.

———. 1989a. *Hobbes*. Paris: Presses Universitaires de France.

———. 1989b. L'apport de l'aristotelisme à la pensée de Hobbes. In *Thomas Hobbes. De la métaphysique à la politique*, ed. Martin Bertman and Michel Malherbe, 9–15. Paris: Vrin.

———. 1990a. L'œuvre de Hobbes en optique et en théorie de la vision. In *Hobbes Oggi*, ed. Bernard Willms et al., 245–268. Milan: Franco Angeli.

———. 1990b. Grandeur, substance et accident: une difficulté du *De corpore*. In *Thomas Hobbes. Philosophie première, théorie de la science et politique*, ed. Yves-Charles Zarka and Jean Bernhardt, 39–46. Paris: Presses Universitaires de France.

———. 1993. La question du vide chez Hobbes. *Revue d'histoire des sciences* 46: 225–232.

Bernstein, Howard R. 1980. Conatus, Hobbes, and the Young Leibniz. *Studies in History and Philosophy of Science* 11: 25–37.

Berthier, Jauffrey. 2015. Prudence juridique et prudence politique chez Bacon et Hobbes (I) : Le philosophe, le juriste et l'homme politique chez Francis Bacon. *Philosophical Enquiries:*

Revue des philosophies anglophones 4: 93–126. http://www.philosophicalenquiries.com/numero4article5Berthier.html.

Bertman, Martin. 2001. *Conatus* in Hobbes' De Corpore. *Hobbes Studies* 14: 25–39.

Bianchi, Luca. 1997. La struttura del cosmo. In *La filosofia nelle Università: secoli XIII-XIV*, ed. Luca Bianchi, 269–303. Florence: La Nuova Italia.

———. 2003. *Studi sull'aristotelismo del Rinascimento*. Padua: Il Poligrafo.

Bianchi, Luca, and Eugenio Randi. 1990. *Le verità dissonanti*. Rome/Bari: Laterza.

Biener, Zvi. 2004. Galileo's First New Science: The Science of Matter. *Perspectives on Science* 12 (3): 262–287.

Bitpol-Hespériès, Annie. 2005. *L'Homme* de Descartes et le *De Homine* de Hobbes. In *Hobbes, Descartes et la métaphysique*, ed. Dominique Weber, 155–186. Paris: Vrin.

Blay, Michel. 1994. Mersenne expérimentateur: les études sur les mouvements des fluides jusqu'en 1644. In *Études sur Marin Mersenne*, Special issue of *Les Études Philosophiques* 68 (Iss. 1–2), ed. Jean-Robert Armogathe and Michel Blay, 69–86.

Bloch, Olivier René. 1971. *La philosophie de Gassendi: Nominalisme, matérialisme et métaphysique*. The Hague: Martinus Nijhoff.

Blumenberg, Hans. 1979. *Die Lesbarkeit der Welt*. Berlin: Surkhamp.

Boas Hall, M. 1952. The Establishment of the Mechanical Philosophy. *Osiris* 10: 412–541.

Bortolotti, Emilio. 1989. L'oeuvre géométrique d'Évangeliste Torricelli. In *L'œuvre de Torricelli: science galiléenne et nouvelle géométrie*, ed. François De Gandt, 111–146. Paris: Les Belles Lettres.

Brandt, Frithiof. 1928. *Thomas Hobbes' Mechanical Conception of Nature*. Copenhagen/London: Levin & Mungsgaard-Librairie Hachette (or. Ed. 1921).

Breidert, Wolfgang. 1979. Les mathématiques et la méthode mathématique chez Hobbes. *Revue internationale de philosophie* 129: 414–431.

Brunschvicg, Léon. 1913. *Les étapes de la philosophie mathématique*. Paris: Blanchard.

Bucciantini, Massimo. 2003. *Galileo e Keplero. Filosofia, cosmologia e teologia nell'Età della Controriforma*. Turin: Einaudi.

———. 2007. Descartes, Mersenne e la filosofia invisibile di Galileo. *Giornale critico della filosofia italiana* 86 (1): 38–52.

Buccolini, Claudio. 1997. Il ruolo del sillogismo nelle dimostrazioni geometriche della *Vérité des Sciences* di Marin Mersenne. *Nouvelles de la République des Lettres* 1: 7–36.

———. 1998. Opere di Galileo Galilei provenienti dalla biblioteca di Marin Mersenne: *Il saggiatore* e i *Discorsi e dimostrazioni matematiche intorno a due nuove scienze*. *Nouvelles de la Républiques des Lettres* (2): 139–142.

———. 2002. Mersenne traduttore di Bacon? *Nouvelles de la République des Lettres* (2): 7–19.

———. 2014. Mersenne et la philosophie baconienne en France à l'époque de Descartes. In *Bacon et Descartes. Genèse de la modernité philosophique*, ed. Élodie Cassan, 115–134. Lyon: ENS Éditions.

———. 2015. Il 'mos geometricus' fra usi teologici ed esiti materialistici: le obiezioni di Mersenne contro la metafisica cartesiana. In *La ragione e le sue vie. Saperi e procedure di prova in età moderna*, ed. Carlo Borghero and Claudio Buccolini, 59–81. Florence: Le Lettere.

Bunce, Robin. 2003. Thomas Hobbes' Relationship with Francis Bacon – An Introduction. *Hobbes Studies* 16: 41–83.

Califano, Salvatore. 2010. *Storia della chimica*, 2 Vols. Turin: Bollati Boringhieri.

Camerota, Michele. 2004. *Galileo Galilei e la cultura scientifica nell'età della Controriforma*. Rome: Salerno Ed.

———. 2008. Galileo, Lucrezio e l'atomismo. In *Lucrezio: la natura e la scienza*, ed. Marco Beretta and Francesco Citti, 141–175. Olschki: Florence.

Carraud, Vincent. 1994. Mathématique et métaphysique: les sciences du possible. In *Études sur Marin Mersenne*, Special issue of *Les Études Philosophiques* 68 (Iss. 1-2), ed. Jean-Robert Armogathe and Michel Blay, 145–159.

Carugo, Adriano, and Alistair C. Crombie. 1983. The Jesuits and Galileo's Ideas of Science and of Nature. *Annali dell'Istituto e Museo di Storia della Scienza di Firenze* 8 (2): 3): 3–3):68.

Cassirer, Ernst. 1922–1957. *Das Erkenntnisproblem in der Philosophie und Wissenschaft der neuren Zeit*, 4 Vols. Berlin: Bruno Cassirer Verlag.

Celeyrette, Jean. 2015. From Aristotle to the Classical Age, the Debate Around Indivisibles. In *Seventeenth-Century Indivisibles Revisited*, ed. Vincent Jullien, 19–30. Dordrecht: Springer.

Clavelin, Maurice. 1968. *La philosophie naturelle de Galilée*. Paris: Librairie Armand Colin.

———. 2001. Galilée astronome philosophe. In *Largo campo di filosofare. Eurosymposium Galileo 2001*, ed. José Montesinos and Carlos Solís, 19–39. La Orotava: Fundaciòn Canaria Orotava de Historia de la Ciencia.

Clericuzio, Antonio. 2000. *Elements, Principles and Corpuscles. A Study of Atomism and Chemistry in Seventeenth Century*. Dordrecht: Kluwer.

Clucas, Stephen. 2004. Walter Warner (*c*.1558–1643). In *Oxford Dictionnary of National Biographies*. Oxford: Oxford University Press. http://www.oxforddnb.com/view/article/38108.

Collins, Jeffrey R. 2002. Thomas Hobbes and the Blackloist Conspiracy of 1649. *Historical Journal* 45: 305–331.

———. 2005. *The Allegiance of Thomas Hobbes*. Oxford: Oxford University Press.

Costabel, Pierre. 1964. La roue d'Aristote et les critiques françaises à l'argument de Galilée. *Revue d'histoire des sciences* 17: 385–396.

———. 1994. Mersenne et la cosmologie. In *Quatrième centenaire de la naissance de Marin Mersenne (Actes du colloque)*, ed. Jean-Marie Constant and Anne Fillon, 47–55. Le Mans: Université du Maine.

Cozzi, Gaetano. 1979. *Paolo Sarpi tra Venezia e l'Europa*. Turin: Einaudi.

Cozzoli, Daniele. 2007. Light, Motion, Space and Time in Mersenne's Optics. In *Notions of Space and Time. Early Modern Concepts and Fundamental Theories (Begriffe von Raum and Zeit. Frühneuzeitliche Konzepte und Fundamentale Theorien)*, ed. Frank Linhard and Peter Eisenhardt, 29–43. Frankfurt am Mein: Klostermann.

———. 2010. The Development of Mersenne's Optics. *Perspectives on Science* 18 (1): 9–25.

Crapulli, Giovanni. 1969. *Mathesis Universalis. Genesi di un'idea nel XVI secolo*. Rome: Edizioni dell'Ateneo & Bizzarri.

Crombie, Alistair C. 1953. *Augustine to Galileo. The History of Science A.D. 400–1650*. Cambridge, MA: Harvard University Press.

———. 1975. Sources of Galileo's Early Natural Philosophy. In *Reason, Experiment, and Mysticism in the Scientific Revolution*, ed. Maria Luisa Righini Bonelli and William Shea, 157–175. New York: Science History Publ.

———. 1990. *Science, Optics and Music in Medieval and Early Modern Thought*. London/Ronceverte: The Hambledon press.

Curley, Edwin M. 1995. Hobbes Versus Descartes. In *Descartes and his contemporaries. Meditations, Objections, and Replies*, ed. Roger Ariew and Marjorie Grene, 97–109. Chicago: The University of Chicago Press.

Dal Pra, Mario. 1962. Note sulla logica di Hobbes. *Rivista critica di storia della filosofia* 17: 411–433.

de Buzon, Frédéric. 2013. *La Science cartésienne et son objet. Mathesis et phénomène*. Paris: Honoré Champion.

De Gandt, François. 1989. Les indivisibles de Torricelli. In *L'œuvre de Torricelli: science galiléenne et nouvelle géométrie*, ed. François De Gandt, 147–206. Paris: Les Belles Lettres.

———. 1992. L'evolution de la théorie des indivisibles et l'apport de Torricelli. In *Geometria e atomismo nella scuola galileiana (Atti del convegno)*, ed. Massimo Bucciantini and Maurizio Torrini, 103–118. Florence: Olschki.

Dear, Peter. 1988. *Mersenne and the Learning of the Schools*. Ithaca/London: Cornell University Press.

———. 2001. *Revolutionizing the Sciences. European Knowledge and its Ambition, 1500–1700*. Basingstoke: MacMillan.

————. 2006. *The Intelligibility of Nature.* Chicago/London: The University of Chicago Press.

Dijksterhuis, Eduard J. 1961. *The Mechanization of the World Picture: From Pythagoras to Newton.* Oxford: Clarendon Press.

Drabkin, Israel E. 1950. Aristotle's Wheel: Notes on the History of a Paradox. *Osiris* 9: 161–198.

Drake, Stillman. 1957. *Discoveries and Opinions of Galileo.* New York: Doubleday & Co..

————. 1960. *The Controversy on the Comets.* Philadelphia: University of Pennsylvania Press.

————. 1970. *Galileo Studies. Personality, Tradition, and Revolution.* Ann Arbor: The University of Michigan Press.

————. 1978. *Galileo at Work: His Scientific Biography.* Mineola: Dover Publications, Inc..

Dubos, Nicolas. 2015. Le passage de Bacon à Hobbes: remarques sur la question de l'histoire. *Philosophical Enquiries: Revue des philosophies anglophones* 4: 59–76. http://www.philosophicalenquiries.com/numero4article6Dubos.html.

Ducrocq, Myriam-Isabelle. 2015. Le silence, le secret, la transparence: de l'art du gouvernement à la science civile chez Francis Bacon et Thomas Hobbes. *Philosophical Enquiries: Revue des philosophies anglophones* 4: 77–91. http://www.philosophicalenquiries.com/numero4article7Ducrocq.html.

Duhem, Pierre. 2003. *Sauver les apparences. Sur la notion de théorie physique de Platon à Galilée.* Paris: Vrin (or. Ed. 1908)

Ernst, Germana. 2010. *Tommaso Campanella. Il libro e il corpo della natura.* Rome/Bari: Laterza.

Ernst, Germana, and Eugenio Canone. 1994. Una lettera ritrovata: Campanella a Periesc, 19 giugno 1636. *Rivista di Storia della Filosofia* (2): 353–366.

Fabbri, Natacha. 2003. *Cosmologia e armonia in Kepler e Mersenne. Contrappunto a due voci sul tema dell'*Harmonice Mundi. Florence: Olschki.

————. 2005. Echi della condanna di Galilei in Marin Mersenne. In *I primi Lincei e il Sant'Uffizio. Questioni di scienza e di fede. (Atti del convegno, Roma, 12 e 13 giugno 2003),* 449–480. Rome: Bardi Editore.

Fattori, Marta. 2000. Fortin de la Hoguette tra Francis Bacon e Marin Mersenne: intorno all'edizione francese del *De Augmentis Scientiarum* (1624). In *Linguaggio e filosofia nel Seicento europeo,* ed. Marta Fattori, 385–411. Florence: Olschki.

————. 2012. *Études sur Francis Bacon.* Paris: Presses Universitaires de France.

Favaro, Antonio. 1966. *Galileo Galilei e lo studio di Padova,* 2 Vols. Padua: Antenore (or. Ed. 1883).

Favino, Federica. 1997. A proposito dell'atomismo di Galileo: da una lettera di Tommaso Campanella a uno scritto di Giovanni Ciampoli. *Bruniana & Campanelliana* 3: 265–282.

————. 2015. *La filosofia naturale di Giovanni Ciampoli.* Florence: Olschki.

Feingold, Mordechai. 1984a. Galileo in England: The First Phase. In *Novità celesti e crisi del sapere. Atti del Convegno di studi,* ed. Paolo Galluzzi, 411–420. Florence: Giunti-Barbera.

————. 1984b. *The Mathematicians' Apprenticeship.* Cambridge: Cambridge University Press.

————. 1985. A Friend of Hobbes and an Early Translator of Galileo: Robert Payne of Oxford. In *The Light of Nature: Essays in History and Philosophy of Science Presented to A. C. Crombie,* ed. John D. North and John J. Roche, 265–280. Dordrecht/Boston/London: Martinus Nijhoff Publisher.

Festa, Egidio. 1992. Quelques aspects de la controverse sur les indivisibles. In *Geometria e atomismo nella scuola galileiana (Atti del convegno),* ed. Massimo Bucciantini and Maurizio Torrini, 193–206. Florence: Olschki.

————. 1999. Le galiléisme de Gassendi. In *Géométrie, atomisme et vide dans l'école galiléenne,* ed. Egidio Festa, Vincent Jullien, and Maurizio Torrini, 213–227. Fontenay St.-Cloud: ENS Éditions.

————. 2007. *Galileo. La lotta per la scienza.* Rome/Bari: Laterza.

Finocchiaro, Maurice A. 2008. *The Essential Galileo.* Indianapolis/Cambridge: Hackett Publishing Company.

Fisher, Saul. 2005. *Pierre Gassendi's Philosophy and Science.* Leiden/Boston: Brill.

————. 2008. Gassendi et l'hypothèse dans la méthode scientifique. In *Gassendi et la modernité,* ed. Sylvie Taussig, 399–425. Turnhout: Brepols.

Frajese, Vittorio. 1994. *Sarpi scettico*. Bologna: Il Mulino.

Fumaroli, Marc. 2015. *La République des Lettres*. Paris: Gallimard.

Funkenstein, Amos. 1986. *Theology and the Scientific Imagination from the Middle Ages to the Seventeenth Century*. Princeton: Princeton University Press.

Gabbey, Alan. 2001. Disciplinary Transformation in the Age of Newton: The Case of Mathematics. In *Between Leibniz, Newton, and Kant. Philosophy and Science in the Eighteenth Century*, ed. Wolfgang Lefèvre. Dordrecht: Kluwer.

———. 2004. What Was 'Mechanical' About 'The Mechanical Philosophy'. In *The Reception of Galilean Science of Motion in Seventeenth-Century Europe*, ed. Carla Rita Palmerino and J.M.M.H. Thijssen, 11–23. Dordrecht/Boston/London: Kluwer.

Galluzzi, Paolo. 1973. Il 'Platonismo' del tardo Cinquecento e la filosofia di Galileo. In *Ricerche sulla cultura dell'Italia moderna*, ed. Paola Zambelli, 39–79. Rome/Bari: Laterza.

———. 1979a. *Momento. Studi galileiani*. Rome: Edizioni dell'Ateneo & Bizzarri.

———. 1979b. Il tema dell''ordine' in Galileo. In *Ordo. Atti del II colloquio internazionale del Lessico Intellettuale Europeo*, ed. Marta Fattori and Massimo L. Bianchi, 235–277. Rome: Ateneo & Bizzarri.

———. 1993. Gassendi e *l'affaire Galilée* delle leggi del moto. *Giornale critico della filosofia italiana* 72 (1): 86–119.

———. 1994. Ratio/Ragione in Galileo. Del dialogo tra la ragione e l'esperienza. In *Ratio (VII Colloquio Internazionale del Lessico Intellettuale Europeo)*, ed. Marta Fattori and Massimo L. Bianchi, 379–401. Florence: Olschki.

———. 2011. *Tra atomi e indivisibili. La materia ambigua di Galileo*. Florence: Olschki.

Garber, Daniel. 2000. A Different Descartes. Descartes and the Programme for a Mathematical Physics in His Correspondence. In *Descartes' Natural Philosophy*, ed. Stephen Gaukroger, John Schuster, and John Sutton, 113–130. London-New York: Routledge.

———. 2004. On the Frontlines of Scientific Revolution: How Mersenne Learned to Love Galileo. *Perspectives on Science* 12 (2): 135–163.

———. 2009. *Descartes Embodied*. Cambridge: Cambridge University Press.

———. 2010. Philosophia, Historia, Mathematica: Shifting Sands in the Disciplinary Geography of the Seventeenth Century. In *Scientia in Early Modern Philosophy. Seventeenth-Century Thinkers on Demonstrative Knowledge from First Principles*, ed. Sorell Tom, G.A.J. Rogers, and Jill Kraye, 1–17. Dordrecht: Springer.

———. 2013. Remarks on the Pre-History of Mechanical Philosophy. In *The Mechanization of Natural Philosophy*, ed. Daniel Garber and Sophie Roux, 3–26. Dordrecht: Springer.

Gargani, Aldo G. 1971. *Hobbes e la scienza*. Einaudi: Turin.

Garin, Eugenio. 1961. *La cultura filosofica del Rinascimento italiano*. Florence: Sansoni.

Gaukroger, Stephen. 2001. *Francis Bacon and the Transformation of Early-Modern Philosophy*. Cambridge: Cambridge University Press.

———. 2002. *Descartes' System of Natural Philosophy*. Cambridge: Cambridge University Press.

———. 2006. *The Emergence of a Scientific Culture*. Oxford: Oxford University Press.

———. 2010. The Unity of Natural Philosophy and the End of *Scientia*. In *Scientia in Early Modern Philosophy. Seventeenth-Century Thinkers on Demonstrative Knowledge from First Principles*, ed. Sorell Tom, G.A.J. Rogers, and Jill Kraye, 18–33. Dordrecht: Springer.

Gert, Bernard. 1989. Hobbes's Account of Reason and the Passions. In *Thomas Hobbes: de la métaphysique à la politique*, ed. Martin Bertman and Michel Malherbe, 83–92. Paris: Vrin.

———. 1996. Hobbes's Psychology. In *The Cambridge Companion to Hobbes*, ed. Tom Sorell, 157–174. Cambridge: Cambridge University Press.

Geymonat, Lodovico. 1957. *Galileo Galilei*. Turin: Einaudi.

Gilbert, Neal W. 1963. Galileo and the School of Padua. *Journal of History of Philosophy* 1: 223–231.

Giorello, Giulio. 1990. Pratica geometrica e immagine della matematica in Thomas Hobbes. In *Hobbes Oggi*, ed. Bernard Willms et al., 215–255. Milan: Franco Angeli.

Giovannozzi, Delfina, and Marco Veneziani, eds. 2011. *Materia. XIII Colloquio Internazionale (Lessico Intellettuale Europeo) Roma, 7–9 gennaio 2010.* Florence: Olschki.

Giudice, Franco. 1996. Teoria della luce e struttura della materia nello *Short Tract on First Principles* di Thomas Hobbes. *Nuncius* 2: 545–561.

———. 1997. Thomas Hobbes and Atomism: A Reappraisal. *Nuncius* 12: 471–485.

———. 1999. *Luce e visione. Thomas Hobbes e la scienza dell'ottica.* Florence: Olschki.

———. 2011. Echi del caso Galileo nell'Inghilterra del XVII secolo. In *Il caso Galileo. Una rilettura storica, filosofica, teologica. Convegno internazionale di studi. Firenze, 26–30 maggio 2009*, ed. Massimo Bucciantini, Michele Camerota, and Franco Giudice, 277–287. Florence: Olschki.

———. 2015. The Most Curious of Science: Hobbes's Optics. In *The Oxford Handbook of Hobbes*, ed. Aloysius P. Martinich and Kinch Hoekstra, 147–165. Oxford: Oxford University Press.

———. 2016. Optics in Hobbes's Natural Philosophy. *Hobbes Studies* 29 (1): 86–102.

Giusti, Enrico. 1980. *Bonaventura Cavalieri and the Theory of Indivisibles.* Bologna: Edizioni Cremonese.

———. 1995. Il ruolo della matematica nella meccanica di Galileo. In *Galileo Galilei e la cultura veneziana*, 321–337. Venice: Istituto Veneto di Lettere e Arti.

Gómez Lopez, Susana. 2001. Galileo y la naturaleza de la luz. In *Largo campo di filosofare*, ed. José Montesinos and Carlos Solis, 403–418. La Orotava: Fundaciòn Canaria Orotava de Historia de la Ciencia.

———. 2008. The Mechanization of Light in Galilean Science. *Galilaeana* 5: 207–244.

Grant, Edward. 1981. *Much Ado About Nothing: Theories of Space and Vacuum from Middle Ages to the Scientific Revolution.* Cambridge: Cambridge University Press.

———. 1996a. *The Foundation of Modern Science in the Middle Ages.* Cambridge: Cambridge University Press.

Grant, Hardy. 1996b. Hobbes and Mathematics. In *The Cambridge Companion to Hobbes*, ed. Tom Sorell, 108–128. Cambridge: Cambridge University Press.

Gregory, Tullio. 2000. *Genèse de la raison classique de Charron à Descartes.* Paris: Presses Universitaires de France.

Guenancia, Pierre. 1990. Hobbes—Descartes: le nom et la chose. In *Thomas Hobbes. Philosophie première, théorie de la science et politique*, ed. Yves-Charles Zarka and Jean Bernhardt, 67–79. Paris: Presses Universitaires de France.

Hall, Alfred R., and Marie Boas Hall. 1964. *A Brief History of Science.* New York: The New American Library of World Literature Inc.

Halliwell, James O. 1841. *Letters on the Progress of Science in England from Queen Elizabeth to Charles II.* London: Taylor.

Hanson, Donald W. 1991. Reconsidering Hobbes's Conventionalism. *The Review of Politics* 53 (4): 627–651.

Heilbron, John L. 2010. *Galileo.* Oxford: Clarendon Press.

Henry, John. 2011. Galileo and the Scientific Revolution: The Importance of His Kinematics. *Galilaeana* 8: 3–36.

———. 2016. Hobbes, Galileo and the Physics of Simple Circular Motion. *Hobbes Studies* 29: 9–38.

Hine, William L. 1973. Mersenne and Copernicanism. *Isis* 64: 18–32.

Holden, Thomas. 2004. *The Architecture of Matter. Galileo to Kant.* Oxford: Oxford University Press.

Hooper, Wallace. 1996. Inertial Problems in Galileo's Preinertial Framework. In *The Cambridge Companion to Galileo*, ed. Peter Machamer, 146–171. Cambridge: Cambridge University Press.

Horstmann, Frank. 1998. Ein Baustein zur Kepler-Rezeption: Thomas Hobbes'. *Physica Coelestis Studia Leibnitiana* 30 (2): 135–160.

———. 2000. Hobbes und Sinusgesetz der Refraktion. *Annals of Science* 57: 415–440.

———. 2001. Hobbes on Hypotheses in Natural Philosophy. *The Monist* 84 (4): 487–501.

Iori, Luca. 2015. *Thucydides Anglicus. Gli Eight Bookes di Thomas Hobbes e la Ricezione Inglese delle Storie di Tucicide (1450–1642)*. Roma: Edizioni di Storia e Letteratura.

Jacquet, Chantal. 2015. Des mots aux choses: le problème du langage chez Bacon et Hobbes. *Philosophical Enquiries: Revue des philosophies anglophones* 4: 27–39. http://www.philosophicalenquiries.com/numero4article3Jaquet.html.

Jacquot, Jean. 1952a. Notes on an Unpublished Work of Thomas Hobbes. *Notes and Records of the Royal Society of London* 9: 188–195.

———. 1952b. Sir Charles Cavendish and His Learned Friends: A Contribution to the History of Scientific Relation between England and the Continent in the Earlier Part of the 17th Century, *Annals of Science* 8 (1): 13–27; and. 8: (2): 175–191.

Jacquot, Jean, and Harold Whitmore Jones. 1973. Introduction. In *Thomas Hobbes, Critique du De Mundo de Thomas White*, 9–102. Paris: Vrin.

Jesseph, Douglas M. 1993. Of Analytics and Indivisibles: Hobbes on the Modern Mathematics. *Revue d'histoire des sciences* 46: 153–193.

———. 1996. Hobbes and Method of Natural Science. In *The Cambridge Companion to Hobbes*, ed. Tom Sorell, 86–107. Cambridge: Cambridge University Press.

———. 1999. *Squaring the Circle. The War Between Hobbes and Wallis*. Chicago/London: University of Chicago Press.

———. 2004. Galileo, Hobbes and the Book of Nature. *Perspectives on Science* 12 (2): 191–211.

———. 2006. Hobbesian Mechanics. *Oxford Studies in Early Modern Philosophy* 3: 119–152.

———. 2010. *Scientia* in Hobbes. In *Scientia in Early Modern Philosophy. Seventeenth-Century Thinkers on Demonstrative Knowledge from First Principles*, ed. Sorell Tom, G.A.J. Rogers, and Jill Kraye, 117–127. Dordrecht: Springer.

———. 2013. Logic and Demonstrative Knowledge. In *The Oxford Handbook of British Philosophy in the Seventeenth Century*, ed. Peter Anstey, 373–392. Oxford: Oxford University Press.

———. 2016. Hobbes on 'Conatus': A Study in the Foundations of Hobbesian Philosophy. *Hobbes Studies* 29 (1): 66–85.

Jullien, Vincent. 2006. Gassendi, Roberval à l'académie de Mersenne. Lieux et occasions de contact entre ces deux auteurs. *Dix-septième siècle* 233 (4): 601–613.

———. 2015. Indivisibles in the Work of Galileo. In *Seventeenth-Century Indivisibles Revisited*, ed. Vincent Jullien, 87–103. Dordrecht: Springer.

Kargon, Robert H. 1966. *Atomism in England from Hariot to Newton*. Oxford: Clarendon Press.

Koyré, Alexandre. 1966. *Études galiléennes*. Paris: Hermann (or. Ed. 1939).

———. 1981. *Études d'histoire de la pensée philosophique*. Paris: Gallimard.

Kuhn, Thomas S. 1962. *The Structure of Scientific Revolutions*. Chicago: University of Chicago Press.

Laird, Walter Roy. 1997. Galileo and the Mixed Sciences. In *Method and Order in Renaissance Philosophy of Nature (The Aristotle Commentary Tradition)*, ed. Daniel A. Di Liscia, Eckhard Kessler, and Charlotte Methuen, 252–270. Ashgate: Aldershot.

Laird, Walter Roy, and Sophie Roux, eds. 2008. *Mechanics and Natural Philosophy Before the Scientific Revolution*. Dordrecht: Springer.

Lasswitz, Kurd. 1963. *Geschichte der Atomistik von Mittelalter bis Newton*, 2 Vols. Hildesheim: Georg Olms, (or. Ed. 1890).

Laudan, Larry. 1981. *Science and Hypothesis. Historical Essays on Scientific Methodology*. Dordrecht: Springer.

Le Grand, Homer E. 1978. Galileo's Matter Theory. In *New Perspectives on Galileo*, ed. Robert E. Butts and Joseph C. Pitt, 197–208. Dordrecht/Boston: Reidel Publishing Company.

Leijenhorst, Cees. 1997. Motion, Monks and Golden Mountains: Campanella, and Hobbes on Perception and Cognition. *Bruniana & Campanelliana* 3 (1): 93–121.

———. 2002. *The Mechanisation of Aristotelianism. The Late Aristotelian Setting of Thomas Hobbes' Natural Philosophy*. Leiden/Boston/Köln: Brill.

———. 2004a. Hobbes' Corporeal Deity. In *Nuove prospettive critiche sul Leviatano di Hobbes*, ed. Luc Foisneau and George Wright, 73–95. Milan: Franco Angeli.

———. 2004b. Hobbes and the Galilean Law of Free Fall. In *The Reception of the Galilean Science of Motion in Seventeenth-Century Europe*, ed. Carla Rita Palmerino and J.M.M.H. Thijssen, 165–184. Dordrecht/Boston/London: Kluwer.

———. 2005. La causalité chez Hobbes et Descartes. In *Hobbes, Descartes et la métaphysique*, ed. Dominique Weber, 79–119. Paris: Vrin.

———. 2007. Sense and Nonsense About Sense: Hobbes and the Aristotelians About Sense Perception and Imagination. In *The Cambridge Companion to Hobbes's Leviathan*, ed. Patricia Springborg, 82–108. Cambridge: Cambridge University Press.

Lenoble, Robert. 1943. *Mersenne ou la naissance du mécanisme*. Paris: Vrin.

Lerner, Michel-Pierre. 2001. La réception de la condamnation de Galilée en France. In *Largo campo di filosofare. Eurosymposium Galileo 2001*, ed. José Montesinos and Carlos Solis, 513–547. La Orotava: Fundaciòn Canaria Orotava de Historia de la Ciencia.

Lewis, John. 2006. *Galileo in France. French Reactions to the Theories and Trial of Galileo*. New York: Peter Lang.

Lindberg, David C. 1976. *Theories of Visions from Al-Kindi to Kepler*. Chicago: University of Chicago Press.

Lupoli, Agostino. 1999. 'Fluidismo' e *corporeal deity* nella filosofia naturale di Thomas Hobbes: A proposito dell'hobbesiano 'Dio delle cause'. *Rivista di Storia della filosofia* (4): 573–609.

———. 2001. Power (Conatus-Endeavour) in the 'Kinetic Actualism' and in the 'Inertial' Psychology of Thomas Hobbes. *Hobbes Studies* 14: 83–103.

———. 2004. Hobbes e Sanchez. In *Nuove prospettive critiche sul Leviatano di Hobbes*, ed. Luc Foisneau and George Wright, 263–301. Milan: Franco Angeli.

———. 2006. *Nei limiti della materia. Hobbes e Boyle: materialismo epistemologico, filosofia corpuscolare e Dio corporeo*. Milan: Baldini, Castoldi, Dalai.

Machamer, Peter. 1978. Galileo and the Causes. In *New Perspectives on Galileo*, ed. Robert E. Butts and Joseph C. Pitt, 161–180. Dordrecht/Boston: Reidel Publishing Company.

Magnard, Pierre. 1997. La mathématique mystique de Pierre Gassendi. In *Gassendi et l'Europe*, ed. Sylvia Murr, 21–29. Paris: Vrin.

Maier, Annaliese. 1968. *Zwei Grundprobleme der Scholastischen Naturphilosophie*. Rome: Edizioni di Storia e Letteratura.

———. 1984. *Scienza e filosofia nel Medioevo. Saggi sui secoli XIII e XIV*. Milan: Jaca Book.

Malcolm, Noel. 1990. Hobbes's Science of Politics and His Theory of Science. In *Hobbes Oggi*, ed. Bernard Willms et al., 145–159. Milan: Franco Angeli.

———. 1996. A Summary Biography of Hobbes. In *The Cambridge Companion to Hobbes*, ed. Tom Sorell, 13–44. Cambridge: Cambridge University Press.

———. 2002. *Aspects of Hobbes*. Oxford: Clarendon Press.

———. 2005. Hobbes, the Latin Optical Manuscript, and the Parisian Scribe. *English Manuscript Studies 1100–1700* 12: 210–232.

Malet, Antoni. 1996. *From Indivisibles to Infinitesimals*. Bellaterra: Universitad Autònoma de Barcelona.

———. 2001. The Power of Images: Mathematics and Metaphysics in Hobbes's Optics. *Studies in History and Philosophy of Science* 32 (2): 303–333.

Malherbe, Michel. 1984. *Thomas Hobbes ou l'oeuvre de la raison*. Paris: Vrin.

———. 1989. Hobbes et la fondation de la philosophie première. In *Thomas Hobbes: de la métaphysique à la politique*, ed. Martin Bertman and Michel Malherbe, 17–32. Paris: Vrin.

Mancosu, Paolo, and Ezio Vailati. 1991. Torricelli's Infinitely Long Solid and Its Philosophical Receptions in the Seventeenth Century. *Isis* 82: 52–70.

Marion, Jean-Luc. 1994. Le concept de métaphysique selon Mersenne. In *Études sur Marin Mersenne*, Special issue of *Les Études Philosophiques* 68 (Iss. 1–2), ed. Jean-Robert Armogathe and Michel Blay, 129–143.

———. 2005. Hobbes et Descartes: l'étant comme corps. In *Hobbes, Descartes et la métaphysique*, ed. Dominique Weber, 59–77. Paris: Vrin.

Marquer, Éric. 2005. Ce que sa polémique avec Descartes a modifié dans la pensée de Hobbes. In *Hobbes, Descartes et la métaphysique*, ed. Dominique Weber, 15–32. Paris: Vrin.

———. 2015. Analyse du langage et science politique selon Bacon et Hobbes. *Philosophical Enquiries: Revue des philosophies anglophones* 4: 41–57. http://www.philosophicalenquiries. com/numero4article4Marquer.html.

Martinich, Aloysius P. 1999. *Hobbes. A Biography*. Cambridge: Cambridge University Press.

Maury, Jean-Pierre. 2003. *À l'origine de la recherche scientifique: Mersenne*. Paris: Vuibert.

McCord Adams, Marilyn. 1987. *William Ockham*, 2 Vols. Notre Dame: Notre Dame University Press.

McLaughlin, Peter. 2000. Force, Determination and Impact. In *Descartes' Natural Philosophy*, ed. Stephen Gaukroger, John Schuster, and John Sutton, 81–112. London/New York: Routledge.

Médina, José. 1997. Nature de la lumière et science de l'optique chez Hobbes. In *Le siècle de la lumière 1600–1715*, ed. Christian Biet and Vincent Jullien, 33–48. Paris: ENS Éditions Fonteany-St. Cloud.

———. 2013a. Mathématique et philosophie chez Thomas Hobbes: une pensée du mouvement en mouvement. In *Lectures de Hobbes*, ed. Jauffrey Berthier, Nicolas Dubos, Arnaud Milanese, and Jean Terrel, 85–132. Paris: Ellipses.

———. 2013b. Physiologie mécaniste et mouvement cardiaque: Hobbes, Harvey et Descartes. In *Lectures de Hobbes*, ed. Jauffrey Berthier, Nicolas Dubos, Arnaud Milanese, and Jean Terrel, 133–162. Paris: Ellipses.

———. 2015a. La philosophie de l'optique du *De Homine*. In *Thomas Hobbes, De l'Homme*, ed. Jean Terrel et al., 31–82. Paris: Vrin.

———. 2015b. Le traité d'optique. In *Thomas Hobbes, De l'Homme*, ed. Jean Terrel et al., 87–146. Paris: Vrin.

———. 2016a. Matière, mémoire et mouvement chez Hobbes. In *Hobbes et le matérialisme*, ed. Jauffrey Berthier and Aranud Milanese, 33–74. Paris: Éditions matériologiques.

———. 2016b. Hobbes's Geometrical Optics. *Hobbes Studies* 29 (1): 39–65.

Messeri, Marco. 1985. *Causa e spiegazione: la fisica di Pierre Gassendi*. Milan: Franco Angeli.

Milanese, Arnaud. 2011. *Principes de la philosophie chez Hobbes*. Paris: Classiques Garnier.

———. 2013. Philosophie première et philosophie de la nature. In *Lectures de Hobbes*, ed. Jauffrey Berthier, Nicolas Dubos, Arnaud Milanese, and Jean Terrel, 35–62. Paris: Ellipses.

———. 2015. Sur le passage de Bacon à Hobbes: un système et ses tensions. *Philosophical Enquiries: Revue des philosophies anglophones* 4: 7–26. http://www.philosophicalenquiries. com/numero4article2Milanese.html.

———. 2016. Le matérialisme de Hobbes en question: la critique du matérialisme et la réception de Hobbes depuis l'École de Marbourg. In *Hobbes et le matérialisme*, ed. Jauffrey Berthier and Arnaud Milanese, 91–109. Paris: Éditions matériologiques.

Minerbi Belgrado, Anna. 1993. *Linguaggio e mondo in Hobbes*. Rome: Editori Riuniti.

Mintz, Samuel I. 1952. Galileo, Hobbes and the Circle of Perfection. *Isis* 43: 98–100.

Moody, Ernest A. 1951. Galileo and Avempace, The Dynamics of the Leaning Tower Experiment. *Journal of the History of Ideas* 12 (3): 375–422.

Moreau, Pierre-François. 1989. *Hobbes. Philosophie, Science, Religion*. Paris: Presses Universitaires de France.

Mori, Gianluca. 2010a. Hobbes, Cartesio e le idee: un dibattito segreto. *Rivista di storia della filosofia* (2): 229–246.

———. 2010b. *Cartesio*. Rome: Carocci.

Morris, Katherine. 2000. Bêtes-machines. In *Descartes' Natural Philosophy*, ed. Stephen Gaukroger, John Schuster, and John Sutton, 401–419. London/New York: Routledge.

Murdoch, John E. 1982. Infinity and Continuity. In *The Cambridge History of Later Medieval Philosophy*, ed. Norman Kretzmann, Anthony Kenny, and Jan Pinborg, 564–591. Cambridge: Cambridge University Press.

———. 2009. Beyond Aristotle: Indivisibles and Infinite Divisibility in the Later Middle Ages. In *Atomism in Late Medieval Philosophy and Theology*, ed. Christophe Grellard and Aurélien Robert, 15–38. Leiden/Boston: Brill.

Napoli, Andrea. 1990. Hobbes e lo *'Short Tract'*. *Rivista di Storia della Filosofia* (3): 539–569.
Nardi, Antonio. 1994. Théorème de Torricelli ou Théorème de Mersenne. In *Études sur Marin Mersenne*, Special issue of *Les Études Philosophiques* 68 (Iss. 1–2), ed. Jean-Robert Armogathe and Michel Blay, 87–118.
Naylor, Ron. 2007. Galileo's Tidal Theory. *Isis* 98: 1–22.
———. 2014. Paolo Sarpi and the First Copernican Tidal's Theory. *British Journal for the History of Science* 47 (4): 661–675.
Nonnoi, Giancarlo. 2000. Galileo Galilei, Quale atomismo? In *Atomismo e continuo nel XVII secolo*, ed. Egidio Festa and Romano Gatto, 275–319. Naples: Vivarium.
Osler, Margaret J. 1994. *Divine Will and Mechanical Philosophy. Gassendi and Descartes on Contingency and Necessity in the Created World.* Cambridge: Cambridge University Press.
Pacchi, Arrigo. 1964. Il *'De motu, loco et tempore'* e un inedito hobbesiano. *Rivista critica di storia della filosofia* 19: 159–168.
———. 1965. *Convenzione e ipotesi nella filosofia naturale di Thomas Hobbes*. Florence: La Nuova Italia.
———. 1971. *Introduzione a Hobbes*. Rome/Bari: Laterza.
———. 1978. Hobbes e l'epicureismo. *Rivista di storia della filosofia* (1): 54–71.
———. 1984. Hobbes e il Dio delle cause. In *La storia della filosofia come sapere critico. Studi offerti a Mario Dal Pra*, ed. Nicola Badaloni et al., 295–307. Milan: Franco Angeli.
———. 1987. Hobbes and the Passions. *Topoi* 6: 111–119.
———. 1988. Hobbes and the Problem of God. In *Perspectives on Thomas Hobbes*, ed. G.A.J. Rogers and Alan Ryan, 171–189. Oxford: Clarendon Press.
———. 1990. Hobbes e la teologia. In *Hobbes Oggi*, ed. Bernard Willms et al., 101–121. Milan: Franco Angeli.
———. 1998. *Scritti hobbesiani (1978–1990)*, ed. Agostino Lupoli. Milan: Franco Angeli.
Paganini, Gianni. 1990. Hobbes, Gassendi e la psicologia del meccanicismo. In *Hobbes Oggi*, ed. Bernard Willms et al., 351–445. Milan: Franco Angeli.
———. 1999. Thomas Hobbes e Lorenzo Valla. Critica umanistica e filosofia moderna. *Rinascimento* 39: 515–568.
———. 2003. Hobbes Among Ancient and Modern Sceptics: Phenomena and Bodies. In *The Return of Scepticism from Hobbes and Descartes to Bayle*, ed. Gianni Paganini, 3–35. Dordrecht/Boston/Leiden: Kluwer.
———. 2004a. Hobbes e lo scetticismo continentale. In *Nuove prospettive critiche sul Leviatano di Hobbes*, ed. Luc Foisneau and George Wright, 303–328. Milan: Franco Angeli.
———. 2004b. Hobbes, Gassendi und die Hypothese der Weltvernichtung. In *Konstellationsforschung*, ed. Martin Mulsow and Marcelo Stamm, 258–339. Frankfurt am Mein: Surkhamp.
———. 2006a. Hobbes, Gassendi e l'ipotesi annichilatoria. *Giornale critico della filosofia italiana* 85 (1): 55–81.
———. 2006b. Le lieu du néant. Gassendi et l'hypothèse de l'annihilatio mundi. *Dix-septième siècle* 58: 587–600.
———. 2006c. Alle origini del 'mortal God': Hobbes, Lipsius e il *Corpus Hermeticum*. *Rivista di Storia della Filosofia* (3): 509–532.
———. 2007. Hobbes's Critique of Essence and Its Source. In *The Cambridge Companion to Hobbes's Leviathan*, ed. Patricia Springborg, 337–357. Cambridge: Cambridge University Press.
———. 2008a. Hobbes alla ricerca del primo motore. Il De motu, loco et tempore. *Rinascimento* 48: 527–541.
———. 2008b. Le néant et le vide: les parcours croisés de Gassendi et Hobbes. In *Gassendi et la modernité*, ed. Sylvie Taussig, 177–214. Turnhout: Brepols.
———. 2008c. *Skepsis. Le débat des modernes sur le scepticisme*. Paris: Vrin.
———. 2010a. Introduzione. In *Moto, luogo e tempo*, ed. Thomas Hobbes, 9–104. Turin: Utet.
———. 2010b. Hobbes e l'umanesimo. In *L'umanesimo scientifico dal Rinascimento all'Illuminismo*, ed. Lorenzo Bianchi and Gianni Paganini, 135–158. Naples: Liguori.

———. 2011. Il piacere dell'amicizia: Hobbes, Gassendi e il circolo neo-epicureo dell'Accademia di Montmor. *Rivista di Storia della Filosofia* (1): 23–38.

———. 2015a. Hobbes and the French Skeptics. In *Skepticism and Political Thought in Seventeenth and Eighteenth Centuries*, ed. John Christian Laursen and Gianni Paganini, 55–82. n. p: University of Toronto Press.

———. 2015b. Hobbes's Galilean Project. Its Philosophical and Theological Implications. *Oxford Studies in Early Modern Philosophy* 7: 1–46.

Palmerino, Carla Rita. 1999. Infinite Degrees of Speed: Marin Mersenne and the Debate Over Galileo's Law of Free Fall. *Early Science and Medicine* 4 (4): 269–328.

———. 2000. Una nuova scienza della materia per la *Scienza Nova* del moto. La discussione dei paradossi dell'infinito nella prima giornata dei *Discorsi* galileiani. In *Atomismo e continuo nel XVII secolo*, ed. Egidio Festa and Romano Gatto, 275–319. Naples: Vivarium.

———. 2001. Galileo's and Gassendi's Solutions to the *Rota Aristotelis* Paradox: A Bridge Between Matter and Motion Theories. In *Late Medieval and Early Modern Corpuscolar Matter Theories*, ed. Cristopher Lüthy, John E. Murdoch, and William R. Newman, 381–422. Leiden/Boston/Köln: Brill.

———. 2002. The Mathematical Characters of Galileo's Book of Nature. In *The Book of Nature in Early and Modern History*, ed. Klaas Van Berkel and Arjo Vanderjagt, 27–44. Peeters: Leuvens.

———. 2004. Galileo's Theories of Free Fall and Projectile Motion as Interpreted by Pierre Gassendi. In *The Reception of the Galilean Science of Motion in Seventeenth-Century Europe*, ed. Carla Rita Palmerino and J.M.M.H. Thijssen, 137–164. Dordrecht/Boston/London: Kluwer.

———. 2008. Une force invisible à l'œuvre: le rôle de la *vis attrahens* dans la physique de Gassendi. In *Gassendi et la modernité*, ed. Sylvie Taussig, 141–176. Turnhout: Brepols.

———. 2010. The Geometrization of Motion: Galileo's Triangle of Speed and Its Various Transformations. *Early Science and Medicine* 15: 410–447.

———. 2011. The Isomorphism of Space, Time and Matter in Seventeenth-Century Natural Philosophy. *Early Science and Medicine* 16: 296–330.

Pantin, Isabelle. 2011. Premières répercussions de l'affaire Galilée en France chez les philosophes et les libertins. In *Il caso Galileo. Una rilettura storica, filosofica, teologica. Convegno internazionale di studi. Firenze, 26–30 maggio 2009*, ed. Massimo Bucciantini, Michele Camerota, and Franco Giudice, 237–257. Florence: Olschki.

Parodi, Massimo. 1981. *Tempo e spazio nel medioevo*. Turin: Loescher.

———. 1984. Misura, analogia e peso. Un'analogia del XII secolo. In *La storia della filosofia come sapere critico. Studi offerti a Mario Dal Pra*, ed. Nicola Badaloni et al., 52–71. Milan: Franco Angeli.

Pastore, Stocchi. 1976–1986. Manlio. *Il periodo veneto di Galileo Galilei*. In *Storia della cultura veneta*, ed. Girolamo Arnaldi and Manlio Pastore Stocchi, vol. 4, T. 2, 37–66. Neri Pozza: Vicenza.

Pécharman, Martine. 1992. Le vocabulaire de l'être dans la philosophie première: *ens, esse, essentia*. In *Hobbes et son vocabulaire*, ed. Yves-Charles Zarka, 31–59. Paris: Vrin.

Piccolino, Marco. 2005. *Lo zufolo e la cicala. Divagazioni galileiane tra la scienza e la sua storia*. Bollati Boringhieri: Turin.

Pietarinen, Juhani. 2001. Conatus as Active Power in Hobbes. *Hobbes Studies* 14: 71–82.

Popkin, Richard H. 1957. Father Mersenne's War Against Pyrrhonism. *The Modern Schoolman* 34: 61–78.

———. 1979. *The History of Scepticism from Erasmus to Spinoza*. Berkeley/Los Angeles/London: University of California Press (or. Ed. 1960).

———. 1982. Hobbes and Scepticism. In *History of Philosophy in the Making. A Symposium of Essays to Honor Professor James D. Collins on His 65th Birthday*, ed. Linus J. Thro, 133–148. Washington, DC: University Press of America. Re-published in 1992, with an appendix in *The Third Force in 17th Century Thought*, ed. by Richard H. Popkin, 9–26 and 2–49. Leiden/New York/København/Köln: Brill, 1992.

Prins, Jan. 1990. Hobbes and the School of Padua: Two Incompatible Approaches of Science. *Archiv für Geschichte der Philosophie* 72: 26–46.

———. 1993. Ward's Polemic with Hobbes on the Sources of His Optical Theories. *Revue d'histoire des sciences* 46: 195–224.

———. 1996. Hobbes on Light and Vision. In *The Cambridge Companion to Hobbes*, ed. Tom Sorell, 129–156. Cambridge: Cambridge University Press.

Radelet-De Grave, Patricia. 2015. Kepler, Cavalieri, Guldin. Polemics with the Departed. In *Seventeenth-Century Indivisibles Revisited*, ed. Vincent Jullien, 57–86. Dordrecht: Springer.

Randall, John H. 1961. *The School of Padua and the Emergence of Modern Science*. Padua: Antenore.

Raphael, Renée. 2008. Galileo's *Discorsi* and Mersenne's *Nouvelles Pensées*: Mersenne as a Reader of Galilean 'Experience'. *Nuncius* 23: 7–36.

Raylor, Timothy. 2001. Hobbes, Payne and *A Short Tract on First Principles*. *The Historical Journal* 44: 29–58.

———. 2005. The Date and Script of Hobbes's Latin Optical Manuscript. *English Manuscript Studies 1100–1700* 12: 201–209.

———. 2018. *Philosophy, Rhetoric, and Thomas Hobbes*. Oxford: Oxford University Press.

Redondi, Pietro. 1982. Galilée aux prises avec les théories aristotéliciennes de la lumière (1610–1640). *Dix-Septième Siècle* 136 (3): 267–283.

———. 1985. Atomi, indivisibili e dogma. *Quaderni storici* 59 (Iss. 2): 529–571.

———. 1997. I fondamenti metafisici della fisica di Galileo. *Nuncius* 12: 267–289.

———. 2009. *Galileo eretico*. Rome/Bari: Laterza (or. Ed. 1983).

Robertson Croom, George. 1886. *Hobbes*. Edinburgh/London: William Blackwood and Sons. (Reprint: Thoemmes Press, 1993).

Robinet, Alexandre. 1990. Hobbes: structure et nature du *conatus*. In *Thomas Hobbes: Philosophie première, théorie de la science et politique*, ed. Yves-Charles Zarka and Jean Bernhardt, 127–138. Paris: Presses Universitaires de France.

Rochot, Bernard. 1957. Gassendi et les mathématiques. *Revue d'histoire des sciences* 10 (1): 69–78.

Rossi, Mario Manlio. 1942. *Alle fonti del deismo e del materialismo moderno*. Florence: La Nuova Italia.

Rossi, Paolo. 1974. *Francesco Bacone. Dalla magia alla scienza*. Turin: Einaudi (or. Ed. 1957).

Roux, Sophie. 1997. La nature de la lumière selon Descartes. In *Le siècle de la lumière 1600–1715*, ed. Christian Biet and Vincent Jullien, 49–66. Paris: ENS Éditions Fonteany-St. Cloud.

———. 1998. Le scepticisme et les hypothèses de la physique. *Revue de Synthèse* 4e s (2–3): 211–255.

———. 2004. Cartesian Mechanics. In *The Reception of Galilean Science of Motion in Seventeenth-Century Europe*, ed. Carla Rita Palmerino and J.M.M.H. Thijssen, 24–66. Dordrecht/Boston/London: Kluwer.

Sacksteder, William. 1980. The Art of the Geometricians. *Journal of the History of philosophy* 18: 131–146.

———. 1992. Three Diverse Sciences in Hobbes: First Philosophy, Geometry, and Physics. *The Review of Metaphysics* 45 (4): 739–772.

Sarasohn, Lisa T. 1985. Pierre Gassendi, Thomas Hobbes and the Mechanical World-View. *Journal of the History of Ideas* 46 (3): 363–379.

Schmitt, Charles B. 1985. *Problemi dell'aristotelismo rinascimentale*. Naples: Bibliopolis.

———. 2001. *Filosofia e scienza nel Rinascimento*. Florence: La Nuova Italia.

Schuhmann, Karl. 1985. Rapidità del pensiero e ascensione al cielo: alcuni motivi ermetici in Hobbes. *Rivista di Storia della Filosofia* 2: 203–227.

———. 1990. Hobbes and Reinassance Philosophy. In *Hobbes Oggi*, ed. Bernard Willms et alii, 331–349. Milan: Franco Angeli.

———. 1992. Le vocabulaire de l'espace. In *Hobbes et son vocabulaire*, ed. Yves-Charles Zarka, 61–82. Paris: Vrin.

————. 1995a. Hobbes dans les publications de Mersenne en 1644. *Archives de Philosophie* 58 (2): 2–7.

————. 1995b. Le *Short Tract*. Première oeuvre philosophique de Hobbes. *Hobbes Studies* 8: 3–36.

————. 1998. *Hobbes. Une chronique*. Paris: Vrin.

Sergio, Emilio. 2001. *Contro il Leviatano. Hobbes e le controversie scientifiche 1650–1665*. Soveria Mannelli: Rubbettino.

————. 2003. Bacon, Hobbes e l'idea di '*Philosophia naturalis*': due modelli di riforma della scienza. *Nuncius* 18 (2): 515–547.

————. 2006a. *Verità matematiche e forme della natura da Galileo a Newton*. Rome: Aracne.

————. 2006b. Hobbes, Campanella e il Cavendish Circle. *Filologia antica e moderna* 16: 173–189.

————. 2007. Campanella e Galileo in un 'english play' del circolo di Newcastle: 'Wit's triumvirate, or the philosopher' (1633–1635). *Giornale critico della filosofia italiana* 86 (2): 298–315.

————. 2016. Hobbes lecteur de Campanella: autour des sources cachées du matérialisme hobbesien. In *Hobbes et le matérialisme*, ed. Jauffrey Berthicr and Arnaud Milanese, 75–89. Paris: Éditions matérilogiques.

Shapin, Steven. 1998. *Scientific Revolution*. Chicago: University of Chicago Press. (or. Ed. 1996).

Shapin, Steven, and Simon Schaffer. 1985. *Leviathan and the Air-Pump*. Princeton: Princeton University Press.

Shapiro, Alan E. 1973. Kinematics Optics: A Study of the Wave Theory of Light in the Seventeenth Century. *Archive for History of Exact Sciences* 11: 134–266.

Shea, William R. 1970. Galileo's Atomic Hypothesis. *Ambix* 17: 13–27.

————. 1974. *La rivoluzione intellettuale di Galileo*. Florence: Sansoni (or. Ed. 1972).

———— 1977. Marin Mersenne: Galileo's 'traduttore-traditore'. *Annali dell'Istituto e Museo di Storia della Scienza di Firenze* A. 2 (1): 55–70.

————. 1978. *Descartes as Critic of Galileo*. In *New Perspectives on Galileo*, ed. Robert E. Butts and Joseph C. Pitt, 139–159. Dordrecht/Boston: Reidel Publishing Company.

————. 1991. *The Magic of Numbers and Motion*. Canton: Science History Publication.

Skinner, Quentin. 1996. *Reason and Rhetoric in the Philosophy of Hobbes*. Cambridge: Cambridge University Press.

————. 2002. *Visions of Politics*, 3 Vols. Cambridge: Cambridge University Press.

Sommerville, Johann P. 1992. *Thomas Hobbes: Political Ideas in Historical Context*. New York: Plagrave MacMillan.

Sorell, Tom. 1986. *Hobbes*. London/New York: Routledge.

————. 1993. Hobbes Without Doubt. *History of Philosophy Quarterly* 10: 121–135.

Sorell, Tom, G.A.J. Rogers, and Jill Kraye, eds. 2010. *Scientia in Early Modern Philosophy. Seventeenth-Century Thinkers on Demonstrative Knowledge from First Principles*. Dordrecht: Springer.

Sosio, Libero. 1995. Galileo Galilei e Paolo Sarpi. In *Galileo Galilei e la cultura veneziana. Atti del convegno di studio*, 269–311. Venice: Istituto Veneto di Lettere ed Arti.

Southgate, Beverley C. 1993. *Covetous of Truth.' The Life and Work of Thomas White, 1593–1676*. Dordrecht/Boston/London: Kluwer.

Spragens, Thomas A. 1973. *The Politics of Motion. The World of Thomas Hobbes*. London: Croom Helm.

Stabile, Giorgio. 2002. Il concetto di esperienza in Galilei e nella scuola galileiana. In *Experientia. X Colloquio del Lessico Intellettuale Europeo*, ed. Marco Veneziani, 217–241. Florence: Olschki.

Strauss, Leo. 1952. *The Political Philosophy of Hobbes. Its Basis and Its Genesis*. Chicago/London: University of Chicago Press (or. Ed. 1948).

Stroud, Elaine C. 1983. *Thomas Hobbes' A Minute or First Draught of the Optiques: A Critical Edition*. PhD Dissertation. Madison: University of Winsconsin-Madison.

————. 1990. Light and Vision: Two Complementary Aspects of Optics in Hobbes' Unpublished Manuscript *A Minute or First Draught of the Optiques*. In *Hobbes Oggi*, ed. Bernard Willms et al., 269–277. Milan: Franco Angeli.

Talaska, Richard A. 2013. *The Hardwick Library and Hobbes's Early Intellectual Development*. Charlottesville: Philosophy Documentation Center.

Taton, René. 1994. Le P. Marin Mersenne et la communauté scientifique au XVIIe siècle. In *Quatrième centenaire de la naissance de Marin Mersenne (Actes du colloque)*, ed. Jean-Marie Constant and Anne Fillon, 13–25. Le Mans: Université du Maine.

Terrel, Jean. 1994. *Hobbes. Matérialisme et politique*. Paris: Vrin.

————. 2008. *Hobbes: vies d'un philosophe*. Rennes: Presses Universitaires de Rennes.

————. 2009. Hobbes et Boyle: enjeux d'une polémique. In *La philosophie naturelle de Robert Boyle*, ed. Myriam Dennehy and Charles Ramond, 277–293. Paris: Vrin.

————. 2015. Hobbes et Bacon: une relation decisive. *Philosophical Enquiries. Revue des philosophies anglophones* 4: 1–6. http://www.philosophicalenquiries.com/numero4article1Terrel.html.

————. 2013. Hobbes: définition et rôle de l'expérience. In *Lectures de Hobbes*, ed. Jauffrey Berthier, Nicolas Dubos, Arnaud Milanese, and Jean Terrel, 63–83. Paris: Ellipses.

————. 2016. Le matérialisme de Hobbes dans les *Troisièmes Objections*. In *Hobbes et le matérialisme*, ed. Jauffrey Berthier and Arnaud Milanese, 13–31. Paris: Éditions Matériologiques.

Tönnies, Ferdinand. 1879. Anmerkungen über die Philosophie des Hobbes. *Vierteljahrsschrift für Wissenschaftliche Philosophie* 3: 453–466.

————. 1925. *Thomas Hobbes. Leben und Lehre*. Stuttgart: Fromman Verlag (Faksimile-Neudruck: Stuttgart-Bad Cannstatt: Fromman Holzboog, 1971).

Torrini, Maurizio. 1993. Galileo copernicano. *Giornale critico della filosofia italiana* 13 (A. 72, Iss. 1, January–April): 26–42.

————. 2013. Galilei, Galileo. In *Il Contributo italiano alla storia del pensiero*. Roma: Treccani. http://www.treccani.it/enciclopedia/galileo-galilei_%28Il-Contributo-italiano-alla-storia-del-Pensiero:-Scienze%29/.

Tricaud, François. 1985. Éclaircissements sur les six premières biographies de Hobbes. *Archives de Philosophie* 48: 277–286.

————. 1992. Le vocabulaire de la passion. In *Hobbes et son vocabulaire*, ed. Yves-Charles Zarka, 139–154. Paris: Vrin.

Tuck, Richard. 1988a. Hobbes and Descartes. In *Perspectives on Thomas Hobbes*, ed. G.A.J. Rogers and Alan Ryan, 11–41. Oxford: Oxford University Press.

————. 1988b. Optics and Sceptics: The Philosophical Foundation of Hobbes's political Thought. In *Conscience and Casuistry in Early Modern Europe*, ed. Edmund Leites, 235–263. Cambridge: Cambridge University Press.

————. 1989. *Hobbes. A Very Short Introduction*. Oxford: Oxford University Press.

Watkins, John W.N. 1965. *Hobbes's System of Ideas*. London: Hutchinson.

Westfall, Richard S. 1971. *The Construction of Modern Science. Mechanisms and Mechanics*. Cambridge: Cambridge University Press.

Wilson, Margaret D. 1993. Descartes on Perception of Primary Qualities. In *Essays on Philosophy and Science of René Descartes*, ed. Stephen Voss, 162–176. New York/Oxford: Oxford University Press.

Wisan, Winifred L. 1978. Galileo's Scientific Method: A Reexamination. In *New Perspectives on Galileo*, ed. Robert E. Butts and Joseph C. Pitt, 1–57. Dordrecht/Boston: Reidel Publishing Company.

Wootton, David. 1983. *Paolo Sarpi. Between Renaissance and Enlightment*. Cambridge: Cambridge University Press.

Zagorin, Perez. 1993. Hobbes's Early Philosophical Development. *Journal of the History of Ideas* 54: 505–518.

Zarka, Yves-Charles. 1985. Empirisme, nominalisme et matérialisme. *Archives de philosophie* 48: 177–233.

————. 1988. La matière et la représentation: Hobbes lecteur de *La Dioptrique* de Descartes. In *Problématique et réception du Discours de la méthode*, ed. Henri Méchoulan, 81–98. Paris: Vrin.

————. 1989. Aspects sémantiques, syntaxiques et pragmatiques dans la pensée de Hobbes. In *Thomas Hobbes: de la métaphysique à la politique*, ed. Martin Bertman and Michel Malherbe, 33–46. Paris: Vrin.

————. 1992. Le vocabulaire de l'apparaître: le champ sémantique de la notion de *phantasma*. In *Hobbes et son vocabulaire*, ed. Yves-Charles Zarka, 13–29. Paris: Vrin.

————. 1999. *La décision métaphysique de Hobbes*. Paris: Vrin (1st Ed.: 1987).

————. 2004. Liberté, nécessité, hasard: la théorie générale de l'événement chez Hobbes. In *Nuove prospettive critiche sul Leviatano di Hobbes*, ed. Luc Foisneau and George Wright, 249–262. Milan: Franco Angeli.

Index

© Springer Nature Switzerland AG 2020
G. Baldin, *Hobbes and Galileo: Method, Matter and the Science of Motion*,
International Archives of the History of Ideas Archives internationales
d'histoire des idées 230, https://doi.org/10.1007/978-3-030-41414-6

Printed in the United States
by Baker & Taylor Publisher Services

Printed in the United States
by Baker & Taylor Publisher Services